TECHNICAL WRITING

Second Edition

TECHNICAL WRITING

John M. Lannon
Southeastern Massachusetts University

Little, Brown and Company
Boston   Toronto

Library of Congress Catalog Card No. 81-84676

ISBN 0-316-51435-7

9 8 7 6 5 4 3

HAL

*Published simultaneously in Canada
by Little, Brown & Company (Canada) Limited*

Printed in the United States of America

Preface

This text provides a comprehensive and flexible introduction to technical communication. Designed specifically for heterogeneous classes, it speaks to a diversity of student interests. The principles of effective report writing are applied to a broad variety of assignments, from brief memos and summaries to detailed formal reports and proposals. The goals of the exercises parallel the writing demands students will face in college and on the job.

The book is organized in four parts. Part I treats technical writing as a deliberate act of communication for a definite purpose to a specified audience. It defines technical writing by comparing and contrasting it with writing done in traditional composition courses. It also offers guidelines for analyzing specific writing situations and for adapting a message to the needs of a stipulated audience. Finally, it explains how to refine paragraph structure, sentence style, and diction.

Part II covers the strategies of rhetoric, design, and research that a competent writer must master. Specifically, it shows how to achieve economy and precision, how to organize information, how to describe and explain mechanisms and processes, how to design various formats and prepare visual aids, and how to collect, report, and document research data.

Part III applies the earlier concepts and strategies to assorted correspondence and reports appropriate for various writing situations: letters, resumes, memos, prepared-form reports, proposals, formal reports, and oral reports.

Throughout the text, each chapter fully explains the rhetorical purpose and principles of each major assignment. To unify the concepts of adaptation for a particular audience and readability, introduced in Part I, with the strategies in Part II and the applications in Part III, sample reports are preceded by an audience and use analysis.

Finally, appendixes contain a handbook for easy reference, an explanation of brainstorming as a prewriting strategy, and guidelines for eliminating sexist language.

The rationale for the sequence of chapters and their individual development is based on six assumptions.

1. That students need intensive practice and guidance in the *process* of writing: in generating worthwhile content, sensible organization, and readable style, and in revising their writing to ensure the adequacy — or excellence — of these three elements. They should not simply be *told* about this process; they must be *shown* the nitty-gritty of "why" and "how to."

2. That often the only real difference between freshmen or sophomores and juniors or seniors is their respective levels of specialized knowledge. Juniors and seniors generally face the same difficulties in planning, writing, and revising as their younger counterparts.

3. That the proliferation of technical writing courses has led to classes that are grouped heterogeneously. This assortment of people with varied backgrounds calls for explanations that are accessible, examples that are broadly engaging and intelligible, and goals that are rigorous but collectively achievable.

4. That a frequent and most difficult challenge in report writing is translating technical information for a nonspecialized audience.

5. That there are nearly as many approaches to the teaching of technical and professional writing as there are instructors. Some schools regard report writing as a lower-level course, and others upper-level. Flexibility in a textbook is therefore crucial.

6. That class time in a writing course should not be taken up by lectures reiterating information readily found in a textbook. Instead, a writing course ought to *apply* textbook principles through workshops that focus on the texts composed by the students in that course. The workshop approach calls for a text that is as comprehensive as possible. (The Instructor's Manual gives detailed suggestions for workshop design.)

In line with these assumptions, the book follows a pattern of cumulative skills, moving from assignments of summaries and expanded definitions to more complex tasks, ending with the formal report of proposal — an assignment that draws upon most skills developed earlier. Within this structure, however, each chapter is self-contained for greater flexibility in course planning.

Numerous exercises in each chapter offer practical applications at various levels of challenge and complexity. Thus the instructor who wishes to spend more time on certain chapters, such as those on letters or informal reports, will find plentiful resources. Timely examples and models are drawn mostly from student writing in a variety of technical and business fields, and are intelligible to students in all majors.

Individual chapters move from theory to practice by following these six steps:

1. Define each assignment in detail.
2. Explain its purpose and usefulness to students, whose implied question is "Why are we doing this?"
3. Discuss the specific criteria that the completed assignment should embody — with an emphasis on rhetorical purpose.
4. Provide models accompanied by explanations and audience analyses.
5. Give step-by-step guidelines in planning, organizing, drafting, and revising the assignment for a specified audience with specified needs.
6. Provide revision checklists for self- or peer evaluation.

The emphasis throughout is on the *process* of writing. Students are carefully guided in organizing and developing their materials systematically. They learn to move from theory to model to statement of purpose to outline to completed assignment. The importance of an introduction-body-conclusion organization is stressed repeatedly to show that the structure of any piece of writing parallels that of the standard paragraph.

This book is easy to teach and to learn from, whether your approach is basic or accelerated. The Instructor's Manual contains sample syllabi and suggestions for implementing either approach. All materials have been used successfully in both community colleges and universities, with lower- and upper-level students in just about every major. Students using this text come to see writing as a process of applied rhetoric instead of mere mechanical transcription.

This new edition is far more than a cosmetic revision. In addition to offering more concise chapters and sharper exercises, the text now addresses students who have substantial job experience as well as those who do not. Specific improvements in this edition:

– an expanded Chapter 2, on tailoring a report to the audience's needs
– a new Chapter 3, on improving readability by refining paragraph structure, sentence style, and diction
– a new Chapter 16, with a detailed treatment of proposals
– a new appendix, with guidelines for avoiding sexist language
– audience and use analyses preceding most sample reports
– a greater emphasis on writing for audiences outside the classroom
– revision checklists — partitioned into sections on "Content," "Arrangement," and "Style" — for easier use
– greater emphasis in exercises on the *writing situation* (message, purpose, and audience)

– improved examples throughout, especially in Chapters 8 and 9
– greater distinction between levels of headings throughout
– division of the discussion of research into Chapter 12 (Researching Information) and Chapter 13 (Writing and Documenting the Research Report)
– a new section on advanced library research
– expanded research exercises
– greater emphasis on technical and scientific documentation systems
– an expanded and improved section on the job search, addressing both new and experienced job applicants

Many of the improvements outlined above were inspired by rigorous and perceptive reviews from Richard S. Beal of Boston University, Barbara Cambridge of Indiana University at Indianapolis, Rebecca Carosso of Northern Essex Community College, Lynn W. Denton of Auburn University, Joseph Harris of Parkland College, Mead Johnson of California State Polytechnic University, Michael Marcuse of the University of Maryland, Robert Rudolf of the University of Toledo, Helen Schwartz of Oakland University, Dennis Vannatta of the University of Arkansas at Little Rock, Edith Weinstein of the University of Akron, and John Woodcock of Indiana University. My sincere thanks to you all.

At Southeastern Massachusetts University, Shaleen Barnes, reference librarian, was invaluable in helping revise the research section. Raymond Dumont, Jr., provided samples and helped me throughout. Special thanks to Jean Morgan for just about everything.

At Little, Brown and Company, Jan Young, Jeff Denham, and Molly Faulkner nursed me along — and kept me well fed. Tim Kenslea's eye for detail kept me honest.

At home, Chega and Daniel kept me going.

Contents

PART I
COMMUNICATING WITH A SPECIFIED AUDIENCE   1

1
Introduction to Technical Writing   2

Definition   3
Technical versus Nontechnical Writing   3
The Value of Technical Writing Skills   6
Specific Uses of Technical Writing Skills   7
Audience Needs   8
Chapter Summary   8
Exercises   9

2
Writing for Readers   10

Definition   11
The Concept of "Audience"   11
Adapting the Message to Your Audience   13
Focusing on Your Reader's Needs   18
Sample Situations   20
Chapter Summary   23
Exercises   23

3
Making It Readable   26

Definition   28
Writing Effective Paragraphs   28
Writing Effective Sentences   39
Choosing the Right Words   50
Exercises   59

PART II
STRATEGIES FOR TECHNICAL REPORTING   65

4
Summarizing Information   66

Definition   67
Purpose of Summaries   67
Elements of an Effective Summary   69
Writing the Summary   71
Applying the Steps   72
Writing the Descriptive Abstract   79
Placing Summaries and Abstracts in Your Report   79
Chapter Summary   80
Revision Checklist   80
Exercises   81

5
Defining Your Terms   86

Definition   87
Purpose of Definitions   87
Using Definitions Selectively   88
Elements of an Effective Definition   89
Choosing the Best Type of Definition   90
Expanding Your Definition   94
Applying the Steps   97
Placing Definitions in Your Report   101
Chapter Summary   102
Revision Checklist   103
Exercises   104

6
Dividing and Organizing   110

Definitions   111
Using Partition and Classification   111
Guidelines for Partition   112
Guidelines for Classification   117
Applying the Techniques of Division   123
Chapter Summary   124
Revision Checklist   126
Exercises   127

7
Charting Your Course: The Outline   130

Definition   131
The Purpose of Outlining   131
Choosing the Best Type of Outline   132
Elements of an Effective Formal Outline   136
Composing the Formal Outline   140
The Report Design Worksheet   143
Chapter Summary   148
Revision Checklist   149
Exercises   150

8
Describing Objects and Mechanisms   152

Definition   153
Purpose of Description   153
Making Your Description Objective   156
Elements of an Effective Description   159
Organizing and Writing Your Description   161
Applying the Steps   172
Chapter Summary   172
Revision Checklist   179
Exercises   180

9
Explaining a Process   184

Definition   185
The Purpose of Process Explanation   185

Types of Process Explanation   186
Elements of Effective Instructions   187
Organizing and Writing a Set of Instructions   193
Applying the Steps   197
Elements of an Effective Process Narrative   198
Organizing and Writing a Process Narrative   207
Applying the Steps   208
Elements of an Effective Process Analysis   208
Organizing and Writing a Process Analysis   215
Applying the Steps   216
Chapter Summary   216
Revision Checklist   226
Exercises   227

10
Designing an Effective Format and Supplements   230

Definitions   231
Purpose of an Effective Format   232
Purpose of Report Supplements   235
Creating an Effective Format   235
Composing Report Supplements   239
Chapter Summary   256
Exercises   258

11
Visual Aids   260

Definition   261
Purpose of Visual Aids   261
Tables   262
Figures   265
Chapter Summary   278
Revision Checklist   279
Exercises   280

12
Researching Information   284

Definition   285
Purpose of Research   285
Identifying Information Sources   287
Taking Effective Notes   311

Chapter Summary   314
Exercises   315

13
Writing and Documenting the Research Report   318

Definition   319
Planning the Report   319
Gathering Information   321
Writing the First Draft   323
Documenting the Report   324
Writing the Final Draft   331
Chapter Summary   333
Exercises   341

PART III
SPECIFIC APPLICATIONS   345

14
Writing Effective Letters   346

Definition   348
Purpose of Letters   348
Elements of an Effective Letter   349
Writing Various Types of Letters   364
Writing the Résumé and Job Application Letter   372
Supporting Your Application   389
Chapter Summary   393
Revision Checklist   393
Exercises   394

15
Writing Informal Reports   400

Definition   401
Purpose of Informal Reports   401
Choosing the Best Report Form for Your Purpose   402
Composing Various Informal Reports   402
Chapter Summary   419
Revision Checklist   424
Exercises   424

16
Writing a Proposal   428

Definition and Purpose   429
The Proposal Process   429
Types of Proposals   431
Elements of an Effective Proposal   439
Planning and Writing the Proposal   447
Applying the Steps   453
Chapter Summary   470
Revision Checklist   471
Exercises   472

17
Analyzing Data and Writing a Formal Report   476

Definition   477
Purpose of Analysis   478
Typical Analytical Problems   478
Elements of an Effective Analysis   481
Finding, Evaluating, and Interpreting Data   484
Planning and Writing the Formal Report   487
Applying the Steps   496
Chapter Summary   518
Revision Checklist   539
Exercises   540

18
Oral Reporting   544

Definition   545
Purpose of Oral Reports   545
Identifying the Best Type of Formal Report   546
Preparing the Extemporaneous Delivery   547
Delivering the Extemporaneous Report   555
Chapter Summary   558
Revision Checklist   558
Exercises   559

Appendix A
Review of Grammar, Usage, and Mechanics   561

Sentence Parts and Types   561
Common Sentence Errors   567

Effective Punctuation 583
Transitions and Other Connectors 599
Effective Mechanics 603

Appendix B
The Brainstorming Technique 608

Appendix C
Eliminating Sexist Language 611

Index 615

COMMUNICATING
WITH A SPECIFIED
AUDIENCE

1

Introduction to Technical Writing

DEFINITION

TECHNICAL VERSUS NONTECHNICAL
WRITING

THE VALUE OF TECHNICAL WRITING
SKILLS

SPECIFIC USES OF TECHNICAL WRITING
SKILLS

AUDIENCE NEEDS

CHAPTER SUMMARY

EXERCISES

DEFINITION

In technical writing, you report factual information objectively for the practical use of your readers. This information is usually specialized and directed toward a specific group of people. It is based on your experiences, observations, and interpretations in a certain field. The purpose of technical writing is to inform and to persuade by providing facts (and opinions based on facts) that help readers answer a question, solve a problem, make a decision, or perform a task. Such writing does not seek to entertain, to create suspense, or to stimulate emotions. It does seek to give readers the information they need — clearly and objectively.

TECHNICAL VERSUS NONTECHNICAL WRITING

Poetry and fiction do not represent technical writing because they are expressions of imagination and feelings.[1] Poets or novelists write about *their* view of a subject. While the feelings and opinions expressed might be shared by many readers, they are not based on demonstrable fact. Thus not all readers might agree with the poet's view. Consider, for example, this poem by Tennyson.

THE EAGLE: A FRAGMENT

He clasps the crag with crooked hands;
Close to the sun in lonely lands
Ringed with the azure world, he stands.

---

[1] The ideas developed in this section were inspired by Patrick M. Kelley and Roger E. Masse, "A Definition of Technical Writing." *The Technical Writing Teacher* 4 (Spring 1977): 94–97.

> The wrinkled sea beneath him crawls;
> He watches from his mountain walls
> And like a thunderbolt he falls.

Although the eagle is often the subject of scientific and technical studies, here it is not described factually. Instead, it is described metaphorically (in emotionally or imaginatively suggestive terms): claws become "crooked hands"; the blue sky becomes "the azure world"; the sea is "wrinkled" and it "crawls"; "like a thunderbolt he falls" in his rapid dive for prey. Clearly, the poet is sharing *his own* imaginative perception of the eagle. He is obviously impressed by the bird's majestic and solitary nature ("Close to the sun in lonely lands"). He writes the poem to evoke the same awe in readers. Such writing is not technical writing because it is not an objective recording of facts. Other poets could describe the eagle in countless other ways, depending on their own feelings.

The nonfiction essay is also not technical writing because an essay, by definition, presents the writer's personal opinions. Consider, for instance, this single-paragraph essay about the eagle:

> The eagle is the *most noble* bird. This large and *impressive* creature perches on the highest cliffs, scanning the earth below. Against the light of the sun he presents a *dignified* silhouette, commanding *full sway* over his world from his solitary perch. On sighting his prey he dives with *swift* and *deadly accuracy*, *sure* and *self-reliant*. The eagle's *majestic demeanor*, *independence*, *pride*, and *invincible spirit* all symbolize basic American values. Thus it is no surprise that this *awesome* bird was chosen as the national emblem of the United States.

The paragraph above is highly subjective; again, the writer describes his own perception of the bird. Notice the number of judgmental terms (in italics). Any factual information contained here is overshadowed by the writer's opinions. Because many of his opinions might not be shared by other people, this writer's description is not technical. (Another writer, for example, might see the eagle as cold and brutal, or as ugly and dumb, or as just another bird to be hunted.)

In contrast to these nontechnical versions, here is a technical description of the eagle:

> **Eagle,** a large diurnal bird of prey that has been a symbol of power and courage since ancient times. The eagle is found throughout the world except in Antarctica and on a few remote oceanic islands. The bird is characterized by stout legs, strong feet with sharp talons, and a strongly hooked bill that is nearly as long as its head. The eagle has large, broad, strong wings, with ragged rounded tips, and a broad tail that in flight is spread like a fan. The female is larger than the male. The bird's eyes are highly developed, and

there are two focusing points (fovea) in each eye, so that it has binocular vision for distance sighting. The various species prey on mammals, reptiles, or fish. The pursuit flight is either a swooping down or direct diving on the prey. The eagle kills with its talons and dismembers with its bill. The nest, a huge bulky structure of sticks lined with grass or moss, is usually built on a cliff, but some species nest in trees.

The name eagle is derived from the Roman name for the golden eagle, *Aquila chrysaëtos.* This species, a typical eagle, was formerly found throughout most of the northern hemisphere. It is nearly extinct in most of the British Isles, and is now very rare east of the Rocky Mountains in North America. The adult golden eagle is blackish, with a golden wash on the back of the neck, and white at the base of the tail and in the wings. The legs and feet are feathered, but the toes are not. The golden eagle is 30 to 40 inches long, and its wingspan is 6 to 7 feet. It feeds chiefly on such small mammals as rodents, but a group will attack animals as large as antelopes. When living prey is scarce, the golden eagle feeds on carrion. A variety of golden eagle, *Aquila chrysaëtos canadensis,* is still found in remote mountain areas, foothills, and plains from northern Canada and Alaska south to the Gulf states and northern Mexico.

The bald eagle, *Haliaeetus leucocephala,* is named for its snow-white head. One of the sea eagles, it nests along fresh or salt waters in polar regions of the northern hemisphere, throughout most of the United States, and south into Mexico. In recent years the number of bald eagles has been much reduced, and they are now most numerous in Alaska. The adult is blackish brown, with a snow-white head and tail. The bald eagle has unfeathered feet and toes. It is 30 to 40 inches long, and it has a wingspan of 6 to 8 feet. It feeds mainly on fish; however, it catches very few itself, either pirating its food from other birds or picking up dead fish on the shore. In 1782, Congress adopted a design displaying the bird for the Great Seal of the United States, and the bald eagle became the national bird.[2]

This version is technical writing because it contains only factual information, presented objectively. All data can be verified and would not change unless new findings were made. As an expression of fact, this version gives readers dependable information to support later opinions, interpretations, conclusions, and recommendations. Thus the singular purpose of all technical writing is to provide facts, or opinions based on facts; the earlier versions in this chapter provided opinions without factual support.

We have just seen that a subject such as the eagle can be the theme of poetry, essay writing, or technical writing. Technical writing, therefore, is defined not by its subject but by the author's treatment of it and by the reader's

---

[2] Reprinted with permission from *Collier's Encyclopedia.* © 1971, Crowell-Collier Educational Corporation.

need for useful information. A technical subject is specialized and usually mechanical or scientific (e.g., a description of screw-thread gauging techniques), but many technical subjects — like the eagle — can be discussed from nontechnical points of view. Your responsibility as a technical writer is to observe, interpret, and report from a technical point of view, that is, on the basis of facts and solid evidence.

Examples of technical writing are found in most specialized textbooks. Other examples include assembly instructions for a stereo system, a car's service manual, a recipe for clam chowder, a police description of an accident, or the curricula listed in your college catalogue. Each gives objective information for the reader's practical use.

THE VALUE OF TECHNICAL WRITING SKILLS

Your ability to communicate useful information increases your chances for success. As jobs become more complex and specialized, the demand for effective communication grows. In fact, much of the communication in the working world is carried out in writing. Therefore, few of us can afford to write poorly.

Top corporation administrators consistently rank communications skills among their highest priorities in evaluating employees and candidates. Here are some of their comments:

> – Certainly it is not necessary that every man we hire be a finished public speaker or writer, but it is necessary that he be able to communicate. . . . Some of the reports that I have had occasion to read over the years would curl your hair, and as for oral presentation — many of them can charitably be called atrocious.
> – One of the chief weaknesses of many college graduates is the inability to express themselves well. Even though technically qualified, they will not advance far with such a handicap.
> – The ability to read and comprehend what one reads and the ability to translate orally are essential to communication. Communication is essential to controlling and directing people, and people (with the help of machines, but, I repeat, *people*) get the job done. . . . A man who can use good, plain, understandable English is worth more to me than a specialist.

Many employers are convinced that you can be trained on the job to perform specific tasks but, after college, you cannot be trained to communicate well. One company official even concluded that a candidate who lacks speaking and writing skills is "a lost cause." [3]

---

[3] These quotations are from Linwood E. Orange, *English: The Pre-Professional Major,* 2nd ed. (New York: MLA, 1973), pp. 4–5.

The message in these brief quotes is clear: technical expertise, motivation, and creativity alone are not enough. At the very least, you need to be a "part-time" technical writer.

Employers first judge your writing skills by the quality of your application letter and résumé. If you join a large organization, your retention and promotion may depend on decisions by executives you have never met. In this case, the quality of your letters, memos, progress reports, work orders, requisitions, recommendations, and instructions will be regarded as an indicator of the overall quality of your work. This is hardly the time to hide your competence beneath carelessly written reports. Good writing skills give you an advantage in any field. And as you advance, your ability to communicate increases in importance while your reliance on your technical background may correspondingly decrease. The higher your professional goals, the better communications skills you need. Your value to any organization depends on how well you can convey to others what you know.

SPECIFIC USES OF TECHNICAL WRITING SKILLS

> Because modern society is becoming increasingly technical, scientists, doctors, engineers, and a wide variety of other technical persons must keep informed of new advances being made by others working in their fields. A discovery made by one scientist may directly affect a problem on which a scientist in another state or country might be working. Instruction manuals must be developed to explain the operation and maintenance of the increasing number of machines in industry, business, and homes. New developments in technical fields are constantly affecting modern living and must be reported and explained to the businessman affected by them and to the general public. People are interested in learning about new inventions, drugs, or equipment which will affect their lives.[4]

Writing skills are important in just about any field, including the health and social sciences, agriculture, law enforcement, engineering, and business. Almost anyone in a responsible position writes accounts of his or her activities and findings, often daily.

Police and fire personnel write detailed incident or investigation reports that must be clear enough to serve as evidence in court. Nurses and medical technicians keep daily records that are crucial to patient welfare and, increasingly,

---

[4] Quoted from "Writer, Technical," *Occupational Brief,* No. 178 (Moravia, New York: Chronicle Guidance Publications, Inc., 1969). This edition was revised in 1974 and is available from Chronicle Guidance Publications, Inc., Moravia, New York 13118 for one dollar a copy.

a basis for litigation. Medical personnel also compile and present research data on various health questions. Executive, medical, and legal secretaries must write clear and precise memos, letters, minutes, and reports. Managers write memos, personnel evaluations, requisitions, and instructions. Contractors and tradespeople write detailed proposals, bids, and specifications for prospective customers. Moreover, they often trace, identify, and report sources of malfunction in structures or machinery. Engineers and architects plan, on paper, the structural details of a project before contracts are awarded and actual construction begins. They must also communicate with members of related fields before presenting a client with a detailed proposal. For example, the architect's plans are reviewed by a structural engineer who certifies that the proposed structure, as designed, is sound. Do not assume that the writing challenge ends with your last composition course. In fact, it only begins there.

AUDIENCE NEEDS

Whenever you write technical reports, you write for a specific audience, translating and interpreting data for the reader's use. Depending on your job and the purpose of your report, your audience may vary. Sometimes you will write for colleagues whose level of technical understanding is higher than or equal to your own. At other times you will write for administrators who are technically informed, but not experts. Much of your reporting may be aimed toward readers outside your organization who have little or no technical understanding. Your report may be read by one person or by dozens. In each case you will have to make your message clear and appropriate for a specific level of technical understanding — sometimes to more than one level. Techniques for reaching your audience are discussed in Chapter 2.

CHAPTER SUMMARY

Technical writing reports factual information objectively for the reader's practical use. The information is often specialized — coming from the writer's experiences, observations, and interpretations within a particular field. Nontechnical writing is subjective, emphasizing soft information (opinion). Technical writing is objective, emphasizing hard information (fact). To write technically means to express a technical point of view — a view based on convincing evidence — toward just about any subject, technical or not.

Much of the communication in the working world is carried out in writing. Therefore, at the very least, you need to be a "part-time" technical writer. In fact, your value to any organization depends mainly on your ability to communicate what you know.

Technical writing skills have application in countless career fields. Your own writing will be based on observable facts instead of unsupported opinion. Whenever you communicate these facts, you write for a specific audience, translating and interpreting data for your reader's use. Depending on your job and the purpose of your report, your reading audience may be experts, technically informed readers, or laypersons. In each case, you need to address a specific level of technical understanding.

EXERCISES

1. *In class:* Politics and religion are emotionally charged subjects that are generally discussed from a nontechnical — that is, subjective — point of view. Identify specific topics within these subjects that might be discussed from a technical point of view.

2. Controversial subjects like abortion, euthanasia, the Equal Rights Amendment, marijuana laws, and the gun-control issue usually elicit discussions that are based on emotions and opinions. How could these subjects be discussed from a technical point of view?

3. In a letter to a friend, describe your dominant impression of your classroom as a learning environment (cheerful, dreary, forbidding, distracting, or the like). Let your reader know how you *feel* about this room. Next, write a description such that a campus visitor could recognize this classroom immediately. In other words, describe the same physical details that anyone else would observe. Finally, explain, in writing, the changes you made in moving from description 1 to description 2 (from subjective to objective).

4. Locate a short example of technical writing in your library. Make a photocopy and bring it to class. Be prepared to explain why your selection can be called technical writing.

5. Interview a person who has a responsible, attractive, and well-paid position. Ask your respondent to describe in detail the kinds of oral and written communication that are part of the job. Write a brief report of your findings.

6. (a) In a brief essay, describe the kinds of "part-time" technical writing you expect to do in your field. Why will you write, and for whom will you write? (To find the answers, you may wish to speak with instructors or workers in your field.) (b) In another brief essay, describe the specific skills you hope to gain from your technical writing course. How do you plan to apply these skills in your career?

2

Writing for Readers

DEFINITION

THE CONCEPT OF "AUDIENCE"

ADAPTING THE MESSAGE TO YOUR
    AUDIENCE

  The Highly Technical Message
  The Semitechnical Message
  The Nontechnical Message
  Primary Versus Secondary Readers

FOCUSING ON YOUR READER'S NEEDS

  Reader Identification
  Purpose of the Request
  The Reader's Technical Background
  The Reader's Knowledge of the Subject
  Appropriate Details and Format
  Due Date

SAMPLE SITUATIONS

CHAPTER SUMMARY

EXERCISES

DEFINITION

All writing (at least all the writing discussed in this book) is for readers who will use your information for some purpose. You might write *to define* something — as to an insurance customer who wants to know what *variable annuity* means. You might write *to describe* something — as to an architectural client who wants to know what a new addition to her home will look like. You might write *to explain* something — as to a fellow stereo technician who wants to know how to eliminate base flutter in your company's new line of speakers. You might write *to persuade* someone — as to your vice-president in charge of marketing who wants to know if it's a good idea to launch an expensive advertising campaign for a new oil additive. In short, as a technical writer, you do not write for yourself. You write to inform others.

In any kind of factual writing, you must connect with the intended readers — sometimes one reader, sometimes a few, sometimes many. Your goal is always to communicate useful information clearly. If your writing confuses readers, alienates them, bores them, or wastes their time, then it has not done its job.

THE CONCEPT OF "AUDIENCE"

Because different readers have different backgrounds, certain information that makes sense to you might confuse others unless it is explained at their level of understanding. Your information, then, is useful only if it makes sense to your audience.

There are numerous examples all around us of the differences in audience knowledge and needs. Watching and discussing a TV football game with ex-players or fans, for instance, is different from watching it with someone who knows nothing about the game. Terms such as *power sweep, blitz,* and *shot-gun formation* are clear to expert or informed viewers, but they have to be defined or illustrated for uninformed audiences. Explaining how to repair a leaking faucet to someone with no mechanical background is different from explaining it to a do-it-yourselfer. Undoubtedly you can think of occasions when you've had to clarify or explain a message for a less specialized audience.

Good writing connects with its readers by recognizing their differences in background and their specific needs. In some cases, a single message may have to be written in several versions for several audiences. A highly technical ar-ticle about a new treatment for heart disease, for instance, might first appear in the *Journal of the American Medical Association,* which will be read by doc-tors and nurses. The same article may later be rewritten less technically for a cardiology textbook, which will be read by medical and nursing students. Finally, a nontechnical version might appear in the *Reader's Digest.* Though the three versions treat the same subject, each is adapted to the needs of its particular audience.

Technical writing is intended to be *used.* Whether you are giving instruc-tions, describing a product, reporting on research, or defining a specialized term, you become the teacher and the reader becomes the student. Because your readers will know less than you do in this instance, you can expect them to have certain questions about your subject:

- "What is it?"
- "What does it look like?"
- "How do I do it?"
- "How did you do it?"
- "Why did it happen?"
- "When will it happen?"
- "Why should we do it?"
- "How much will it cost?"
- "What materials and equipment does it require?"
- "What are the dangers?"

These are just some of the questions that various reports are designed to an-swer. (Later chapters treat these questions in detail.) All readers need answers they can immediately understand and effectively put to use.

Think of your readers' needs as you think of your own. If chapter 1 of your introductory math textbook, for example, covered differential equations (ad-vanced calculus), the author would have ignored your needs. For your background and purposes, the chapter would be useless. How clearly any message answers its readers' questions depends on how precisely it is adapted to their level of understanding.

ADAPTING THE MESSAGE TO YOUR AUDIENCE

When you write for a close acquaintance (friend, fellow camera buff, classmate in experimental psychology, fellow dental technician, engineering colleague, chemistry professor who reads your lab reports), you know a good deal about your particular reader. Therefore, you automatically adapt your report to that particular reader's knowledge and needs. There are times, however, when you have to write for less clearly defined audiences. This is particularly true when dealing with large audiences (when you are writing for a certain magazine or professional journal, or when you are asked to explain your company's stock-options program to members of many different departments). In such situations, where you have only general knowledge about your audience's background, you must decide whether your message should be *highly technical, semitechnical,* or *nontechnical.*

The Highly Technical Message

Readers whose specialized training is similar to your own expect technical data, without long background explanations or detailed interpretations. To illustrate, the following account of emergency treatment given a heart attack victim is a technical message. The writer, a doctor on call in the emergency room, is reporting to the patient's doctor. This reader needs an exact record of his patient's symptoms and treatment.

> Mr. X was brought to the emergency room by ambulance at 1:00 A.M., September 27, 1977. The patient complained of severe chest pains, shortness of breath, and dizziness. Auscultation and electrocardiogram revealed a massive cardiac infarction and pulmonary edema marked by pronounced cyanosis. Vital signs were as follows: blood pressure, 80/40; pulse, 140/min.; respiration, 35/min. Lab tests recorded a wbc count of 20,000, an elevated serum transaminase and a urea nitrogen level of 60 mg%. Urinalysis showed 4+ protein and 4+ granular cast/field, suggesting acute renal failure secondary to the hypotension.
>
> The patient was given 10 mg of morphine stat, subcutaneously, followed by nasal oxygen and a 5% D & W IV. At 1:25 A.M. the cardiac monitor recorded an irregular sinus rhythm, suggesting left ventricular fibrillation. The patient was defibrillated stat and given a 50 mg bolus of Xylocaine IV. A Xylocaine drip was started, and sodium bicarbonate was administered until

a normal heartbeat was established. By 3:00 A.M., the oscilloscope was recording a normal sinus rhythm.

As the heartbeat stabilized and cyanosis diminished, the patient was given 5 cc of Heparin IV, to be repeated every six hours. By 5:00 A.M. the BUN had fallen to 20 mg% and the vital signs had stabilized as follows: blood pressure, 110/60; pulse, 105/min.; respiration, 22/min. The patient was now conscious and responsive.

Written at the highest level of technicality, this report is clear only to the medical expert. The writer correctly assumes that her reader has the specialized knowledge to understand the message. Therefore, she does not define technical terms (pulmonary edema, sinus rhythm, etc.). Nor does she interpret lab findings (4+ protein, elevated serum transaminase, etc.). She uses abbreviations that she knows are familiar to her reader (wbc, BUN, 5% D & W IV). Because her reader knows the reasons for specific treatments and medications (defibrillation, Xylocaine drip, etc.), she includes no theoretical background. Her report is designed to provide concise answers to the main questions she can anticipate from her particular reader: "What happened?" "What course of treatment was given?" "What were the results?"

Other highly technical messages are found in specialized journals. A forestry researcher, for instance, might publish an article in the *Journal of Wood Technology* to describe the results of a new seeding technique. The article might consist mainly of charts, graphs, and equations. Untranslated, these data would mean nothing to the nontechnical reader.

The Semitechnical Message

Readers whose specialized training is somewhat less than your own expect to have the background and interpretations of technical data spelled out. Semitechnical messages may address a broad array of readers. First-year medical students, for example, have a basic technical understanding, but second-, third-, and fourth-year students have increasingly higher levels of understanding. Yet in many cases students in all four groups can be considered semitechnical readers. Therefore, when you write for a semitechnical audience, identify the *lowest* level of understanding among the group, and write to that level. You are wiser to give too much explanation than too little.

Here is a partial version of the earlier medical report. Written at a semitechnical level, it might appear in a textbook for first-year medical or nursing students, in a report for a medical social worker, in a patient history for the medical technology department, or in a monthly report for the hospital administration.

Examination by stethoscope and electrocardiogram revealed a massive failure of the heart muscle along with fluid build-up in the lungs, which produced a cyanotic discoloration of the lips and fingertips from lack of oxygen.

The patient's blood pressure at 80 mm Hg (systolic)/40 mm Hg (diastolic) was dangerously below its normal measure of 130/70. A pulse rate of 140/minute was almost twice the normal rate of 60–80. Respiration at 35/minute was over twice the normal rate of 12–16.

Laboratory blood tests yielded a white blood cell count of 20,000/cu mm (normal values: 5,000–10,000), indicating a severe inflammatory response by the heart muscle. The elevated serum transaminase enzymes (only produced in quantity when the heart muscle fails) confirmed the earlier diagnosis. A blood urea nitrogen level of 60 mg% (normal values: 12–16 mg%) indicated that the kidneys had ceased to filter out metabolic waste products. The 4+ protein and casts reported from the urinalysis (normal values: 0) revealed that the kidney tubules were degenerating as a result of the lowered blood pressure.

The patient was immediately given morphine to ease the chest pain, followed by oxygen to relieve strain on the cardiopulmonary system, and an intravenous solution of dextrose and water to prevent shock.

This version defines, interprets, and explains the significance of the raw data. Exact dosages here are not significant to these readers because they are not treating the patient; therefore, dosages are not mentioned. Normal values of lab tests and vital signs, however, make interpretation easier. (Highly technical readers would know these values, as well as the significance of specific dosages.) Knowing what medications the patient received would be especially important to the lab technician, because some medications affect blood test results. For a nontechnical audience, however, the above message needs even further translation.

The Nontechnical Message

Readers with little or no specialized training in your field expect technical data to be translated into their simplest form. Nontechnical readers are impatient with abstract theories but want enough background to help them make the right decision or take the right action. They are bored by long explanations but frustrated by bare facts not explained or interpreted. They want to understand a message without having to become students of the discipline. They expect a report that is clear on first reading rather than one that requires review or study.

Following is a nontechnical version of the earlier medical report. The attending physician might write this version for the patient's spouse who is on a business trip overseas, or as part of a script for a documentary film about emergency room treatment.

> Both heart sounds and electrical impulses were abnormal, indicating a massive heart attack caused by failure of a large part of the heart muscle. His lungs were swollen with fluid and his lips and fingertips showed a bluish discoloration from lack of oxygen.
>
> The patient's blood pressure was dangerously low, creating the danger of shock. His pulse and respiration were almost twice the normal rate, indicating that the heart and lungs were being overworked in keeping oxygenated blood circulating freely.
>
> Blood tests confirmed the heart attack diagnosis and indicated that waste products usually filtered out by the kidneys were building up in the bloodstream. Urine tests showed that the kidneys were failing as a result of the lowered blood pressure.
>
> The patient was given medication to ease his chest pain, oxygen to ease the strain on his heart and lungs, and an intravenous solution to prevent his blood vessels from collapsing and causing irreversible shock.

This nontechnical version translates all specialized information into simple terms. It mentions no specific medications, lab tests, or normal values because these would mean nothing to the nontechnical reader. The writer simply summarizes the events and explains the causes of the crisis and the reasons for the particular treatment.

In some other situation, however (say, in a jury trial for malpractice), the nontechnical audience might need information about specific medication and treatment. Such a situation would call for a much longer report — a kind of mini-course in emergency room treatment of coronary patients.

Each version of the medical report is useful *only* to readers at a particular level of technical understanding. Doctors and nurses have no need of the explanations in the two latter versions, but they do need the specialized data in the first. Medical students and paramedics might be confused by the first version and bored by the third. Nontechnical readers would find both the first and second versions meaningless.

For another illustration of these differences in delivery consider the hypothetical article about the new forest-seeding technique mentioned previously. A less specialized version might appear as a set of instructions for growing trees, in an ecology, agriculture, or gardening magazine whose readers are informed but not expert. Here the scientific findings in the earlier article would be applied to practical ends — with an emphasis on procedure ("how to") instead of theory. An even more simplified version might be found in a magazine with a general readership, like *Outdoor Life*. Instead of discussing scientific and technical details of the seeding technique, this version might show how the new method will help rejuvenate our depleted forests.

In your own job — as doctor, nurse, medical technician, accountant, dental technician, building contractor, banker, sales manager, or engineer — you will need to connect with audiences at various levels. As a building contractor, for instance, you might order "800 board feet of 1 × 6 no. 2 white pine, rough sawn and kiln dried" from your lumber dealer (a wood expert). For the bank financing the project you might describe the wood as "rough pine interior finish." For the client you might write "new barn-board interior walls." Each version is clear and appropriately detailed for its intended audience.

Primary versus Secondary Readers

When you write the same basic message for readers at different levels of technicality, it helps if you classify those readers as *primary* or *secondary*. The primary readers usually are those who have requested the report and who will probably use it as a basis for decisions or actions. The secondary readers are those who will carry out the project, who will advise the primary reader about his or her decision, or who will somehow be affected by this decision. They will read your report for information that will help them to get the job done, to provide educated advice, or to remain abreast of new developments.

As often as not, the technical background of these two audiences will differ. Primary readers may require highly technical messages, while secondary readers may need semitechnical or nontechnical messages — or vice versa. When you must write for audiences at different levels, follow this rule of thumb:

1. If your report is short (a letter, memo, or anything less than a few pages), rewrite it at various levels for various readers.

2. If your report is longer than several pages, maintain a level of technicality that connects with your primary readers. Then supplement the report with appendixes that address the secondary readers (technical appendixes when secondary readers are technical; or vice versa). Letters of transmittal, informative abstracts, and glossaries are other supplements that help less specialized audiences understand a highly technical report.

The following situation illustrates how certain reports must be adapted to both primary and secondary readers.

You are a metallurgical engineer in a Detroit consulting firm. Your supervisor has asked that you test the fractured rear axle of a 1981 Delphi pickup truck recently involved in a fatal accident. Your job is to determine whether the axle fracture was the cause or a result of the accident.

After testing the hardness and chemical composition of the metal and examining a series of microscopic photographs of the fractured surfaces (fractographs), you conclude that the fracture resulted from stress that developed *during* the accident.

Because your report may serve as evidence in a court case, you must explain your findings in meticulous detail. But your primary readers will be nonspecialists (attorneys, insurance representatives, possibly a judge and a jury), so you will have to translate your report, explaining the principles behind the various tests, defining specialized terms such as "chevron marks," "shrinkage cavities," and "dimpled core," and showing the significance of these features as evidence.

Secondary readers will include your supervisor and outside consulting engineers who will be evaluating your test procedures and assessing the validity of your findings. For these readers you will have to include appendixes that spell out the technical details of your analysis: *how* hardness testing of the axle's case and core indicated that the axle had been properly carburized; *how* chemical analysis ruled out the possibility that the manufacturer had used inferior alloys; *how* light-microscope fractographs revealed that the origin of the fracture, its propagation direction, and the point of final rupture indicated a ductile fast fracture, not one caused by torsional fatigue.

In the above situation, the primary readers need to know *what your findings mean*, whereas the secondary readers need to know *how you arrived at your conclusions*. Unless the needs of each group are served independently, your information will be worthless.

FOCUSING ON YOUR READER'S NEEDS

In addition to choosing the appropriate level of technicality for your message, you need to make other decisions as well. Your choice of *what* you write (content) and *how* you write it (format, arrangement, style) is guided by what you know about your readers beforehand. Learn all you can about your audience *before* you write.

When you write for a particular reader or a small group of readers, you can focus sharply on your audience by asking yourself specific questions:

1. Who wants the report? Who else will read it?
2. Why do they want the report? How will they use it?
3. How much technical background does the primary audience have? The secondary audience?
4. How much does the audience already know about the subject?
5. What exactly does the audience need to know, and in what format?
6. When is the report due?

Follow the suggestions on the next two pages for answering these questions.

Reader Identification

Identify the primary readers by name, job title, and specialty (Martha Jones, Director of Quality Control, B.S. and M.S. in mechanical engineering). Are they superiors, colleagues, or subordinates? Are they inside or outside your organization? Are they apt to agree or disagree with your conclusions and recommendations?

Although your report must satisfy the needs of primary readers, it should not ignore the needs of any secondary readers. Identify those additional readers who are likely to be interested in or affected by your report, or who will affect the primary reader's perception or use of your report.

Purpose of the Request

Find out why your readers want the report and how they will use it. Ask them directly. Do they merely want a record of your activities or progress? Are you expected to supply only raw data or conclusions and recommendations as well? Will your readers take immediate action based on your report? Do they need step-by-step instructions? Are they simply collecting information for later use? Will the report be read and discarded, filed, or published? The more you learn about your audience's exact expectations and needs, the more useful you can make your report.

The Reader's Technical Background

Colleagues on your project speak your technical language and can understand raw data. Supervisors responsible for several technical areas may want interpretations and recommendations. Managers may have only vague technical knowledge and thus will expect definitions and background explanations. Clients who have no technical background will need writing in plain English. Keep in mind, however, that none of these generalizations may apply to your specific situation. Find out for yourself. By assessing your readers' technical background, you can avoid insulting their intelligence or writing over their heads. Make the level of technicality appropriate for your primary audience. For secondary audiences, rewrite the short report or include supplements with the long report.

The Reader's Knowledge of the Subject

Don't waste your time and your audience's by rehashing information that readers already have. If preliminary reports on your subject have been written earlier, a brief summary of their major points will do. Information is not worthwhile unless it covers new ground.

Appropriate Details and Format

In the earlier medical reports we saw that dosages, drug names, and medical terminology are significant to some readers but not to others. The amount and kind of detail in your report will depend on what you have learned about your readers and their purposes. Were you asked to "keep it short" or to "be comprehensive"? Can you summarize certain items or do they all need to be spelled out? Are your primary readers most interested in conclusions and recommendations or do they want a full description of your investigative procedure as well? Have they requested a letter, a memo, a short report, or a long, formal report with supplements (title page, table of contents, appendixes, etc.)? Will visuals (charts, graphs, drawings, photographs, etc.) make certain information more accessible? What level of technicality will connect with your primary readers?

High Technicality	The diesel engine generates 10 BTUs per gallon of fuel, as opposed to the conventional gas engine's 8 BTUs.
Low Technicality	The diesel engine yields 25 percent better gas mileage than its gas-burning counterpart.

Sometimes, in writing for a large audience, you will be unable to identify all its members. In such cases, write for the least specialized members of the audience. It is safer to risk boring some experts than to write over other people's heads.

Due Date

Is there a deadline for your report? Find out immediately. If you do have a deadline, don't procrastinate. Give yourself plenty of time to collect data, to write the report, and to revise it. If possible, ask your primary readers to review the preliminary draft and to suggest improvements. A report dashed off at the eleventh hour is not likely to impress anyone.

SAMPLE SITUATIONS

In most jobs you will have to ask questions about your audience's needs often — perhaps daily. Here are two scenarios in which part-time technical writers have to assess the various needs of their various audiences.

A DAY IN THE LIFE OF A POLICE OFFICER

You are a police officer in the burglary division of an urban precinct.

– A local community college has asked you to lecture on the various fields of investigative police work during its career week. Your audience will be interested laypersons, possibly those considering a career in police work. You know they've been captive audiences at "lectures" for years, so you decide to liven things up with visuals. The police artist agrees to draw up a brightly colored flip chart with illustrations to accompany your summaries of each investigative area.

– Tomorrow in court you will testify for the prosecution of a felon whom you caught red-handed last month. His lawyer is a sly character who is known for making police testimony look foolish, so you are busy reviewing your notes and getting your report in perfect form. Your audience will consist of legal experts (judge, lawyers) as well as laypersons (jury). The key to a convincing testimony will be *clarity, accuracy,* and *precision,* with objective and factual descriptions of *what* happened as well as *when* and *where.*

– You are drafting an article about new fingerprinting techniques for a law enforcement journal. Your readers will be expert (veteran officers) and informed (junior officers). Because they understand "shoptalk" and know about the theory and practice of fingerprinting they won't need extensive background information or definitions of specialized terms. Instead, they will read your article to learn about a new *procedure.* You will have to convince them that your way is better. You decide to discuss the merits of your technique before explaining the process and its results. If your article is well received, you will later write a manual that gives instructions for using this technique.

– Your chief asks you to write a manual of investigative procedures for junior officers in your division. The manual will be bound in pocket size and carried by all junior officers on duty. Your audience is informed, but not expert. They have studied the theory at the police academy, but now they need the nitty-gritty "how to." They will want rapid access to the various instructions for handling various problems, with each step spelled out. Most likely, they will use your instructions as a guide to *immediate* actions and responses. Therefore, warnings and cautions will have to be clearly labeled *before* the steps they apply to. Clarity throughout will be imperative.

– Next month you will speak before the chamber of commerce on "Protecting Your Business Against Burglary." The audience will be highly interested laypersons. Because burglary protection can be costly (alarm systems, guard dogs, etc.), the audience is apt to be most interested in the less expensive precautions they can take. They will want to remember your advice, so you decide to supplement your talk with visuals: specifically, a checklist for business owners that you will discuss item-by-item by using an overhead projector. The division secretary agrees to type the checklist and make copies for distribution at the end of your talk.

A DAY IN THE LIFE OF A CIVIL ENGINEER

You are a civil engineer in a materials-testing lab who is in charge of friction studies on various road surfaces.

— Your vice-president calls to say that your firm is competing for a state contract for a safety study of state highway surfaces. Full reports of your testing procedures, progress, and results are needed for inclusion in your firm's contract bid. As engineer in charge, you are responsible for the quality of your firm's proposal. Your audience consists of experts (state engineers), informed readers (officers in the highway department), and laypersons (members of the state legislature). The legislature ultimately will decide which firm gets the contract, but they will act on the advice of engineers and the highway department. Therefore, you decide to keep your proposal at a level of technicality that will connect with the more specialized audience. The legislative committee will read only conclusions and cost data.

— During another call, the vice-president asks you to write safety instructions for building crews on the new bay bridge. Your instructions will be read by all crew members, from project engineer to manual laborer, but when it comes to safety on this kind of job, even the specialized personnel may be considered laypersons. Therefore, you decide to keep the instructions simple and clear. To increase reader interest, you ask the graphics department to design a brochure for your instructions, including sketches and photographs of the hazardous areas and situations.

— At next month's engineering convention in Dallas you are scheduled to deliver a paper describing a new road-footing technique that reduces frost-heave damage in northern climates. (This paper will soon be published in an engineering journal.) Your audience will be civil engineering colleagues who are highly interested in your new procedure. This expert audience wants to know what it is and how it works — without lengthy background. You decide to compress your data by using visuals (cross-sectional drawings, charts, graphs, maps, and formulas).

— You draft a letter to a colleague you've never met to ask about her new technique for increasing the durability and elasticity of asphalt surfaces. Your expert reader is a busy professional with little time to read someone's life story, but she will need *some* background about your own work in this area, along with clear, specific, and precise questions that she can answer quickly. As a gesture of goodwill, you close your letter with an offer to share *your* findings with her.

— You are putting the finishing touches on an article for a national magazine. The article describes the hazards encountered in building the trans-Alaskan highway. Your audience consists almost entirely of laypersons who are reading for general knowledge (and perhaps even for entertainment). To hold the reader's interest, you decide to focus on twelve dramatic incidents (avalanches, earthquakes, etc.). To further increase dramatic appeal, you decide to relate the factual details of each event in first-person narrative form, as in a journal or diary (e.g., "Friday, March 12: This morning's wind-chill factor lowered the temperature at our excavation site to $-55°F$").

In each situation above, the writers gave first priority to their audience's need for clear, appropriate, and useful information. What they could learn about each audience beforehand guided their choice of what was appropriate and necessary.

Much of your writing in this course — and much of it on the job — will be aimed at general readers, but even when your readers are informed or expert, they will know less than you do about the particular subject of your report. Whereas in college you write for an audience (the professor) who knows more than you and who is *testing* your knowledge, on the job you write for an audience who knows less and who is *using* your knowledge.

CHAPTER SUMMARY

Whenever you write, make your audience your first concern. Assume that readers will use your information for practical purposes. Readers whose specialized training is similar to yours expect a highly technical message. Those who are less specialized expect a semitechnical message. Readers with no technical training expect a nontechnical message.

When you write a single report for readers at different levels of specialization, classify them as *primary* or *secondary*. Rewrite short reports at various levels for various readers. Aim a long report at your primary audience, adding appendixes and other supplements for secondary readers.

Learn all you can about your audience, *before you write,* by answering these questions:

1. Who wants the report? Who else will read it?
2. Why do they want the report? How will they use it?
3. How specialized is the primary audience? The secondary audience?
4. How much does the audience already know about the subject?
5. What exactly does the audience need to know, and in what format?
6. When is the report due?

Ask yourself these questions whenever you have to be a part-time technical writer.

EXERCISES

1. In a short essay, identify and discuss the kinds of writing and speaking assignments you expect in your career. (If necessary, interview someone in your field for the information.) For whom will you be writing (colleagues,

supervisors, clients, etc.)? Will you need to address both primary and secondary audiences? How will your readers use your information? For which audience are you likely to write most often?

2. Review the questions for audience analysis on page 18 and the scenarios on pages 21–22. Then ask a member of your field to describe five or six situations in which he or she has had to be a part-time technical writer. Finally, based on your notes from the interview, compose a scenario for a typical person in your field. Use the sample scenarios as models for your own.

3. Locate a short article from your field. (Or select part of a long article or a section from one of your textbooks for an advanced course.) Choose a piece written at the highest level of technicality that you can understand and then translate the piece for a layperson. Exchange translations with a classmate from a different major. Read your neighbor's translation and write a paragraph evaluating the appropriateness of its level of technicality. Submit to your instructor a copy of the original, your translated version, and your evaluation of your neighbor's translation.

4. Assume that a new employee is taking over your job because you have been promoted to another one. Identify a specific problem in your old job that could cause difficulty for the new employee. Write the employee a set of instructions for avoiding or dealing with the problem. Before writing, perform an audience analysis by posing and answering (on paper) the questions on page 18. Submit to your instructor both your audience analysis and your instructions.

3

Making It Readable

DEFINITION

WRITING EFFECTIVE PARAGRAPHS
Paragraph Length
Paragraph Structure
The Introduction
The Body
The Conclusion
Structural Variation
Paragraph Unity
Paragraph Coherence
Paragraphs Developed in Logical
Sequence
Spatial Sequence
Explanation Sequence
Chronological Sequence
Examples Sequence
Effect-to-Cause Sequence
Cause-to-Effect Sequence
Definition Sequence
Reasons Sequence
Problem-Causes-Solution Sequence
Comparison/Contrast Sequence

WRITING EFFECTIVE SENTENCES
Making Sentences Clear
Omit No Key Words
Avoid Ambiguous Phrasing
Avoid Overstuffing
Unstack the Modifiers
Rearrange Word Order
Use Active and Passive Voice
Selectively
Making Sentences Concise
Eliminate Redundancy
Avoid Needless Repetition
Avoid "There" Sentence Openers
Avoid Certain "It" Sentence Openers
Delete Needless "to Be" Constructions
Avoid Excessive Prepositions
Use "That" and "Which" Sparingly
Fight Noun Addiction
Make Negatives Positive
Clear Out the Clutter Words
Delete Needless Prefaces
Delete Needless Qualifiers
Making Sentences Fluent
Combine Related Ideas
Vary Sentence Construction and
Length
Use Short Sentences for Special
Emphasis

CHOOSING THE RIGHT WORDS
  Keeping the Language Simple
    Deflate the Diction
    Avoid Needless Jargon
  Being Convincing
    Avoid Triteness
    Avoid Overstatement
    Avoid Sweeping Generalizations
    Avoid Euphemisms
  Being Precise
  Being Concrete and Specific
  Remaining Unbiased

EXERCISES

DEFINITION

You may be called upon to write for a broad or a specific audience, or for an expert or nonexpert audience. In every case, your audience's needs will not be served unless your report is easy to follow and understand — a report that is *readable*. Information not clear and appropriate for a particular reader is useless. To make your reports readable, be sure that (1) paragraphs are logically structured; (2) sentences are clear, concise, and fluent; and (3) words are chosen with care.

WRITING EFFECTIVE PARAGRAPHS

A paragraph is an orderly and logical arrangement of sentences designed to express an idea and to explain it with supporting details. Most well-written paragraphs should be able to stand alone as complete and meaningful messages. In fact, a good deal of your job-related writing will likely consist of one- or two-paragraph messages (as shown in the memo examples in Chapter 15).

Effective paragraphs help you stay in control of your writing. They provide individual spatial units for discussing a main idea in detail. For example, if you begin your paragraph by describing a work accident you can easily stay on course. All words, phrases, and sentences within that paragraph can be tailored to tell your readers what they need to know about the accident. Thus, if you find yourself arguing for longer lunch breaks by the third sentence you know that you have strayed.

Effective paragraphs also make your message more readable. Visually, they segment a whole page into smaller units that your reader can easily digest.

Logically, they help your reader follow your reasoning. The indentation on your page provides a breathing space. It is a signal that one "central-idea block" has ended and another is beginning.

Paragraph Length

There is no rule that tells how long a specific paragraph should be. We can only say that a paragraph should be long enough to tell the readers what they need to know and to reflect the content and emphasis the writer wants to achieve. However, a few general guidelines are worth keeping in mind.

1. A series of short paragraphs (one or two sentences each) makes your writing seem choppy and poorly organized. Imagine, for example, the previous section divided into eight or ten paragraphs instead of three. A series of short paragraphs, however, *is* effective in a set of step-by-step instructions.

2. A paragraph that is too long can easily obscure an important idea buried somewhere in the middle. A series of long paragraphs can make your material difficult to follow.

3. A single-sentence paragraph can be a good way of calling your reader's attention to an important idea by setting it off from the rest of your text.

> In summary, we can prevent further damage from mud slides only by building a retaining wall around our #3 construction site immediately.

4. Unless your paragraph is patterned as a list (like this one), in general it should be no longer than fourteen lines.

5. A combination of shorter and longer paragraphs is most effective, if your subject lends itself to this kind of arrangement.

Paragraph Structure

In technical writing, paragraphs usually follow an introduction-body-conclusion pattern. With this pattern you move from the general to the specific by (1) stating your central idea in a topic sentence, (2) supporting and explaining your idea with specific details and examples, and (3) tying your paragraph together with a concluding statement or summary. Each part is discussed in detail below.

The Introduction

Your topic sentence introduces your central idea. It sets boundaries by promising what your paragraph will deliver — no more and no less. Before you can write a good topic sentence, you must identify your purpose, based on what you know of your reader's needs. Then you can tailor your topic sentence to meet those needs.

Assume, for instance, that you're writing about whales for readers whom you'd like to recruit as members of the Save-the-Whales Foundation. First, you must decide exactly what point you want to make about whales. And when that point becomes part of your topic sentence, it must provide enough direction for you to develop a worthwhile paragraph. Avoid topic sentences that lead nowhere:

> Whales are a species of mammal.
> Whales live only in salt water.

Also, the point made in your topic sentence must be focused enough to be covered in one paragraph. Avoid broad and abstract topic sentences like this one:

> Whales are interesting animals.

What is meant by "interesting"? Their breeding habits, their migration patterns, the way they exhibit intelligence, or something else?

Let's say you decide to discuss whale intelligence. What exact point do you want to make about whale intelligence?

> Whales seem to exhibit some intelligence.
> Whales are fairly intelligent.
> Whales are highly intelligent.

You decide that the final sentence above expresses your point most accurately. Now you think of ways to make this topic sentence more informative. After all, your readers will be asking: "Highly intelligent relative to what?" So you decide to relate whales' intelligence to that of other mammals:

> Whales are among the world's most intelligent mammals.

You now have a clear direction for developing support in the body section.

Let's look at some other directions your topic sentence might have taken:

> A good indication of whales' high intelligence is the way they play in game-like patterns.
>
> *or*
>
> Like children, a group of whales can spend hours playing tag.

Depending on your purpose and your reader's needs, you can make any main point more and more specific by focusing on smaller and smaller parts of it. The paragraph should then deliver what the topic sentence promises.

The Body

The body of your paragraph contains the supporting details that explain and expand on your central idea. This support material answers the questions that you can expect readers to have about your topic sentence:

> Says who?
> What proof do you have to support your claim?
> Can you give examples?

To answer these questions, you brainstorm your topic (see Appendix B) and make a list of everything you know about whale intelligence:

1. Whales help wounded fellow whales.
2. They communicate through sonar clicks and pings.
3. Scientists have studied whales.
4. They exhibit complex group behavior.
5. They teach and discipline their young.
6. They use marine vegetation, in addition to animal life, as a food source.
7. Scientists compare whales' intelligence with that of higher primates.
8. They follow fixed patterns of annual migration.
9. They engage in elaborate sexual foreplay.
10. They play in gamelike patterns.
11. They have a sophisticated communications system.

After selecting those facts that support the main idea,[1] you arrange them in related categories. Looking at your list, you notice that items 3, 4, 7, and 11 are major points, while the remaining items provide specific support for these points. You now have three categories of general evidence, which are further supported by specific details. When you combine and arrange this material, your paragraph looks like this:

> Whales are among the world's most intelligent mammals. Scientists study-ing whales rate their intelligence on a level with higher primates because of their complex group behavior. These impressive mammals have been seen teaching and disciplining their young, helping wounded fellow whales, en-gaging in elaborate sexual foreplay, and playing in definite gamelike pat-terns. They are able to coordinate such complex activities because of a highly effective communications system of sonar clicks and pings.

With your topic sentence and supporting details on paper, you are ready to write your conclusion.

The Conclusion

Your concluding statement signals readers that the discussion of the central idea stated in your topic sentence is ending. It usually ties the paragraph to-gether by summarizing the discussion. If the paragraph is part of a longer

---

[1] Items 6 and 8 can be deleted from the list because they are not functions of a whale's intelligence. The former is a function of physiology while the latter is instinct.

report, your conclusion can also prepare readers for a subsequent paragraph. Assume, for example, that your paragraph on whale intelligence is to be followed by one describing abuses in the whaling industry. Here is an effective conclusion for the first paragraph:

> All in all, scientific evidence shows that whales have a high level of social organization. Unfortunately, the whales' intelligence is ignored by an industry that threatens them with extinction.

The final statement in this conclusion serves as a transition to the next major idea.

An introduction-body-conclusion structure should serve most of your paragraph needs in report writing. Begin each paragraph with a solid topic sentence and you will stay on target.

Structural Variation

Sometimes a central idea that needs detailed definition might call for a topic statement of two or more sentences (as shown in the needle-description paragraph, on page 35). On the other hand, your central idea might have several distinct parts, which would result in an excessively long paragraph. In this case you might break up the paragraph, letting your topic statement stand as a brief introductory paragraph and serve the separate subparts, which are set off as independent paragraphs for the readers' convenience.

> **The most common types of strip-mining procedures are open-pit mining, contour mining, and auger mining. The specific type employed will depend on the type of terrain covering the coal.**
>
> Open-pit mining is employed in the relatively flat lands in Western Kentucky, Oklahoma, and Kansas. Here, draglines and scoops operate directly on the coal seams. This process produces long, parallel rows of packed spoil banks, ten to thirty feet high, with steep slopes. Between the spoil banks are large pits which soon fill with water to produce pollution and flood hazards.
>
> Contour mining is most widely practiced in the mountainous terrain of the Cumberland Plateau and Eastern Kentucky. Here, bulldozers and explosives cut and blast the earth and rock covering a coal seam. Wide bands are removed from the mountain's circumference to reach the embedded coal beneath. The cutting and blasting form a shelf along with a man-made cliff some sixty feet high at a right angle to the shelf. The blasted and churned earth is pushed over the shelf to form a massive and unstable spoil bank which creates a danger of mud slides.
>
> Auger mining is employed when the mountain has been cut so thin that it can no longer be stripped. It is also used in other difficult-access terrain. Here, large augers bore parallel rows of holes into the hidden coal seams in

order to extract the embedded coal. Among the three strip-mining processes, auger mining causes the least damage to the surrounding landscape.

Each paragraph begins with a clear statement of the part of the central idea to be discussed.

Paragraph Unity

A paragraph is unified when all its parts work toward the same end — when every word, phrase, and sentence explains, illustrates, and clarifies the central idea expressed in the topic sentence. Paragraph unity is destroyed when you drift away from your stated purpose by adding irrelevant details. Here is how the paragraph on whale intelligence could become disunified:

> A Disunified Paragraph
>
> Whales are among the most intelligent mammals ever to inhabit the earth. Scientists studying whales rate their intelligence on a level with higher primates because of their complex group behavior. For example, these impressive mammals have been seen teaching and disciplining their young, helping wounded fellow whales, engaging in elaborate sexual foreplay, and playing in definite gamelike patterns. **Whales continually need to search for food in order to survive. As fish populations decrease because of overfishing, the whale's quest for food becomes more difficult.**

This paragraph is not unified because it begins with whale intelligence and drifts into food problems. Stay on the track set by your topic sentence.

Paragraph Coherence

A paragraph is coherent when it hangs together and flows smoothly in a clear direction — when all sentences are logically connected like links in a chain, leading toward a definite conclusion.

One way to damage paragraph coherence is to use too many short, choppy sentences. Two other ways to damage coherence are (1) to place sentences in the wrong order and (2) to use insufficient transitions and other connectors (pp. 599–602) to link related ideas. Here is how the paragraph on whaling intelligence could become incoherent:

> An Incoherent Paragraph
>
> Whales are among the most intelligent creatures ever to inhabit the earth. Scientists rate their intelligence on a level with higher primates. Whales exhibit complex group behavior. **The whaling industry ignores the whales'**

intelligence and threatens them with extinction. Whales have been seen teaching and disciplining their young, helping wounded fellow whales, engaging in elaborate sexual foreplay, and playing in definite gamelike patterns. **Scientific evidence shows that whales have a high order of social organization.** They are able to coordinate such complex group activities because of their apparently effective communication system of sonar clicks and pings.

This paragraph is not coherent because it fails to progress logically. The fourth sentence (in boldface) does not follow logically from the first two; it should be the last sentence. Also, the second-to-last sentence should follow the sentence that is now last. Finally, more transitions and connectors should be added to help the sentences flow smoothly from idea to idea. Here is the same paragraph with sentences written in correct sequence and with enough transitional expressions and other connectors (in boldface):

A Coherent Paragraph

Whales are among the most **intelligent** creatures ever to inhabit the earth. Scientists studying **whales** rate **their intelligence** on a level with higher primates **because** of **their** complex group behavior. **For example, these huge and impressive mammals** have been seen teaching and disciplining **their** young, helping wounded fellow whales, engaging in elaborate sexual foreplay, and playing in definite gamelike patterns. **They** are able to coordinate complex group activities **because** of **their** apparently effective communications system of sonar clicks and pings. **All in all,** scientific evidence shows that **whales** have a high order of social organization. **Unfortunately,** the **whales' intelligence** is ignored by an industry that threatens **them** with extinction.

The paragraph is now tighter and moves in a logical sequence. The sequence of development here can be called a *reasons sequence,* because reasons are given to support an observation and to lead to a recommendation. The logical sequence in the paragraph can be expressed like this:

1. a topic sentence about whale intelligence (*general*)
2. scientific documentation to support the thesis (*specific*)
3. specific examples and explanations of the scientific claims (*more specific*)
4. a statement that sums up and interprets the earlier evidence
5. a concluding statement that provides a bridge to the later discussion: abuses by the whaling industry

Paragraphs Developed in Logical Sequence

Once you have identified your reader and purpose, and gathered your supporting details, you will have to arrange these details in a way that makes the most sense. Following a logical sequence within a paragraph simply means that you decide on which idea to discuss first, which second, and so on. The sequence you select for any paragraph will depend on your subject and your purpose. Some possibilities follow.

Spatial Sequence

A spatial order of development begins at one location and ends at another. This order is most useful in a paragraph that describes a physical item or a mechanism. Simply describe the parts in the order in which readers would actually view them: left or right, inside to outside, etc. This writer has chosen a spatial order that proceeds from the needle's base (hub) to its point:

> A hypodermic needle is a slender, hollow, steel instrument used to introduce medication into the body (usually through a vein or muscle). It is a single piece composed of three parts, all considered sterile: the hub, the cannula, and the point. The hub is the lower, larger part of the needle that attaches to the necklike opening on the syringe barrel. Next is the cannula (stem), the smooth and slender central portion. Last is the point, which consists of a beveled (slanted) opening, ending in a sharp tip. The diameter of a needle's cannula is indicated by gauge number; commonly, a 24–25 gauge needle is used for subcutaneous injections. Needle lengths are varied to suit individual needs. Common lengths used for subcutaneous injections are ⅜, ½, ⅝, and ¾ inches. Regardless of length and diameter, all needles have the same functional design.

Explanation Sequence

Some paragraphs can be arranged according to the order of details used to explain the topic sentence (or an earlier paragraph). Here is the paragraph that explains the preceding one. The writer has wisely chosen to move from the point backward because the needle's function begins at the point.

> Functionally, the needle serves as a passage for the flow of medication. First, the firm, sharp point penetrates the vial for withdrawal of medication. Next, the cannula provides a channel from the vial, through the hub attachment, to the syringe. In turn, the sharp point permits rapid penetration of the skin and tissue while the slender cannula provides a smooth surface, reducing the body's resistance to entrance. Finally, the entire needle works as a unit to allow passage of solution from the syringe barrel into the subcutaneous tissue during the injection.

Chronological Sequence

A paragraph describing a series of events or giving instructions is most effective when its details are arranged according to a strict time sequence: first step, second step, etc.

When you have collected all needed ingredients, prepare your cake batter. Begin by sifting 3½ cups of flour. Then measure the flour and sift it again, along with 2½ cups of granulated sugar, 5 teaspoons of baking powder, and 1 teaspoon of salt into a mixing bowl. To this mixture add ¾ cup of shortening, 4 eggs, 1½ cups of milk, and 2 teaspoons of vanilla extract. Finally, blend the ingredients by hand until the batter is mixed well enough to divide.

Examples Sequence

Often, a topic sentence can best be supported by specific examples.

Although strip mining is safer and cheaper than conventional mining, it is highly damaging to the surrounding landscape. Among its effects are scarred mountains, ruined land, and polluted waterways. Strip operations are altering our country's land at the rate of 5,000 acres per week. An estimated 10,500 miles of streams have been poisoned by silt drainage in Appalachia alone. If strip mining continues at its present rate, 16,000 square miles of United States land will eventually be stripped barren.

Effect-to-Cause Sequence

A paragraph that first identifies a problem and then discusses its causes is typically found in problem-solving reports.

Modern whaling techniques have brought the whale population to the threshold of extinction. In the nineteenth century the invention of the steamboat increased hunters' speed and mobility. Shortly afterward, the grenade harpoon was invented so that whales could be killed quickly and easily from the ship's deck. In 1904 a whaling station opened on Georgia Island in South America. This station became the gateway to Antarctic whaling for the nations of the world. In 1924 factory ships were designed that enabled 'round-the-clock whale tracking and processing. These ships could reduce a ninety-foot whale to its by-products in roughly thirty minutes. After World War II, more powerful boats with remote sensing devices gave a final boost to the whaling industry. The number of kills had now increased far above the whales' capacity to reproduce.

Cause-to-Effect Sequence

In a cause-to-effect sequence the topic sentence identifies the cause(s) and the remainder of the paragraph discusses its effects.

The state prisoner furlough program was devised primarily to help exoffenders stay out of jail after their release. It was believed that short periods of societal exposure would help inmates prepare for their eventual reintegration. According to figures compiled by the State Corrections Department, the program is working.

Definition Sequence

For adequate definition, a term may require a full paragraph (as discussed in Chapter 5).

> The prisoner furlough program permits inmates to leave a state or county institution, unescorted, for no less than twelve hours and no more than seven consecutive days. The state corrections commissioner or the administrator of a county jail may grant furloughs to inmates for the following purposes: attending the funeral of a relative; visiting a critically ill relative; obtaining medical, psychiatric, or counseling services when such services are not available within the institution; contacting prospective employers; finding a suitable residence for use on permanent release; or for any other purpose that will help the inmate's reintegration into the community. Inmates released in 1973 after receiving at least one furlough had a one-year recidivism° rate of 17 percent. In contrast, inmates released in the same year after receiving no furloughs had a one-year recidivism rate of 25 percent. These figures are significant because convicts from all crime categories have been furloughed.

Reasons Sequence

A paragraph that provides detailed reasons to support a specific viewpoint or recommendation is often used in job-related writing.

> The Wankel Engine has obvious advantages over the conventional V-8 engine. Its simplicity of design and operation add to its efficiency and adaptability. The Wankel's lack of valves and camshaft — parts found in all conventional engines — increases its breathing abilities making for a smooth running engine on low octane gas. Because of its high power-to-weight ratio the Wankel allows more passenger room in large automobiles, while providing better power and performance in smaller ones. Even the cost of manufacturing and maintaining the Wankel should be lower because of its few moving parts.

Problem-Causes-Solution Sequence

The problem-causes-solution paragraph is commonly used in daily activity or progress reports.

> The unpainted wood exteriors of all waterfront buildings had been severely damaged by the high winds and sandstorms of the previous winter. After repairing the damage we took the following protective steps against further storms. First, all joints, edges, and sashes were treated with water-repellent preservative to protect against water damage. Next, three coats of

---

° For the purpose of the study, a recidivist was defined as an ex-convict who was returned to any federal, state, or county jail for thirty days or more.

nonporous primer were applied to all exterior surfaces to prevent paint blistering and peeling. Finally, two coats of wood-quality latex paint were applied over the nonporous primer. To avoid future separation between coats of paint, the first coat was applied within two weeks of the priming coats and the second within two weeks of the first. As of two weeks after completion, no blistering, peeling, or separation has occurred.

Comparison/Contrast Sequence

A paragraph discussing the similarities or differences (or both) between two or more items is often used in job-related writing.

> The ski industry's quest for a binding that ensures both good performance and safety has led to the development of two basic types. Although both bindings improve performance and increase the safety margin, they have different release and retention mechanisms. The first type consists of two units (one at the toe, another at the heel) that are spring-loaded. These units apply their retention forces directly to the boot sole. Thus the friction of the boot against the ski allows for the kind of ankle movement needed at high speeds over rough terrain, without causing the boot to release. In contrast, the second type has one spring-loaded unit at either the toe or the heel. From this unit extends a boot plate that travels the length of the boot to a fixed receptacle on its opposite end. With this plate binding, the boot plays no part in release or retention. Instead, retention force is applied directly to the boot plate, providing more stability for the recreational skier, but allowing for less ankle and boot movement before releasing. On the whole, the double-unit binding performs better in racing, but the plate binding is safer.

For the comparison and contrast of more specific data on each of these bindings, two lists would be most effective.

> The Salomon 555 offers the following features.
>
> 1. upward release at the heel and lateral release at the toe (thus eliminating 80 percent of all leg injuries)
> 2. lateral antishock capacity of fifteen millimeters with the highest available return-to-center force
> 3. two methods of reentry to the binding: for hard and deep powder conditions
> 4. five adjustments
> 5. maximum hold-down power for racers and experts
> 6. the necessity of boot alteration
> 7. release torque applied to the boot sole, which, under stress of normal use, alters its release characteristics, requiring readjustment and eventual replacement of boots

8. high durability and a combined weight of seventy-four ounces for eighty dollars

9. the most endorsements among alpine racers today

The Americana offers these features:

1. upward release at the toe as well as upward and lateral release at the heel

2. lateral antishock capacity of thirty millimeters with a moderate return-to-center force

3. two methods of reentry to the binding

4. two adjustments, one for boot length and another comprehensive adjustment for all angles of release and elasticity

5. hold-down power that is slightly compromised at the heel to provide lateral release potential

6. a boot plate that eliminates the need for boot alterations, wear on boots, and danger from friction

7. the durability of two moving parts and a combined weight of fifty-six ounces for $54.50

Instead of this block structure (in which one binding is discussed and then the other) the writer might have chosen a point-by-point structure (in which points in common for each item are listed together: e.g., "Reentry Methods"). The point-by-point comparison is particularly favored in feasibility and recommendation reports because it offers readers a meaningful comparison on the basis of common points rather than by cataloguing items separately.

WRITING EFFECTIVE SENTENCES

A paragraph is only as readable as the sentences it contains. Generally, sentences in technical writing should be no longer than twenty-five words. But word or syllable count (page 51) alone are not adequate measures of good sentences. An effective sentence emphasizes relationships, offers density of information, and makes for easy reading. In short, an effective sentence is *clear, concise,* and *fluent.*

Making Sentences Clear

A clear sentence communicates its meaning upon the very first reading. The following guidelines will help you achieve clarity.

Omit No Key Words

By leaving blanks in your information, you force readers to work that much harder to grasp your meaning.

Faulty   The film container is your first step. (Can a container be a step?)

Correct   Opening the film container is your first step.

Faulty   Before painting walls, place drop cloth on floor. (necessary articles omitted)

Correct   Before painting the walls, place a drop cloth on the floor.

Avoid Ambiguous Phrasing

In technical writing, a sentence should have *one* meaning only. An ambiguous message is useless.

Ambiguous   I cannot recommend this candidate too highly.

Correct   This candidate has my highest recommendation.
or
I cannot recommend this candidate highly.

Ambiguous   Our patients are enjoying the warm days while they last.

Correct   While these warm days last, our patients are enjoying them. (or vice versa)

Avoid Overstuffing

A sentence that crams in too many ideas forces readers to struggle over its meaning.

Overstuffed   A smoke-filled room causes not only teary eyes and runny noses but also can alter auditory and visual perception as well, which is irritating but not associated with any serious disease, except for people with heart and lung ailments, who are threatened with major problems from smoke.

Correct   A smoke-filled room not only causes teary eyes and runny noses but can alter auditory and visual perception as well. Although the smoke itself does not produce serious disease, it does pose a threat to people with heart and lung ailments.

Unstack the Modifiers

Too many nouns stacked up as modifiers in front of another noun make for hard reading.

Faulty   Her job involves **fault-analysis systems troubleshooting handbook** preparation.

Correct   In her job, she prepares handbooks for troubleshooting fault-analysis systems.

Faulty   Our **vehicle air conditioner compression cut-off**
device will reduce fuel consumption by 5 percent.

Correct   Our compression cut-off device for vehicle air con-
ditioners will reduce fuel consumption by 5 percent.

Generally, no such problem with readability occurs when adjectives are
stacked.

Correct   He was a **perplexed, angry, confused,** but **dedicated**
employee.

Rearrange Word Order

Just as any paragraph has a key sentence, any sentence has a key word or
phrase. For emphasis, place the key word or phrase at the beginning or end
of the sentence.

Faulty   I expect a **refund** because of your error in my shipment.

Correct   Because of your error in my shipment, I expect a
refund.

Faulty   These days, **"quality"** means little to **workers** in some
industries.

Correct   **"Quality"** in some industries these days means little
to **workers.**

Use Active and Passive Voice Selectively

Most often, the active voice ("I did it") is better than the passive voice ("It
was done by me"). But for certain kinds of emphasis, the passive voice is preferred. Let the following guidelines govern your choice.

The active voice follows an actor-action-recipient pattern:

actor   *action*   *recipient*
Joe   lost   the check
subject   *verb*   *object*

The passive voice reverses the pattern, making the recipient the subject and
either omitting the actor or naming him/her in a prepositional phrase:

recipient	*action*	*actor*
The check	was lost	by Joe
subject	*verb*	*prepositional phrase*

In each version, the word or phrase in the *subject* position receives the emphasis.

Active   **A falling girder** injured the foreman.

Passive   **The foreman** was injured by a falling girder.

In most writing, give preference to the active voice. It is more direct and economical. Use the passive voice only to emphasize the *recipient* rather than the *actor*.

Passive   **Mr. X** was brought to the emergency room by ambulance.

Passive   **The hijacking story** was broadcast by all stations.

Use the passive voice to emphasize events or results when the *actor* is unknown, not apparent, or unimportant.

Passive   **The victim** was badly beaten.

Passive   **The leak in the nuclear-core housing** was finally discovered.

Passive   **Parts of the plane** were found as far as two miles from the crash site.

One danger of passive construction is that it camouflages the person responsible for the action:

Passive   A mistake was made in your shipment. (By whom?)
"Irresponsible"   The manager was kissed.
It was decided not to offer you the job.
These offices were designed poorly.

Don't hide behind the passive unless, of course, the person behind the action has reason for being protected.

Correct   The criminal was positively identified.

In reporting errors or bad news, use the active voice for greater sincerity.
The passive voice is a weaker construction. It creates an impersonal tone.

Weak and   An offer will be made by us next week.
Impersonal

Stronger   We will make you an offer next week.

Ordinarily, use the active voice in giving instructions:

Faulty   The lid should be sealed with wax.

Correct   Seal the lid with wax.

Faulty   Care should be exercised.

Correct   Be careful.

Use the active voice when you want action. Otherwise your statement will have no impact.

Faulty   If the aforementioned claim is not settled immediately, the Better Business Bureau will be contacted, and their advice on legal action will be taken.

This statement is not likely to move readers to action. Who is making the claim? Who should be doing the settling? Who will be contacting the Bureau? Who will take action? There's nobody here!

Correct   If you do not settle my claim by May 15, I will contact the Better Business Bureau. Upon their advice, I will take legal action.

Making Sentences Concise

A concise sentence is brief but informative. It gets right to the point without clutter. A brief but vague message is useless.

Brief but Vague   These structural supports are too heavy.

Brief but   These structural supports weigh 300 pounds, thereby
Informative   exceeding our specified load tolerance by 50 percent.

Avoid wordiness. Use fewer words when fewer will do.

Cluttered   At this point in time I would like to say that we are ready to move ahead with the project.

Concise   We are now ready.

The passage below is both wordy and poorly detailed.

Low-information   The lawn tennis court is bounded at each end by a
Sentences   screen. This *high* steel or wooden fence is placed
                   so that it is at an *appropriate* distance beyond the
                   baseline *of the court. Its function is significant in
                   that* it prevents the ball from bouncing out of the
                   playing area. The screen *is located so as to* mark the
                   distal boundaries of the playing surface. Each segment *of the screen* is supported by *sturdy* poles
                   which are set at *measured lengths.* (82 words)

There is no density here, no solid information. The words in italics either cause clutter or are too vague. Here is a concise version — informative and to the point.

High-information
Sentences

The lawn tennis court is bounded at each end by a screen. This ten-to-twelve-foot high steel or wooden fence is set exactly 21 feet beyond each baseline. The screen prevents the ball from bouncing out of the playing area and marks the rear boundaries of the playing surface. Each segment is supported by four-inch diameter poles set at three-foot intervals. (65 words)

First drafts are rarely concise. Always make revisions to refine parts that are wordy, repetitious, or vague. Trim the fat by getting rid of anything that adds no meaning.

Eliminate Redundancy

Avoid using a phrase when a word will do. Each redundant phrase below can be reduced to a single word — without any loss in meaning.

at a rapid rate	=	rapidly
has the ability to	=	can
in this day and age	=	today
situated toward	=	near
in close proximity	=	near
the majority of	=	most
due to the fact that	=	because
aware of the fact that	=	know
on a personal basis	=	personally
take the place of	=	substitute

Another form of redundancy occurs when the same thing is said twice, in different words. In each sample below, the bracketed word or phrase merely adds clutter because its meaning is contained in the other word.

a [dead] corpse	[totally] monopolize
the [final] conclusion	[very] vital
[utmost] perfection	[past] experience
[mental] awareness	[totally] oblivious
[the month of] August	mix [together]
[good] asset	[viable] alternative
[the color] green	correct [amount of] change
[mutual] cooperation	stands out [the most]
[fellow] colleagues	[the person,] Jim

Avoid Needless Repetition

Much repetition in the following passage can be eliminated when sentences are combined.

Repetitious   Breathing is restored by artificial respiration. Artificial respiration means that breathing is being maintained by artificial means. Techniques of artificial respiration are mouth-to-mouth and mouth-to-nose. Artificial respiration must always be performed when external cardiac massage is being carried out. (43 words)

Revised   Breathing is restored by artificial respiration, either mouth-to-mouth or mouth-to-nose. Always give artificial respiration when performing external cardiac massage. (23 words)

Avoid "There" Sentence Openers

Save words and improve your emphasis by avoiding "There is" and "There are" at the beginning of sentences.

Weak   There are several reasons why Joe left the company.

Revised   Joe left the company for several reasons.

Weak   There is a danger of collapse in number 2 mine shaft.

Revised   Number 2 mine shaft is in danger of collapsing.

Avoid Certain "It" Sentence Openers

Eliminate any "It" that does not refer to something specific.

Wordy   [It was] his negative attitude [that] caused him to be fired.

Wordy   It gives me great pleasure to introduce our new colleague.

Improved   I am pleased to introduce our new colleague.

Delete Needless "to Be" Constructions

Forms of the verb "to be" ("is," "was," "are," etc.) often add clutter without adding meaning.

Wordy   She seems [to be] upset.

Wordy   I find some employees [to be] incompetent.

Avoid Excessive Prepositions

Prepositions (especially "of") can combine with forms of the verb "to be" to make wordy sentences.

Wordy	These are the recommendations of some of the members of the committee.
Revised	Some committee members made these recommendations.
Wordy	Dr. Karloff is a grouchy sort of teacher.
Revised	Dr. Karloff is a grouchy teacher.

Use "That" and "Which" Sparingly

Excessive use of "is" usually drags along a needless "that" or "which."

Wordy	The report [, which is] about consumer attitudes[,] is three hundred pages long.
Wordy	This [is a] problem [that] requires immediate attention.

Fight Noun Addiction

Excessive nouns make sentences awkward and wordy. They are often found with needless prepositions or weak verbs, such as "is," "has," or "make." Break up noun clusters by substituting action verbs.

Wordy	His memo is a request that we conduct a study of the problem.
Revised	His memo requests that we study the problem.
Wordy	I am giving consideration to the possibility of a change in career.
Revised	I am considering changing careers.

Make Negatives Positive

Save words and get to the point by eliminating negative constructions.

Weak and Wordy	I did not gain anything from this course.
Revised	I gained nothing from this course.
Wordy	I don't see any difference between these two models.
Revised	I see no difference between these two models.

Clear Out the Clutter Words

Clutter words stretch a message without adding meaning. Here are some of the most common: "very," "definitely," "quite," "extremely," "rather," "somewhat," "really," "actually," "situation," "aspect," "factor." If you must use them, be sure they serve a purpose in your message.

Cluttered	Actually, one aspect of a business situation that could definitely make me quite happy would be to have a somewhat adventurous partner who really shared my extreme love of taking risks.
Revised	I would like to have an adventurous business partner who enjoys taking risks.

Delete Needless Prefaces

Get to the point. Deliver the pitch without a long wind-up.

Wordy	[I am writing this letter because] I wish to apply for the position of copy editor.
	[The conclusion we can draw is that] writing is hard work.

Delete Needless Qualifiers

Qualifiers are expressions such as "I feel," "it seems," "I believe," "in my opinion," and "I think." Use them only to emphasize that your assertion is one of opinion, not of fact — an assertion subject to change.

Appropriate Qualifier	Despite Frank's poor academic performance last semester, he will, I think, do well in college.

If you are sure of your assertion, however, eliminate the qualifier. Otherwise, it waters down your assertion and sounds as though you are beating around the bush.

Needless Qualifiers	[It seems that] I've wrecked the company car.
	[It would appear that] I've lost your credit card.
	[In my opinion,] you have a valid argument.

Making Sentences Fluent

Fluent sentences are polished, graceful, easy to read — the work of a careful writer. Their variety of length and word order make them free of choppiness and monotony. The following suggestions will help you achieve sentence fluency.

Combine Related Ideas

Use coordination and subordination (pages 572–574) to combine a series of short, choppy sentences.

> Choppy   Jogging can be healthful if you have the right equipment. Most important are well-fitting shoes. These are important because without them you take the chance of injuring your legs. Your knees are especially prone to injury.

> Fluent   Jogging can be healthful if you have the right equipment. Well-fitting shoes are most important because they prevent injuries to your legs, especially your knees.

Most sets of information can be combined in various ways — depending on where the emphasis should be. Consider the following sets of information.

1. Roland James is the third candidate for our engineering position.
2. He has graduated from an excellent program.
3. He has no practical experience.
4. He comes highly recommended.

Assume that you are writing a memo to your colleagues that records your impression of this candidate. If you want to emphasize a negative impression, you might combine the above data this way:

> Although Roland James, the third candidate for our engineering position, has graduated from an excellent program and comes highly recommended, **he has no practical experience.**

The thought in the independent clause is the one that receives the emphasis.

If, on the other hand, you are undecided about this candidate, but are leaning in a negative direction, you might combine the data this way:

> Roland James, the third candidate for our engineering position, has graduated from an excellent program and comes highly recommended, but he has no practical experience.

In the above example, the two independent clauses are joined by a coordinating conjunction ("but"), which suggests that both sides of the issue are of equal importance. Placing the negative feature last, however, gives it slightly more emphasis. If instead you were leaning toward the positive, you could reverse the order, placing the positive feature after the "but."

To emphasize strong support for the candidate, you might use the following combination:

> Although Roland James, the third candidate for our engineering position, has no practical experience, **he has graduated from an excellent program and comes highly recommended.**

In short, our meaning is reflected not only by what we say but also by how we arrange our information.

The following sentences, excerpted from a student pilot's description of "Taking Off," further illustrate how fluent sentences are both more readable and more meaningful:

> First Draft   My mind's eye sees only the center line. My eyes constantly move from the windshield to the instruments. They read and calculate and they miss nothing.

Because the above actions all took place at the same time, the writer decided to emphasize that relationship by combining her information in a single sentence.

> Revised   My mind's eye sees only the center line, while my eyes constantly move from the windshield to the instruments, reading, calculating, missing nothing.

The revised version captures the mood that the writer set out to convey: one of immediacy, of excitement, of all things happening at once.

Here, from the same writer, is another illustration of how the structure of a sentence can emphasize a certain meaning:

> First Draft   Gravity pulls at us. It insists that we are bound to the earth. It insists that we are slaves of its laws.

The idea of pulling deserves the most emphasis here. So the writer combines her information to reflect that sense:

> Revised   Gravity pulls at us, insisting that we are bound to the earth, slaves of its laws.

The lead verb, "pull," is placed in an independent clause, and the two subsequent ideas are subordinated — carried along by the first clause.

Vary Sentence Construction and Length

Too much of anything is boring. An unbroken series of short or long sentences makes for dreary reading, as does a series with identical openings.

> Dreary   There are some drawbacks about diesel engines. They are much noisier than standard engines. They are difficult to start in cold weather. They cause

> considerable vibration. They also give off an un-
> pleasant odor. They cause sulphur dioxide pollution.
> For these reasons, some auto makers are limiting
> their diesel models to light trucks only.

Combine and rephrase for variety:

> Revised   Diesel engines have some drawbacks. Most obvious
> are their noisiness, cold-weather starting difficulties,
> vibration, unpleasant odor, and sulphur dioxide emis-
> sions. Therefore, some auto makers are limiting their
> diesel models to light trucks only.

Opening most sentences with "The," "This," "He," "She," "It" or "They" creates monotony. When you write in the first person, overusing "I" suggests that you are self-centered. Eliminate such repetition by combining ideas and shifting word order.

Use Short Sentences for Special Emphasis

With all this talk about combining ideas, one might conclude that short sentences have no place in effective writing. Wrong. Short sentences (even one-word sentences) provide valuable emphasis. Consider, for example, another part of the student pilot's description of taking off:

> The airspeed increases. We reach the point where the battle begins.

Instead the student might have written: "The airspeed increases and we reach the point where the battle begins," but she wanted to emphasize two, discrete instances here: (1) the acceleration and (2) the critical point of lifting off the ground.

Whereas long sentences combine ideas to clarify certain relationships, short sentences isolate a single idea for special emphasis. They drive home a crucial point. They stick in the reader's mind.

CHOOSING THE RIGHT WORDS

Your choice of words ultimately determines the quality of your writing. Keep your expressions simple, jargon-free, original, convincing, precise, concrete and specific, and unbiased.

Keeping the Language Simple

Overblown diction and needless jargon obscure your message and make your readers work harder.

Deflate the Diction

Say it in plain English. Avoid inflated diction. Don't use three syllables when one will do. Trade for less:

utilize	=	use
approximately	=	about
to be cognizant	=	to know
to endeavor	=	to try
endeavor	=	effort
securing employment	=	finding a job
demonstrate	=	show
phenomenon	=	event
determine	=	find
multiplicity of	=	many
necessary	=	needed
effectuate	=	do
component	=	part
terminate	=	end

Count the syllables. Trim when you can. Don't write like the author of a report from the Federal Aviation Administration who suggested that manufacturers of the DC-10 be directed "to re-evaluate the design of the entire pylon assembly to minimize design factors which are resulting in sensitive and/or critical maintenance and inspection procedures" (25 words, 50 syllables). Here is a plain-English translation: "Redesign the pylons so they are easier to maintain and inspect" (11 words, 18 syllables). Here is another example, this from a United States senator:

Inflated	There are no major concerns as far as the Trident missiles relate to environmental damage. (15 words, 25 syllables)
Revised	Trident missiles pose no major environmental threat. (7 words, 14 syllables)

These are examples of the worst kind of flab: too many words, and words that are bigger than needed. Keep it lean.

Inflated	Acoustical attenuation for the food consumption area is needed.
Revised	The cafeteria needs soundproofing.
Inflated	Replacement of the weak battery should be effectuated.
Revised	Replace the weak battery.

Inflated   Make an improvement in the clerical situation.

Revised   Hire more (or better?) secretaries. (Inflated language also can be ambiguous!)

Avoid Needless Jargon

Various professions have their own "shorthand." Among specialists, certain technical terms do save time. For example, "stat" on page 13 is shorthand for "Drop whatever else you're doing and deal immediately with the emergency." A mechanic's handbook might use "⅜ socket" to refer to a certain wrench. In film production, "takes" are the number of times a scene is filmed until it is correct. There are two ways, however, to use technical language — appropriately and inappropriately — and the latter is what we call needless jargon, meaningless to insiders as well as outsiders.

Needless Jargon   Unless all parties to the contract interface within the same planning framework at an identical point in time, the project will be rendered inoperative.

Revised   Unless we coordinate our efforts, the project will fail.

Needless Jargon   Intercom utilization will be used to initiate substitute teacher operative involvement.

Revised   Teachers who have to substitute will be notified on the intercom.

Needless Jargon   For the obtaining of the X-33 word processor, our firm will have to accomplish the disbursement of funds to the amount of six thousand dollars.

Revised   Our firm will have to pay six thousand dollars for the X-33 word processor.

The practice in commercial jargon of adding "wise" to nouns as shorthand for "with reference to" is unacceptable: "saleswise," "timewise," "businesswise," "costwise," "moneywise," etc. Jargon-ridden writing makes you seem as if you are trying to put something over on your reader or as if you doubt the validity of your own statements.

The following letter, an unfortunate mix of sixty-five-cent words, jargon, and needless use of passive construction, is a version of one published in a newspaper.

In the absence of definitive studies regarding the optimum length of the school day, I can only state my personal opinion based upon observations made by me and upon teacher observations that have been conveyed to me. Considering the length of the present school day, it is my opinion that the day is excessive length-wise for most elementary pupils, certainly for almost all of the primary children.

To find the answer to the problem requires consideration of two ways in which the problem may be viewed. One way focuses upon the needs of the children, while the other focuses upon logistics, transportation, scheduling, and other limits imposed by the educational system. If it is necessary to prioritize these two ideas, it would seem most reasonable to give the first consideration to the primary reason for the very existence of the system, i.e., to meet the educational needs of the children the system is trying to serve.

Here is a plain-English translation:

Although no studies have defined the best length for a school day, my experience and teachers' comments lead me to believe that the school day is too long for most elementary students — especially the primary students.

We can view this problem from the children's point of view (health, psychological welfare, and so on) or from the system's point of view (scheduling, transportation, utilities costs, and so on). If we consider our most important goals, the children should come first, because the system exists to serve their needs.

Being Convincing

Observe the following guidelines to give your writing the qualities of original expression and convincing tone.

Avoid Triteness

Writers who rely on tired old phrases (clichés) seem either too lazy or careless to find convincing or exact ways to say what they mean. The expressions below are just a few of the thousands that have become worn out through overuse.

first and foremost	not by a long shot
in-depth study	last but not least
in the final analysis	in a manner of speaking
consensus of opinion	the bottom line
it is interesting to note	welcome aboard
it has come to my attention	over the hill
needless to say	water under the bridge
as a matter of fact	take it in stride
to all intents and purposes	holding the bag
close the deal	bite the bullet

If it sounds like something you've heard before, don't write it.

Avoid Overstatement

Overstatement is a sure way for writers to lose their credibility. Resist the temptation to exaggerate in order to make your point.

Overstated	If you try skiing, you will find it to be one of the most memorable experiences of your life.
Revised	If you try skiing, you will enjoy it.
Overstated	If you hire me, I will be the best worker you've ever had.
Revised	If you hire me, I will do my best.

Avoid Sweeping Generalizations

Sweeping generalizations harm a writer's credibility because they lack specific factual support.

Sweeping	Television is rotting everyone's brain.
Revised	Many authorities argue that television is one cause of declining literacy.
Sweeping	Democracy in America is collapsing.
Revised	American democracy is threatened by fanatical organizations — both from the left and right.

Avoid Euphemisms

Euphemisms are expressions that aim at politeness, that make unpleasant or delicate subjects seem less offensive. Thus, one "powders one's nose" instead of using the toilet; "passes on," "passes away," "finds eternal rest," or "goes to meet one's maker" instead of dying. To the extent that euphemisms save hurt feelings or embarrassment, they are by all means legitimate.

One danger with euphemisms, however, is that they can sugar-coat the hard truth when only the truth will serve: criminals become "offenders"; illiterates become "disadvantaged students"; failing students become "underachievers"; political corruption becomes "lobbying"; rape becomes "sexual assault"; wars become "conflicts"; the nuclear disaster at Three Mile Island becomes an "incident"; and the prospect of nuclear holocaust is translated by one United States senator into "in the event of a nuclear exchange."

But as benign as conflicts, nuclear incidents, and nuclear exchanges may sound, they kill people just as dead as do wars, nuclear disasters, and a nuclear holocaust. Plain talk is better.

Being Precise

Be sure that what you say is what you mean. Even words listed as synonyms contain a different shade of meaning. Do you mean to say, "I'm slender; you're thin; he's lean; and she's scrawny"? The wrong choice could be disastrous. Is your nose reacting to a "smell," "essence," "odor," "scent," "fragrance," "stench," or "stink"? Do not use one word when you mean another. Don't, for instance, write to apply for a job with a statement like this:

> Another attractive feature of *X* corporation is its **adequate** training program.

The sense conveyed by "adequate" clashes with the whole notion of "attractive." The above writer later explained his choice of *adequate:* the program, he said, was not highly ranked, so he tried to choose a word that would be sincere. But his choice has two liabilities:

1. *Adequate,* in this context, is insulting. After all, who likes to be called adequate?
2. The word suggests that the writer has placed himself in a position of judgment, overstepping his bounds as an applicant.

Any of several alternatives (*solid, respectable, growing*) retains sincerity without being offensive.

Precision is not only important to the tone of your writing but also to its informative value as well:

Imprecise	When a plug is pushed in the outlet, it connects the cord to the power supply. ("Push" is imprecise. Do we "push" a plug as we push a baby carriage?)
Revised	When a plug is inserted into the outlet, it connects the cord to the power supply.
Imprecise	The diesel is not without fault and skepticism. (Can an engine be skeptical?)
Revised	Certain faults in the diesel make some people skeptical.

Learn the differences between words that sound alike. Here are some of the most commonly confused:

affect/effect	healthy/healthful
continual/continuous	imply/infer
fearful/fearsome	sensual/sensuous
formally/formerly	worse/worst

Do not write "Skiing is healthy" when you mean that skiing is good for one's health (healthful). Healthful things help keep us healthy.

Avoid *due to* when you mean *because*. (A train can be due to arrive!) Even worse is *due to the fact that* (imprecision burdened by flab). Don't write about *getting used to* something when you mean *becoming accustomed*. Tools *get used to* do a job; cars *are used to* travel, but you *become accustomed to* new circumstance.

Be on the lookout for imprecise (and therefore illogical) comparisons.

Imprecise   Your rate of interest is higher than the First National Bank. (Can a rate be higher than a bank?)

Revised   Your rate of interest is higher than that of the First National Bank.

Being Concrete and Specific

Informative writing *shows* as well as *tells*. Consider an abstract and general assertion like this one:

Pedestrians crossing the street in front of our office place their lives in danger.

You need to support the assertion — to *show*, with concrete and specific examples. Do not say simply,

For example, **an office worker** was **injured** there by a **vehicle recently.**

Because they are much too general, the boldface words only *tell*. Instead, say,

Alan Hill was hit by a speeding truck last Tuesday and suffered a broken leg.

Notice how the informative value of the following terms increases as we move down the scale.

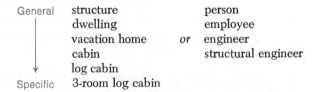

General	structure		person
	dwelling		employee
	vacation home	*or*	engineer
	cabin		structural engineer
	log cabin		
Specific	3-room log cabin		

Choose high-information words that show exactly what you mean. Don't write "thing" when you mean "lever," "switch," "micrometer," or "compass." Instead of evaluating an employee as "swell," "great," and "terrific" or "lousy," "terrible," and "awful," use informative modifiers such as "reliable," "skillful," and "competent" or "dishonest," "irritable," and "incompetent." The earlier modifiers reveal your attitude, but nothing about the person. Readers have to know what we mean.

Abstract   Professor Jones's office looks like a disaster area.

Concrete   Professor Jones's office has a floor strewn with books, a desk buried beneath a mountain of uncorrected papers, and ashtrays overflowing with cigarette butts.

General   Industrial emissions are causing lakes to die.

Specific   Sulphur dioxide emissions from coal-burning plants combine with atmospheric water to produce sulfuric acid. The resultant "acid rain" increases the acidity of lakes to the point where they can no longer support life.

When applicable, provide solid numbers and statistics that get your point across:

General   In 1972, thousands of people were killed or injured on America's highways. Many families had at least one relative who was a casualty. After the speed limit was lowered to 55 miles per hour in late 1972, the death toll began to drop.

Specific   In 1972, 56,000 people died on America's highways; 200,000 were injured; 15,000 children were orphaned. In that year, if you were a member of a family of five, chances are that someone related to you by blood or law was killed or injured in an auto accident. After the speed limit was lowered to 55 miles per hour in late 1972, the death toll dropped steadily to 41,000 deaths in 1975.

Concrete and specific information is not only more informative; it is more convincing as well.

Remaining Unbiased

Your job as a technical writer is to report objectively without injecting "loaded" words that reflect personal attitudes. If you *are* asked to include your interpretations and conclusions, base them on the facts you have discussed. Even controversial subjects deserve objective treatment. For instance, imagine that you have been sent to investigate the causes of an employee-management confrontation at your company's Omaha branch. Your initial report, written for the New York central office, is intended simply to describe the incident. Here is how an unbiased description might read:

> At 9:00 A.M. on Tuesday, January 21, eighty women employees entered the executive offices of our Omaha branch and remained for six hours, bringing business to a virtual halt. The group issued a formal statement of protest, claiming that their working conditions were repressive, their salary scale unfair, and their promotional opportunities limited. The women demanded affirmative action, insisting that the company's hiring and promotional policies and wage scales be revised. The demonstration ended when Garvin Tate, vice-president in charge of personnel, promised to appoint a committee to investigate the group's claims and to correct any inequities.

Notice the absence of implied judgments; the facts are presented objectively. A less objective version of the event, from the women's side, might read as follows:

> Last Tuesday, sisters struck another blow against male supremacy when eighty women employees paralyzed the pin-striped world of solidly entrenched sexism for more than six hours. The timely and articulate protest was aimed against degrading working conditions, unfair salary scales, and lack of promotional opportunities. Stunned executives watched helplessly as the group occupied their offices. The women were determined to continue their occupation until their demands for equal rights were met. Embarrassed company officials soon perceived the magnitude of this protest action and agreed to study the group's demands and to revise the company's discriminatory policies. The success of this long overdue confrontation serves as an inspiration to oppressed women employees everywhere.

Notice how the use of judgmental words and qualifiers ("male supremacy," "degrading," "paralyzed," "articulate," "stunned," "discriminatory," etc.) injects the writer's personal attitude even though it's not called for. In contrast to the above bias, the following version defends the status quo:

> Our Omaha branch was the scene of an amusing battle of the sexes last Tuesday, when a Women's Lib group, eighty strong, staged a six-hour sit-in at the company's executive offices. The protest was lodged against supposed inequities in hiring, wages, working conditions, and promotion for women in our company. The libbers threatened to remain in the building until their demands for "equal rights" were met. Bemused company officials responded to this carnival demonstration with patience and dignity, assuring the militants that their claims and demands — however inaccurate and immoderate — would receive just consideration.

Again, the use of qualifying adjectives and superlatives slants the tone of the report. Let your facts speak for themselves.

EXERCISES

1. The following topic sentences either provide no direction, are un-focused, or are not sufficiently informative. Revise them.

Examples

A. Our firm employs twelve people. (leads nowhere)
   Revised: Working in a small architectural firm has helped me appreciate the importance of being versatile.
B. Writing is a complex skill. (unfocused)
   Revised: Writing is a process that involves a deliberate series of deliberate decisions about audience needs, content, arrangement, and style.
C. A Mercedes Benz is a great all-around car. (not sufficiently informative)
   Revised: A Mercedes Benz offers safety, performance, durability, and luxury.

a. Women have changed radically in the last decade.
b. My colleague's name is Bill.
c. My job is (awful, great).
d. Professor Jones is unfair to students.
e. Nursing is a popular profession.
f. My company hires more men than women.

2. The following sentences are unclear because of missing key words, ambiguous phrasing, or overstuffing. Revise them so that their meaning is clear. (For those sentences that suggest two meanings, write two versions.)

a. State law requires that restaurants serve food with a sanitation certificate.
b. Bring dictionary to exam as reference.
c. Making the shelves look neater should be another one of our priorities for increasing sales, because if the merchandise is not always neatly arranged, customers will not have a good impression, whereas if it is neat they probably will return.
d. His constant aggravation will someday cause a nervous breakdown.
e. Along with losing weight, swimming tones muscles and improves lung capacity.
f. While camping, we live in different surroundings from Friday to Sunday.

3. Improve clarity and emphasis in the following sentences by unstacking the modifiers or by rearranging word order.

a. Our students are tested by an incoming freshman mandatory writing proficiency exam.
b. Education enables us to recognize excellence and to achieve it.

   c. The new diesel engine trailer truck driver training school is now enrolling students.

   d. In a business relationship, trust makes it work.

   e. A densely packed several-inch-thick layer of pine needles covered the ground.

4. Some of the following sentences need to be rewritten in the active or passive voice for better emphasis, less awkwardness, more directness, or greater economy. Make the necessary changes and be prepared to give reasons for each. Mark an E by the sentences that are already effective.

   a. It is believed by us that this contract is faulty.

   b. The tall model wore the fifty-thousand-dollar mink coat.

   c. Joe has been fired.

   d. Hard hats should be worn at all times on this job.

   e. A tornado destroyed our brand new tractor.

   f. It was decided not to accept your invitation.

   g. A check for full payment will be sent next week.

   h. This package should be kept cool.

   i. A rockslide buried the mine entrance.

   j. Searchers found the victim almost dead.

   k. It is my hope that you succeed.

5. Get rid of clutter in the following sentences by eliminating redundancies and needless repetition.

   a. He is a man who works hard.

   b. This report is the most informative report I've read in months.

   c. I am aware of the fact that Sam is a trustworthy person.

   d. On previous occasions, we have worked together.

   e. I have admiration for Professor Jones.

   f. I'll meet you in the downtown area of the city.

   g. She has the talent to restore old furniture back to life.

   h. In the event that you need help, call me.

   i. Sally is an associate of mine.

   j. Stretching is very vital before exercising.

6. Make the following sentences more concise by eliminating "There is" and "There are" sentence openers and the needless use of "it," "to be," "is," "of," "that," and "which."

   a. I consider George to be a talented technician.

   b. The sales volume for last month, which was dreadfully low, disappointed everyone.

   c. This step must be practiced in order for it to become effective.

   d. There are certain people whom I enjoy traveling with.

   e. Another reason the job is attractive is because the salary is excellent.

   f. Smoking of cigarettes is considered by many people to be the worst habit of all habits of human beings.

g. Our summer house, which is located on Cape Cod, is for sale.
h. Many of the jobs that I have held have been interesting.
i. There are many employees who always do excellent work.
j. The static electricity that is generated by the human body is measurable.

7. Improve the economy and directness of the following sentences by replacing nouns with verbs, by changing negatives to positives, and by clearing out clutter words, needless prefaces, and needless qualifiers.

a. We request the formation of a committee of students for the review of grading discrepancies.
b. I am not unappreciative of your help.
c. I tend to disagree with you.
d. It seems that I've made a mistake in your order.
e. Bill made the suggestion that we hire an additional salesperson.
f. Her quick wit is an extremely impressive aspect of her personality.
g. I find Boris to be an industrious and competent employee.
h. Actually, I am very definitely interested in this position.
i. Igor doesn't have any friends at the office.
j. In my opinion, George is a responsible employee.

8. Use two ways of combining each group of sentences below into a single, fluent sentence.

a. Sarah Fields is a structural engineer.
b. She has been assigned to the Northern Ontario Hydroelectric Project.
c. She is completing an inspection of the Winisk River dam site.

a. Calvin and Calvin Associates are an innovative architectural firm.
b. They have designed outstanding homes.
c. They sometimes do work for the city.
d. They lost the urban-renewal contract.

a. Boats with excessive horsepower should be banned from our lakes and ponds.
b. This would be one way to decrease noise and water pollution.
c. It also would save precious fuel.

a. I was employed by the Food Mart supermarket.
b. I held the position of service clerk.
c. It was my job to operate a cash register.
d. Also, I priced items and stocked shelves.

9. Rewrite the following paragraph to make it more concise and fluent.

There are two methods that may be used in glazing pottery. The best method is to use underglazes and glazes. Three coats of each are applied. The underglazes designate the color; the glazes give the pottery a shiny finish and a semitransparent color. The underglazes are put on first. They are usually put on with a fine paint brush. The glazes are put on next. A larger brush is usually used for this.

The glazes are patted on, whereas the underglazes are brushed on. When glazes and underglazes are used, the pottery must be fired again. This method produces a shiny effect.

10. Rewrite the following statements in plain English.

    a. This writer desires to be considered for a position with your company.
    b. At this point in time we cannot agree to your terms.
    c. Please refund our full purchase expenditure in view of the fact that the microscope is defective.
    d. No decisions will be made until next week as far as the contract is concerned.
    e. There are several banks that can be contacted in terms of obtaining a business loan.
    f. I can wish you no better luck than that you find this job as enjoyable as I have.
    g. Prior to this time we have had no such equipment failure.
    h. In relation to your job, I would like to say that we can no longer offer you employment.
    i. This report is useless as far as I am concerned.
    j. Further interviews appear to be a necessity before we can identify the best qualified candidate.
    k. I suggest that you might want to consider shipping your lobsters by air transport.
    l. I suggest that you reduce the amount of cigarettes that you consume daily.
    m. A good writer is cognizant of how to utilize grammar in a correct fashion.

11. Revise the following sentences by eliminating the overstatements and euphemisms.

    a. I was less than candid.
    b. This employee is poorly motivated.
    c. It was decided to terminate your employment.
    d. She expropriated company funds.
    f. When the grenade exploded, his arm was traumatically amputated.
    g. Quitting your job at this stage would be an act of self-destruction.
    h. Cigarette smoking destroys your health.
    i. You're the world's best boss!
    j. The business world follows the law of survival of the fittest.
    k. Today's students are nothing but illiterates.

12. Revise the following sentences to eliminate triteness and needless jargon.

    a. The rising prime rate is sending shock waves through the business world.

    b. The preparation of this report has been facilitated by the assistance of Professor Jones.

    c. We will be appreciative of your input on this plan.

    d. Timewise, this construction schedule is not viable.

    e. In order to optimize our financial return, we should prioritize our investments.

    f. There's never a dull moment on this job.

13. Revise the following sentences to make them precise.

    a. Our outlet does more business than San Francisco.

    b. Low-fat foods are healthy.

    c. Due to hours of studying, I received an "A" on the exam.

    d. Anaerobic fermentation is used in this report.

    e. Prices hope to be held down by building a smaller engine.

    f. The diesel can meet antipollution standards by installing exotic hardware.

14. Revise the following sentences to make them more concrete and specific.

Example
A storm damaged the building.
Revised: A tornado tore the roof off our number 2 warehouse.

    a. He received an excellent job offer.

    b. The group presented its demands.

    c. She repaired the machine quickly.

    d. This thing bothers me.

    e. His performance was awful.

    f. The crew damaged a piece of furniture.

    g. My new employee is disappointing.

    h. They discussed the problem.

    i. She claimed that he had never phoned her about the deal.

    j. An animal injured a person at the construction site.

STRATEGIES
FOR TECHNICAL
REPORTING

4

Summarizing Information

DEFINITION

PURPOSE OF SUMMARIES

ELEMENTS OF AN EFFECTIVE SUMMARY
  Essential Message
  Nontechnical Style
  Independent Meaning
  No New Data
  Introduction-Body-Conclusion Structure
  Conciseness

WRITING THE SUMMARY

APPLYING THE STEPS

WRITING THE DESCRIPTIVE ABSTRACT

PLACING SUMMARIES AND ABSTRACTS
     IN YOUR REPORT

CHAPTER SUMMARY

REVISION CHECKLIST

EXERCISES

DEFINITION

A summary (or informative abstract) is a short version of a longer message that expresses the substance (meaning, emphasis, organization) of the original in a condensed form. Summaries are an economical way to communicate.

PURPOSE OF SUMMARIES

Every effective statement conveys a message — makes one or more points. Anyone who listens to or reads the statement carefully can extract its important ideas without memorizing the original. For instance, your notes of a college lecture do not include the lecturer's every word; instead they summarize the mean points of the message in a way that you will understand them later. Likewise, in studying a textbook for an examination, you extract the main ideas, deciding which information is most important. Accordingly, the best students usually are those who take effective notes and who "know what to study" — those who summarize accurately.

Outside of school you summarize information daily, whether relating an anecdote or describing a recent magazine article you want a friend to read. On the job you have to write concisely about your work. Perhaps you will record the minutes of a meeting, summarize a conference lecture, news article, or report, or write summaries of your progress on a project. Or you might write proposals for new projects, bids for contracts, or summaries of your research. Also, you will include summaries and abstracts with any long reports you write. When you apply for jobs, your letter and résumé summarize your personal qualities and qualifications.

Whether you summarize someone else's information or your own, your job is to communicate the *essential message* — to express the most important information accurately in the fewest words. The principle is simple: include what your readers need and omit what they don't.

The essential message in any well-written piece is easy enough to identify if we read carefully. Consider this example:

> The lack of technical knowledge among owners of television sets leads to their suspicions about the honesty of TV repair technicians. Although TV owners might be fairly knowledgeable about most repairs made to their automobiles, they rarely understand the nature and extent of specialized electronic TV repairs. For example, the function and importance of the automatic transmission in an automobile is generally well known; however, the average TV owner knows nothing about the function of the flyback transformer in a TV set. The repair charge for a flyback transformer failure is roughly $150 — a large amount to a consumer who lacks even a simple understanding of what the repairs accomplished. In contrast, a $450 repair charge for the transmission on the family car, though distressing, is more readily understood and accepted.

Three significant ideas comprise the essential message here: (1) TV owners lack technical knowledge and are suspicious of repair technicians. (2) An owner usually understands even the most expensive automobile repairs. (3) Owners do not understand or accept expenses for repairs and specialized parts needed for their TV sets. A summary of the paragraph might read like this:

> Because TV owners lack technical knowledge of their sets, they are often suspicious of repair technicians. Although consumers may understand expensive automobile repairs, they rarely understand or accept repair and parts expenses for their TVs.

This summary is almost 30 percent of the original length because the original itself is short. With a longer original, a summary might be as short as 5 percent or less. Length, however, is secondary to the need to present all important information from the original.

Summaries are vital whenever people have no time to read in detail everything that crosses their desks.[1] Of course, only by reducing length without distorting the original message can a summary be effective.

---

[1] It is said that one of our recent United States presidents required all significant world information for the last twenty-four hours to be compressed into one typed page and placed on his desk the first thing each morning. And another ex-president had a writer who summarized articles from more than two dozen major magazines.

ELEMENTS OF AN EFFECTIVE SUMMARY

The following 235-word summary (or informative abstract) of a 5000-word report (pages 489–495) is a good distillation. Although less than 5 percent of the original, the summary contains all the significant points.

SUMMARY: SURVIVAL PROBLEMS
OF TELEVISION SERVICE BUSINESSES

The high rate of business failure among qualified independent TV repair technicians (second only to service station failures) is rooted in a tradition of unsound business management and inadequate communication with customers. Soon after World War II, many radio-mechanic veterans opened radio shops, which were to form the basis of today's TV repair shops. The ensuing rapid technological progress saw these veterans and the newer technicians swept up in a frenzy of too much work and too little time to learn effective business methods. Moreover, the repair technicians' somewhat secretive ways of dealing with customers, coupled with the advent of television and its phenomenal growth, led to customer suspicions that helped form the "TV repairman syndrome." Even the best of today's technicians have difficulty in shaking this image. Accounts of dishonesty and ineptitude, prevalent in the early years, linger on. Expensive repairs to electronic equipment are difficult to explain to the nontechnical consumer who feels that he is the victim of a supertechnology that requires elaborate servicing. Today's technician or shop owner is still hesitant to adopt sound management and collection methods, as well as to inform his beleaguered customers about the services he offers. Only when technicians begin to learn improved business methods and sponsor collective advertising to improve their image and better inform the consumer will their chances for survival increase.

The elements that make this summary effective are discussed below.

Essential Message

A good summary answers the reader's implied question: "What point(s) is the original making?" We have just seen that the essential message is the minimum needed for the reader to understand an issue. It is the sum of the significant points — and *only* the significant points — extracted from the original. Significant points include controlling ideas (thesis statements and topic sentences); major findings and interpretations; important names, dates, statistics, and measurements; and major conclusions or recommendations. They do not include background discussions; the author's personal comments, digressions, or conjectures; introductions, explanations, lengthy examples, graphic illustrations, long definitions, or data of questionable accuracy. (These distinctions are illustrated on pages 73–78).

Nontechnical Style

More people are likely to read your summary than the entire original, so write at the lowest level of technicality. Translate technical terms and complex data into plain English; for example, if the original states: "For twenty-four hours, the patient's serum glucose measured a consistent 240 mg%," you might rephrase: "For twenty-four hours, the patient's blood-sugar level remained critically high." When you do know specifically the kinds of people who will read the report, keep these people in mind. If they are expert or informed, you won't need to simplify as much (as discussed in Chapter 2). It is safer, however, to risk oversimplifying than to risk confusing your reader.

Independent Meaning

In meaning, as well as in style, your summary should be clear; a complete message that stands alone. Readers should have to read the original only for further details, examples, or illustrations — not to make sense out of your message.

No New Data

Your job is to represent the original faithfully. Avoid personal comments or judgments ("This interesting report . . ." or "I strongly agree with this last point," etc.). In short, add nothing to the original.

Introduction-Body-Conclusion Structure

Most good writing has an introduction, a body, and a conclusion; so should your summary of that writing.

1. Begin with a clear statement of the controlling idea.
2. Present the supporting details in the same order as in the original.
3. Close with the conclusions and recommendations.

To improve coherence, use transitional words ("however," "in addition," "while," "therefore," "although," "in contrast," etc.).

Conciseness

A summary, above all, is concise. Because the contents of messages differ greatly, however, we cannot set a rule for summary length. Unless a length is specified as part of the job, all we can say is that the summary must be short enough to be economical and long enough to be clear and complete. A long, clear summary is always better than a short, foggy one.

WRITING THE SUMMARY

Summarize your own work only after completing the original. Follow these instructions to pare down and to polish any longer piece — your own or someone else's.

1. *Read the entire original.* When summarizing another's work, read the whole thing before writing a word. You will then get a complete picture. You have to understand the original fully before you can summarize it effectively.

2. *Reread and underline.* Reread the original two or three times, underlining significant points (usually found in the topic sentences of individual paragraphs). If the piece is in a book, journal, or magazine that belongs to someone else, write the points on a separate sheet of paper instead of underlining.

Identify the key (thesis) sentence, which states the controlling idea. Omit all minor supporting details such as introductions, explanations, illustrations, examples, and definitions.

3. *Edit the underlined data.* Reread the underlined material and cross out needless words. Leave only phrases that you can later rewrite in your own words, combining them into sentences.

4. *Rewrite in your own words.* Rewrite the edited, underlined material in your own words, following the original order. Include all important data in the first draft, even if you use too many words; you can always trim later. Avoid judgments ("The author is correct in assuming . . .") and add no outside data.

5. *Edit your own version.* When you are sure you have everything readers need, edit your version for conciseness.

   a. Cross out all your own needless words without compromising clarity. (See Appendix A.) Do not delete "a," "an," or "the" from any of your writing. Use complete sentences.

> The summer internship program in journalism gives the ~~journalism~~ student first-hand experience ~~at what goes on within the system of a real~~ newspaper staff.

   b. Cross out needless prefaces such as "The writer argues . . ." or "The researchers discovered . . ." or "Also discussed is. . . ."

   c. Use numerals for numbers, except when beginning a sentence. (See Appendix A.)

d. Combine related ideas through subordination within longer sentences. (See Appendix A.)

Choppy Sentences

The occupational outlook for journalists is good. There was a 53 percent increase in journalism jobs between 1947 and 1972. The national job increase was only 41 percent. Indications point to a continuation of this trend. This is partly due to an increase in weekly newspapers.

Revised

The occupational outlook for journalists is good, as evidenced by the 53 percent increase in journalism jobs between 1947 and 1972, in contrast with a national job increase of only 41 percent. Indications point to a continuation of this trend, which is partly due to an increase in weekly newspapers.

Five short sentences are combined into two longer ones, thus emphasizing relationships only implied in the original.

6. *Check your version against the original.* When your own version is tightened and refined, check it against the original to make sure that you have preserved the essential message, followed the original order, and added no extraneous comments or data.

7. *Rewrite your edited version.* Rewrite, following an introduction-body-conclusion structure. Add transitional words and phrases to reinforce the logical connection between related ideas ("*X* therefore *Y*" implies that *X* is related to *Y* in a cause-and-effect relationship, and so on).

8. *Document your source.* If you are summarizing another's work, identify the source in a bibliographical note immediately following the summary, and place directly quoted statements within quotation marks. (See Chapter 13 for documentation format.)

When summarizing your own information, eliminate steps 1 and 8. Otherwise the procedure is identical. Use this technique in preparing for essay examinations.

In a sense, a good summary functions like a. digital clock: it saves mental operations ("The big hand is on the 2 and the little on the 11; therefore it is 11:10") by giving a direct reading. When summarizing your own work, remember that, although the abstract is written last by the writer, it is read *first* by the reader. Though you may be tired, take the time to do a good job.

APPLYING THE STEPS

Let's apply the above steps to an actual summarizing process. Steps 1, 2, and 3 have been completed on the following article:

BRIGHTER PROSPECTS FOR WOMEN
IN ENGINEERING

One of the major <u>deterrents to women</u> considering engineering as a career is the <u>all-male image.</u> This barrier <u>is rapidly disappearing as the engineering image changes from that of a hard-hat roustabout at a construction site to that of a thoughtful, logical individual</u> who is genuinely <u>interested in solving the engineering and social problems</u> which face us today. True, she may still show up at a construction site in her hard hat, but her time is more apt to be spent at a desk working on new solutions. A female engineer — unlikely? Not quite.

<u>Although women make up</u> an unimpressive <u>1% of the engineering population, their ranks have been growing.</u> The latest Society of Women Engineers survey of schools accredited by the Engineering Council for Professional Development shows that female engineering enrollment increased from 1,035 during 1959–60 to 3,905 during the 1972–73 school year. This increase may not be as large as it appears on the surface. Only 128 schools replied to the 1959–60 survey. But with the advent of the Civil Rights Act of 1964 and, more recently, the implementation of the federal affirmative action program, as well as an increased awareness on the part of the schools, 201 responded last year. However, since the <u>number of female engineering students</u> per school <u>has increased, even as the number of males</u> enrolled at these schools <u>has decreased</u>, there is little doubt that the percentage of women enrolled in engineering undergraduate programs is growing.

JOBS COME FAST, PROMOTIONS SLOWLY

What happens when <u>the newly</u> minted <u>female engineer</u> tries to enter the field? Initially, she <u>is sought after</u> by almost every employer in sight. Once she is <u>on the job, however,</u> things change. On the average, <u>promotions do not come as rapidly</u> for women as they do for men.

Discrimination can be a double-edged sword, however, for unlike her male counterpart, <u>the female engineer is highly visible</u>, and <u>if she does an outstanding job, she may</u> very well <u>be rewarded faster than a man</u> would be. <u>If her performance is average</u>

(margin notes)

Combine as key sentence (thesis).

Delete example.

Include significant statistic.

Delete data of questionable accuracy.

Include significant finding.

Delete explanation repeating above finding.

Include significant finding.

Delete author's personal comment.

Include significant finding.

or slightly below average, she may be judged in
terms of a number of myths. Perhaps chief among
them is the notion that men (and women) don't like
to work for women. In my personal experience, I
have found that people who enjoy their work get it
done without any thought to whether their supervisor is a man or a woman.

> Delete author's
> personal comment.

Some echoes of other misconceptions about women
are still heard among engineers, and undoubtedly
contribute to the lag in promoting women to top
management ranks. Examples of these myths are:
(a) a company's public image will suffer if a woman
takes over a top management position, because men
have traditionally been the corporate leaders; (b)
a woman won't travel on sales trips, to plant inspections, to professional conferences and so forth; (c)
women don't want to accept responsibility; (d) a
woman's family will always take precedence over her
career. (One must ask, why shouldn't it take precedence over a man's as well?)

> Include factual
> interpretation and
> continuation of
> above finding.

> Delete rhetorical
> question.

It has also been argued that promotion policies
don't favor women because companies prefer long
tenure for those elevated to executive positions, and
they believe that turnover rates are greater for
women. But, not only do government figures show
that professional women have working careers comparable in length to those of men, it is also clear that
promotions generally accrue to men regardless of age
and experience. Over 20% of all male engineers are
in management, as opposed to an estimated 3% of
female engineers.

> Include continuation
> of finding.

> Include significant
> finding.

Admittedly, because the number of women in the
profession is small, the above figure is open to sampling error. Indeed, as many as 40% of the women
surveyed in 1972 by the Society of Women Engineers stated that they supervised groups which
ranged in size from teams to major organizations. It
must be noted, however, that members of SWE (and
engineering societies in general) are probably among
the more qualified and professionally active engineers.

> Delete statistics
> of questionable
> accuracy, along
> with the related
> explanation.

ATTITUDES VARY

A questionnaire on discrimination was included
in the 1972 SWE survey. In a classic case of
"which-came-first-the-chicken-or-the-egg?" the results

> Include significant
> date.

showed that those <u>women</u> who were <u>very successful</u> in terms of salary, responsibility, and years of experience <u>felt they had not encountered any discrimination.</u> Those women who were <u>moderately successful</u> <u>indicated</u> that there was <u>no discrimination</u> encountered <u>from</u> their <u>immediate superiors.</u> They felt, <u>however</u>, that people in the <u>upper</u> levels of <u>management</u> hierarchy <u>did discriminate</u> and that there was some evidence of discrimination by <u>coworkers.</u>

Include significant finding.

Include continuation of finding.

<u>Those</u> women <u>on</u> the <u>low side</u> of the average in terms of salary and responsibility <u>indicated</u> that they had encountered <u>discrimination at all levels.</u> It can be argued that these women have less ability than their male cohorts, and use "discrimination" as an excuse for their lack of advancement.

Delete obvious explanation.

SALARIES

<u>All women encounter discrimination</u>, perhaps not intentional or even conscious, from their male colleagues. This contention is borne out by the results of the SWE salary survey, compared with the results of the survey of Engineers Joint Council for the profession as a whole. For engineers <u>with 11 years'</u> <u>experience</u> (the median for women), <u>the median</u> <u>salary for female engineers is $14,200 per year, while</u> <u>that for all engineers with 11 years' experience is</u> <u>$16,700</u>, according to the EJC. Both surveys were completed <u>in 1972.</u> The disparity may be even greater because, again, SWE members are more professionally active than all engineers taken as a group.

Include factual interpretation.

Delete background information.

Include significant statistics and date.

Delete author's conjecture.

Of course, the engineering profession is not alone in this disparity in salaries. In the federal civil service, men average $14,328 per year, and women only $8,578. This is not because there are separate pay scales for women, but rather because women employees are heavily concentrated in lower-grade jobs.

Delete author's digression.

All is not bleak, <u>however. In 1973, the average</u> <u>starting salary</u> offered to <u>women engineering gradu-</u> <u>ates</u> at the <u>bachelor's degree</u> level was <u>$936 per</u> <u>month — $15 a month more than</u> the average <u>for</u> <u>men</u>, according to the College Placement Council. This represents a closing of the gap when <u>compared</u> <u>with 1971, when women were offered $8 a month</u> <u>less than men. Engineering</u> — the profession offering the highest starting pay for those with bachelor de-

Include significant finding, supporting statistics, and date.

grees, remains the only profession where salary offers are higher for women than for men.

    If one considers salary offers from private industry only, the salary gap between male and female engineers was even greater than the averages indicate, and favored women. But the federal government, which offered significantly lower salaries to entry-level female engineers than to males, dragged the overall averages closer together.

Include significant conclusion.

Delete explanation.

THE FUTURE?

    The current energy crisis and materials shortage indicate that this country is fast moving from a state of have to have not. The only way to maintain our current standard of living is through technology, which means that engineers will continue to be in great demand. It also means that the image of engineering will continue to change as attention is focused on sociological-technological problems. Consequently, we can expect women to enter the engineering profession in greater numbers.[2]

Include significant conclusion and restatement of thesis.

With the reading, rereading and underlining, and editing completed, we can move on to step 4: rewriting. Here is how the first rewrite of the underlined material might read; ideas are simply listed in order, without concern for length or subordination:

FIRST REWRITE

The all-male image has deterred women from engineering careers. This image of a hard-hat worker is giving way to that of a thoughtful individual working on today's engineering and social problems. Women comprise only 1% of the engineering population, but their ranks are growing. Female engineering students have increased in number while males have decreased. The new female engineer easily finds work but is not rapidly promoted. Because of high visibility, she may be promoted faster than a man if her performance is outstanding. If performance is average or below, she may be judged in terms of several myths: that people don't like to work for a woman; that a woman in top management harms a company's public image; that a woman

    [2] Naomi J. McAfee, "Brighter Prospects for Woman in Engineering," *Engineering Education* 64, no. 7 (April 1974): 502–504. Reprinted with permission from *Engineering Education* © 1974 The American Society for Engineering Education.

won't travel on business; that women won't accept responsibility; that her family takes precedence over her career; and that turnover rates are higher for women. But careers of professional women are as long as those of men. Men usually receive the promotions, regardless of age or experience.

In 1972, highly successful women engineers reported no discrimination. Moderately successful women sensed no discrimination from immediate superiors, but felt that higher management and coworkers did discriminate. Those with minimal success felt discrimination at all levels. All women do encounter salary discrimination. In 1972, the median salary for female engineers with 11 years' experience was $14,200 yearly, as opposed to $16,700 for all engineers with equal experience. However, in 1973, the average starting salary for women engineers with bachelor degrees was $936 per month — $15 more than for men. This figure contrasts with 1971 figures when women were offered $8 a month less than men. Engineering is the only profession where salary offers are higher for women than for men.

We now have a shortage of energy and materials. We can only maintain our living standard through technology. Demand for engineers will continue to grow. Their image will continue to change with new emphasis on sociological-technological problems. Thus, women are expected to enter the engineering field in greater numbers.

This version has all significant information, without regard for coherence. In the final draft, transitional terms and punctuation signals (see below) will be added, and related ideas combined to tighten the whole structure. Also, the length (roughly 330 words or 25 percent of the 1400-word original) will be further reduced. Word length, however, is less important than accurate emphasis and faithful representation. This version preserves the original emphasis by recording the brighter prospects as well as the continuing problems. Factual statements are faithfully represented because they are *fully* expressed. Imagine the distortion if, to save space, the statement, "The new female engineer easily finds work but is not rapidly promoted," were only partially expressed as, "The new female engineer easily finds work."

Here is the final draft with steps 5, 6, 7, and 8 completed. Transitional terms and connecting devices, including punctuation signals, are underlined:

SUMMARY OF "BRIGHTER PROSPECTS
FOR WOMEN IN ENGINEERING"

The all-male, hard-hat image, which has deterred women from engineering, is changing to that of a thoughtful individual working on today's engineering and social problems. Although women comprise only 1% of engineers, their growing ranks are evidenced by an increase in female engineering students, contrasted with a decrease in male students. The graduating female easily finds work but no rapid promotion unless her performance is outstanding.

An average or below-average performance may be judged in terms of several myths: that people dislike working for women; that a woman in top management harms a company's image; that she won't travel or accept responsibility; that her family takes precedence over her career; and that female turnover rates are higher. In fact, women's professional careers are as long as men's, but men usually receive the promotions.

In 1972, highly successful women engineers reported no discrimination. Moderately successful women sensed none from immediate superiors, but felt that higher management and coworkers did discriminate. Marginally successful women claimed discrimination at all levels. Women do encounter salary discrimination: the 1972 median salary for female engineers with 11 years' experience was $14,200 yearly, as opposed to $16,700 for all equally experienced engineers. However, the 1973 average starting salary for women graduates was $936 monthly — $15 more than for men. By contrast, in 1971, women received $8 less than men. Thus, only in engineering are women receiving higher offers.

Current energy and materials shortages increase our reliance on technology to maintain living standards; consequently, the demand for engineers with a sociological-technological commitment should attract more women. (Naomi J. McAfee, "Brighter Prospects for Women in Engineering," *Engineering Education* 64 (April 1974): 502–504.)

This version is trimmed, edited, and tightened: word count is reduced to roughly 20 percent of the original. A summary of this length will serve well in many situations, but in others you might want a substantially briefer and more compressed summary — say, 125 to 150 words, or about 10 percent of the original:

> The all-male image of engineering is changing, and although women comprise only 1% of engineers, their ranks are growing. Female students are increasing while males decrease. Although female graduates easily find work, only the outstanding are rapidly promoted, with average or lower performance often judged according to conventional myths about women in "male" professions.
>
> In 1972 no discrimination was reported by the highly successful women engineers; selective discrimination, by the moderately successful; and general discrimination, by the marginal. Women *do* encounter salary discrimination: the 1972 median salary for experienced females was $14,200, compared with $16,700 for all equally experienced engineers. However, in 1973 women graduates commanded starting salaries of $15 more monthly than men, whereas in 1971 they had received $8 less. Only in engineering are women receiving higher offers.
>
> As resource shortages increase reliance on technology to maintain living standards, the demand for "sociological-technological" engineers should attract more women.

Notice that the essential message is still intact; related ideas are again combined and fewer supporting details are included. Clearly, length is adjustable according to your audience and purpose.

WRITING THE DESCRIPTIVE ABSTRACT

A summary (or informative abstract) reflects *what the original contains,* whereas a descriptive abstract reflects *what the original is about.* These differences are best clarified with an analogy. Imagine that you are describing your recent summer travels to a friend; you have two options: (1) You might simply mention the places you visited in chronological order. This catalogue of major areas would convey the basic nature of your trip. (2) In addition to describing your itinerary, you might describe the significant experiences you had in each area you visited. Option 1 is like a roadmap, an overview of the areas covered in your journey. This option is analogous to a descriptive abstract. Option 2, on the other hand, is expanded to include the main points within each area. This second, more detailed, option, is analogous to a summary (or informative abstract).

A descriptive abstract, then, presents a table of contents (a list of major topics) in related-sentence form. Whereas the summary contains the meat of the original, the descriptive abstract contains only its skeletal structure; in effect, a descriptive abstract is "a summary of a summary," as shown below:

ABSTRACT OF "BRIGHTER PROSPECTS FOR WOMEN IN ENGINEERING"

As the all-male image of engineering changes, the number of women engineers increases. Although persisting sexist myths affect women's chances for promotion, women's salaries are increasing. Growing demands for sociologically oriented engineers promise to attract more women to the field.

Because a descriptive abstract simply previews the focus of the original, it is always brief — usually no longer than a short paragraph. Abstracts one or two sentences long often accompany article titles in journal and magazine tables of contents; they give readers a bird's-eye view.

PLACING SUMMARIES AND ABSTRACTS IN YOUR REPORT

If your reader asks for a descriptive abstract of your report, place it in front, on a separate page, right after your table of contents (or on the title page). It is usually single-spaced. With a descriptive abstract in front, your summary (or informative abstract) will go in the conclusion section of your report. Sometimes you will be asked to place your summary in front, instead of writing a descriptive abstract.

CHAPTER SUMMARY

A summary (or informative abstract) — like this one — is an economical way to communicate because it compresses a longer message into essentials. An effective summary extracts *only* the major points from the original and (usually) presents them in a nontechnical style. The summary stands independently as a complete message and adds nothing to the original. Like most good writing, it has an introduction-body-conclusion structure. The key word in summary writing is *conciseness;* that is, the summary must be brief but also clear and complete. It is better to make it a bit long than to omit some key point and distort the original. You can summarize your own work only after you have written the original version.

Follow these steps in writing your summary:

1. Read the entire original.
2. Reread and underline (or copy) the major points.
3. Edit the underlined or copied data to cut out needless words.
4. Rewrite the material in your own words.
5. Edit your version by crossing out needless words and prefaces, using digits for numbers, and combining related ideas through subordination.
6. Check your version against the original for accuracy.
7. Rewrite your version in an introduction-body-conclusion structure with all needed transitions.
8. Document your source if you have summarized another's work.

Analyze your audience and purpose carefully in order to adjust the length of your summary to the demands of the situation.

Whereas a summary (or informative abstract) reflects what the original contains, a descriptive abstract reflects what the original is about (a summary of a summary). Descriptive abstracts are short and simply give a bird's-eye view.

A descriptive abstract always belongs in front of a report, but the summary (or informative abstract) may be in the conclusion or in front, as readers request.

REVISION CHECKLIST

Use this checklist as a guide to refining the content, arrangement, and style of your summaries.

Content

1. Does the summary contain only the essential message (controlling idea; major findings and interpretations; important names, dates, statistics, measurements; conclusions and recommendations)?
2. Does it make sense as an independent piece?
3. Is the summary accurate (checked against the original)?
4. Is it free from personal comments or other additions to the original?
5. Is it free from needless details?
6. Is it short enough to be economical and long enough to be clear and complete?
7. Is the source documented?
8. Is the descriptive abstract an effective "summary of a summary" in that it clearly expresses what the original is about?

Arrangement

1. Is the summary coherent?
2. Are there enough transitions between related ideas?
3. Does it follow an introduction-body-conclusion structure?
4. Does it follow the order of the original?
5. Have you placed the summary (informative abstract) or descriptive abstract at the proper location in your report?

Style

1. Is it written in a style that all readers can readily understand?
2. Is it free from needless words?
3. Are all sentences clear, concise, and fluent (pages 39–50)?
4. Is it written in correct English (mechanics and usage; see Appendix A)?

Now list those features of your summary that need improvement.

EXERCISES

1. In a unified and coherent paragraph, describe the differences between a summary and an abstract in enough detail to give general readers a clear understanding of the distinction.

2. *In class:* Organize into groups of four or five and choose a topic for group discussion: a social problem, a political issue, a campus problem, plans for an event, suggestions for individual energy conservation, etc. Discuss the topic for one full class period, taking notes of significant points. Afterward, organize and edit your notes in line with the directions for "Writing the Summary." Next, write a unified and coherent individual summary of the group discussion in no more than 200 words. Finally, as a group, compare your individual summaries for accuracy, emphasis, conciseness, and clarity.

3. In one or two paragraphs, discuss the specific kinds and frequency of summary-writing assignments you expect to encounter in your occupation. Will your reading audience be mainly colleagues, superiors, or customers, clients, or other general readers? (If you don't know, ask someone in the field.)

4. Read each of the following paragraphs and make lists of the significant ideas comprising the essential message in each. Next, write a unified and coherent summary of each paragraph.

In recent years, ski-binding manufacturers, in line with consumer demand, have redesigned their bindings several times in an effort to achieve a noncompromising synthesis between performance and safety. Such a synthesis depends on what appear to be divergent goals: Performance, in essence, is a function of the binding's ability to hold the boot firmly to the ski, thus enabling the skier to change rapidly the position of his skis without being hampered by a loose or wobbling connection. Safety, on the other hand, is a function of the binding's ability both to release the boot when the skier falls, and to retain the boot when subjected to the normal shocks of skiing. If achieved, this synthesis of performance and safety will greatly increase skiing pleasure while decreasing accidents.

Contrary to public belief, sewage-treatment plants do not fully purify sewage. The product that leaves the plant to be dumped into the leaching (sievelike drainage) fields is secondary sewage containing toxic contaminants such as phosphates, nitrates, chloride, and heavy metals. As the secondary sewage filters into the ground, this conglomeration is carried along. Under the leaching area develops a contaminated mound through which ground water flows, spreading the waste products over great distances. If this leachate reaches the outer limits of a well's drawing radius, the water supply becomes polluted. Furthermore, because all water flows essentially toward the sea, more pollution is added to the coastal regions by this secondary sewage.

5. In 500 words or less, discuss your reasons for applying for a certain job or to a certain school.

6. Attend a campus lecture on a topic of interest to you and take notes of the significant points. Write a clear and organized summary of the lecture's essential message.

7. Use your own modified technique of the steps in "Writing the Summary" as a study aid in preparing for an examination in one of your courses. After taking the exam, write one or two paragraphs evaluating this technique for putting yourself in control of a large and diverse body of information.

8. Find three examples of abstracts or summaries from journals and magazines in your school library and bring them to class. As a group, analyze selected examples on an overhead projector for clarity and meaning.

9. Find an article in the library pertaining to your major field or area of interest and write both an abstract and a summary of the article.

10. Select a long paper you have written for one of your courses; write an abstract and a summary of the paper.

11. Read the following article and write both a descriptive abstract and a summary (or informative abstract) of it, using the steps under "Writing the Summary" as a guide. Bring your summary to class and exchange it with a classmate for proofreading according to the revision checklist. When your proofread copy is returned, revise as necessary before submitting it to your instructor.

EPA SETS STANDARDS FOR NEW MOTORCYCLES
AND MOTORCYCLE REPLACEMENT EXHAUST SYSTEMS

The U.S. Environmental Protection Agency has issued standards which limit the noise from newly manufactured motorcycles and motorcycle replacement exhaust systems.

The standards will be phased in over a two- to five-year period beginning in 1983. Mopeds are considered motorcycles by EPA and will be covered, but will have only one standard to meet — also imposed in 1983. No existing motorcycles, or any built before 1983, will be affected.

Some 93 million people are daily affected by traffic noise. Motorcycles are an integral and important part of the traffic stream.

The motorcycle manufacturing industry has been greatly concerned about potential restrictions on commerce as a result of being required to produce new motorcycles that will comply with a multiplicity of differing state and local noise standards. This regulation will preempt state and local noise standards for newly manufactured motorcycles, thereby providing national uniformity of treatment.

Motorcycles are the source of more annoyance and adverse community response than any other single traffic noise source. EPA realizes that much of this negative response comes about because of excessively loud, exhaust-modified motorcycles. Because of this, the Agency believes that both the noise from newly manufactured motorcycles and from modified motorcycles must be controlled if the public health and welfare benefits Congress expected when it passed the Noise Control Act are to be realized. To control the noise from exhaust-modified motorcycles, the combined efforts of the federal government and state and local governments are essential. In addition to providing the labeling and antitampering provisions of the regulation EPA will assist state and local governments in establishing complementary noise control programs including ordinances that will prohibit the use of noisy exhaust systems.

EPA has set 80 decibels (dB) as the most stringent noise standard for street motorcycles and small off-road motorcycles. Although the standards are less stringent than those that were proposed, EPA anticipates that these standards will, on the average, reduce the noise from new street motorcycles by 5 dB and by 2 to 7 dB on new offroad motorcycles by 1986. The exhaust system regulation and the "antitampering" and labeling provisions of the motorcycle regulation in combination with strong complementary state and local programs, should help reduce exhaust-modified motorcycles to between one-half and one-fourth their current numbers.

These reductions are expected to result in a 55 to 75 percent decrease in interferences with human activities (such as sleeping, conversation), depending on the extent to which state and local governments are able to contribute to reducing the numbers of exhaust-modified motorcycles. Likewise, these reductions are expected to result in a 7 to 11 percent decrease in the severity and extent of overall traffic noise impact, again depending on in-use enforcement.

The standards and effective dates applicable to new motorcycles and to new motorcycle replacement exhaust systems, are as follows:

1. Street motorcycles; 83 dB; January 1, 1983; 80 dB; January 1, 1986.
2. Moped type street motorcycles; 70 dB; January 1, 1983.
3. Off-road motorcycles (displacement 170 cc and below); 83 dB; January 1, 1983; 80 dB; January 1, 1986; (displacement more than 170 cc); 86 dB; January 1, 1983; 82 dB; January 1, 1986.

EPA expects the costs of compliance to be reflected in increased purchase prices for motorcycles and exhaust systems. For street motorcycles, increases will average approximately 2% (or $36.00). The estimated purchase price increase for off-road motorcycles will average 2% (or $21.00). For replacement exhaust systems, the estimated purchase price increase will average 25% (or $30.00).

Although higher retail prices could result in some initial lost sales, total industry sales (in terms of both units and dollars) are projected to significantly expand in the next decade. Furthermore, because all mopeds that the Agency has tested, which are being sold in the United States, already comply with the 70 dB level set for these vehicles, EPA foresees no impact on moped prices and consequently on moped sales.

This is the fourth noise control regulation EPA has issued to limit traffic noise. Regulations have been issued for interstate motor carriers (October 19, 1974), newly manufactured medium and heavy trucks (April 13, 1976), and newly manufactured garbage trucks (October 1, 1979). On the same day that the motorcycle noise standard was issued, EPA proposed an amendment to the testing requirement of the regulations.

The proposed amendment would require manufacturers to take one additional step in their testing program over and above what is now required of them as a result of the final regulations. Specifically, under the proposed amendment, manufacturers would be required to remove all easily removable components from their exhaust systems before conducting the tests necessary to show compliance with applicable standards. The Agency believes that this amendment will enhance the effectiveness of the regulation since the control of motorcycle noise is dependent on exhaust systems retaining their noise suppression performance beyond the time of sale. These amendments are expected to encourage manufacturers to design exhaust systems in ways which will reduce the incidence of tampering by consumers.[3]

[3] U.S. Environmental Protection Agency, *EPA Environment News*, Boston, February 1981, pp. 12–13.

5

Defining
Your Terms

DEFINITION

PURPOSE OF DEFINITIONS

USING DEFINITIONS SELECTIVELY

ELEMENTS OF AN EFFECTIVE
    DEFINITION
  Plain English
  Basic Properties
  Objectivity

CHOOSING THE BEST TYPE OF
    DEFINITION
  Parenthetical Definition
  Sentence Definition
    Classifying the Term
    Differentiating the Term
  Expanded Definition

EXPANDING YOUR DEFINITION
  Etymology
  History and Background
  Example
  Graphic Illustration
  Analysis of Parts
  Comparison and Contrast
  Basic Operating Principle
  Specific Materials or Conditions
    Required

APPLYING THE STEPS

PLACING DEFINITIONS IN YOUR
    REPORT
  Parenthetical Definitions
  Sentence Definitions
  Expanded Definitions

CHAPTER SUMMARY

REVISION CHECKLIST

EXERCISES

DEFINITION

To define a term is to give its precise meaning. As we have said earlier, *clarity* is the most important element in your writing. Clear writing begins with clear thinking; clear thinking begins with a clear understanding of what all the terms mean. Therefore, clear writing begins with a careful definition that both reader and writer understand. Always define something before you discuss it.

PURPOSE OF DEFINITIONS

Virtually every specialty has its own "language," its technical terms. Engineers, architects, and builders talk about "prestressed concrete," "tolerances," or "trusses"; psychologists, social workers, counselors, and police officers may use terms like "manic-depressive psychosis," "sociopathic behavior," or "repression"; lawyers, real estate brokers, and investment counselors discuss "easements," "liens," "amortization," or "escrow accounts" — and so on. Any of these terms is likely to be unfamiliar to nonspecialists. In your own writing, keep your reader's needs in mind as you identify the terms that require definition.

When writing to colleagues you rarely have to define specialized terms used in your field (unless the term is new), but reports are often written for the layperson — the client or some other general reader. When you write for these people, think about their needs. Don't force readers to consult a dictionary or encyclopedia to make sense of your message. Assume that your readers know less than you — general readers, much less! Make your meaning clear with good definitions.

Most of the specialized terms mentioned above are concrete and specific. Once a term such as "truss" has been defined in enough detail to suit the reader's purpose, its meaning will not be appreciably different in another context. And when a term is highly technical it is easy enough to figure out that it should be defined for certain readers. Anyone who isn't a specialist knows that he or she has no idea what "prestressed concrete" or "diffraction" means. However, your readers are considerably less likely to be aware that more familiar terms like "disability," "guarantee," "tenant," "lease," or "mortgage" acquire very specialized meanings in specialized contexts. This is where definition (by all parties) becomes crucial. What "guarantee" means in one situation is not necessarily what it will mean in another. That is why a legal contract is a detailed definition of the subject of the contract.

Let's assume that you're shopping for disability insurance to provide a steady income in case injury or illness should cause you to lose your job. Besides comparing the prices of various policies, you will want each company to define "physical disability." Although Company A offers the least expensive policy, it might define physical disability as your inability to work at any job whatsoever. Therefore, should a neurological disease prevent you from continuing your work as a designer of delicate, transistorized electronic devices, without disabling you for work as a salesperson or clerk, you might not qualify as "disabled," according to Company A's definition. In contrast, Company B's policy, which is more expensive, might define physical disability as your inability to work at your specific job. Thus, although all companies use the term "physical disability," they may not mean the same thing by it.

Similar problems in definition arise with certain purchases. Should you plan to purchase a condominium, you will have to obtain the developer's full definition of "condominium." Otherwise, your failure to read the "fine print" might be disastrous. The same is true for the terms of "warranty" on a new automobile. Because you are legally responsible for all documents carrying your signature, you must understand the importance and technique of clear definition.

USING DEFINITIONS SELECTIVELY

The growth of technology and specialization will continue to make definitions an important part of communication. However, use definitions only when the audience needs them. The point is to know for whom you're writing, and why. Reports in *Psychology Today* (with a general readership), for instance, define many terms not defined in reports to psychologists. It stands to reason that the expert or informed reader will require fewer definitions. If you can't pinpoint your audience, assume a general readership and define generously.

Depending on your subject, purpose, and audience, individual definitions can vary greatly in length. As a rule, make your definition long enough to be understood by a general reader. Often you will need only a *parenthetical definition* —

a few words or a synonym placed in parentheses after the term. Sometimes your definition will require one or more complete sentences. Certain terms require a definition that extends to hundreds of words — an *expanded definition*. (Each type will be discussed later.)

Your choice of parenthetical, sentence, or expanded definition depends on the amount of information your readers need, and that, in turn, depends on why they need it. For instance, "carburetor" could be defined in a single sentence (as shown on page 91) telling the reader what it is and how, in general, it works. This definition, however, should be greatly expanded for the student mechanic who needs to know where the word *carburetor* comes from, how the device was developed and perfected, what it looks like, how it is used, and how its parts work together.

In most cases, abstract and general terms ("loan," "partnership," etc.) will need expanded definitions.

ELEMENTS OF AN EFFECTIVE DEFINITION

For all definitions, use these guidelines:

Plain English

Your purpose is to clarify meaning, not muddy it. Use simple language.

Incorrect	A tumor is a neoplasm.
Correct	A tumor is a growth of cells that occurs in the body, grows independently of surrounding tissue, and serves no useful function.
Incorrect	A solenoid is an inductance coil that serves as a tractive electromagnet. (*This definition might be appropriate for an electrical engineering manual, but is too specialized for the general reader.*)
Correct	A solenoid is an electrically energized coil that converts electrical energy to magnetic energy capable of performing various mechanical functions.

Basic Properties

Any single thing has characteristics that make it different from all others. Its definition should express clearly these basic properties. Thus, a thermometer can be defined in terms of its singular function: it measures temperature; this is the primary information that your reader needs. All other data about a thermometer, such as types, special uses, materials used in construction, and cost are secondary. On the other hand, a book cannot primarily be defined in terms of its function, because books can have several functions. A book can be used to write in or to display pictures (if the pages are blank), to record financial transactions, to read (if the pages are printed or written), and so on. Also, other items — individual sheets of paper, posters, newspapers, picture frames, etc. — serve the same functions. The basic property of a book is physical: it is a bound volume of pages. This is the feature an item must possess in order to be called a book; it is what your reader would have to know *first* in order to understand what a book is. Comments about types of books, uses, sizes, contents, etc. provide only secondary information.

Objectivity

Make your definitions objective. Tell your reader what the item is, not what you think of it. Personal comments and interpretations should come only *after* your data and only at the specific request of your reader. "Bomb," for instance, may be defined properly as "an explosive weapon that is detonated by impact, proximity to an object, a timing mechanism, or other predetermined means." If instead you define a bomb as "a weapon devised and perfected by hawkish idiots to eventually destroy themselves and the world," you are editorializing; furthermore, you are not presenting a bomb's basic property.

Likewise, in defining something like "diesel engine," simply tell your reader what it is and how it works. You might think that diesels are too noisy and sluggish for use in automobiles, but you should reserve these judgments until *after* your definition — that is, if your reader has asked for such comments (as in a comparative study of small buses for your town's public transportation system). Otherwise, let the data speak for themselves.

CHOOSING THE BEST TYPE OF DEFINITION

After deciding to define a term in your report, choose the most appropriate type of definition: parenthetical, sentence, or expanded.

Parenthetical Definition

A parenthetical definition is the simplest type. It explains the term in a word or phrase and often consists of a synonym in parentheses immediately following the term:

> The effervescent (bubbling) mixture was quickly discarded.
> The leaching field (sievelike drainage area) needs fifteen inches of crushed stone.

Another option is to express your definition as a clarifying phrase:

> The trees on the site are mostly deciduous; that is, they shed their foliage at season's end.

Use parenthetical definitions to give readers a general understanding of specialized terms so they can easily follow the discussion where these terms are used. A parenthetical definition of "leaching field" might be adequate in a progress report to a client whose house you are building. A town report titled "Groundwater Contamination from Leaching Fields," however, would call for an expanded definition.

Sentence Definition

Often, a clear definition requires more than just a word or phrase in parentheses. A sentence definition (which may be stated in more than one sentence) follows a fixed structure: (1) the name of the item to be defined, (2) the class (specific group) to which the item belongs, and (3) the features that differentiate the item from all other items in its class.

Term	Class	Distinguishing Features
polygraph	a measuring instrument	that simultaneously records changes in pulse, blood pressure, and respiration, and is often used in lie detection
carburetor	a mixing device	in gasoline engines which blends air and fuel into a vaporized mixture for combustion within the cylinders
transit	a surveying instrument	that measures horizontal and vertical angles
diabetes	a metabolic disease	caused by a disorder of the pituitary gland or pancreas, and characterized by excessive urination, persistent thirst, and, often, an inability to metabolize sugar
liberalism	a political concept	based on belief in progress, the essential goodness of man, the autonomy of the individual, and standing for the protection of political and civil liberties
brief	a legal document	containing all the facts and points of law pertinent to a specific case, and filed by an attorney before arguing the case in court
stress	an applied force	that tends to strain or deform a body

In their presentation, these elements are combined into one or more complete sentences.

> Diabetes is a metabolic disease caused by a disorder of the pituitary gland or pancreas. This disease is often characterized by excessive urination, persistent thirst, and often an inability to metabolize sugar.

Sentence definition is especially useful if you need to stipulate the precise working definition of a term that has several possible meanings. In a construction, banking, or real estate report, for example, a term such as "qualified buyer" could have different meanings for different readers. The same is true for "compact car" in a report comparing various brands of cars for use in the company fleet.

State your working definitions at the beginning of your report, as in the following example:

> Throughout this report, the term "disadvantaged student" is taken to mean. . . .

Classifying the Term

Be precise in your classification. The item's class will reflect its similarities to all other items with common attributes (as discussed in Chapter 6). The narrower your class, the more specific your meaning. For example, "transit" is correctly classified as a "surveying instrument," not as a "thing" or simply as an "instrument." Likewise, "stress" is correctly classified as "an applied force"; to say that stress "is what . . ." or "takes place when . . ." or "is something that . . ." is incorrect; these formulas are not words of classification. Also, select the most accurate terms of classification: "Diabetes" is accurately classified as "a metabolic disease," not as "a medical term."

Differentiating the Term

Differentiate the term by separating the item it names from every other item in its class. If the distinguishing features can be applied to more than one item, your definition is imprecise. Make these features narrow enough to pinpoint the item's unique identity and meaning, yet broad enough to be inclusive. For example, a definition of "brief" as "a legal document introduced in a courtroom" is not narrow enough because the definition doesn't differentiate "brief" from all other legal documents. Conversely, a differentiation of "carburetor" as "a mixing device used in automobile engines" is too narrow because it fails to indicate the carburetor's use in all other gasoline engines.

Also, avoid circular definitions (repeating, as part of the distinguishing features, the word you are defining). Thus "stress" should not be defined as "an applied force that places stress on a body." In short, the class and distinguishing features must express the item's basic property.

Expanded Definition

The sentence definition of "solenoid" on page 89 would be good for a general reader who simply needs to know what a solenoid is. An instruction manual for mechanics or mechanical engineers, however, would define this item in great detail (as on pages 97–99); these readers need to know what a solenoid is, how it works, and how it can be used. So the choice of length and detail in a definition of a concrete and specific term depends on the purpose of the definition and the needs of the audience.

The problem with defining an abstract and general word, such as "condominium" or "bodily injury," is different. "Condominium," for example, is a vaguer term than "solenoid" (solenoid A is pretty much like solenoid B) because the former refers to a wide range of ownership agreements; therefore, its meaning is much more variable and needs to be spelled out.

Concrete, specific terms such as "diabetes," "transit," and "solenoid" often can be defined by a sentence, and will require an expanded definition only according to particular audience needs. Terms such as "disability" and "condominium," however, will almost always require expanded definition. The more general and abstract the term, the more likely the need for an expanded definition.

An expanded definition may be a single paragraph (as in defining a simple tool) or may extend to many pages (as in defining a new aerospace navigational device); sometimes the definition itself comprises the whole report.

The following excerpt from an automobile insurance policy defines the coverage for "bodily injury to others." Its style and detail make this definition clear to general readers. Instead of the fine print "legalese" seen in many policies, this definition is written in plain English.

PART 1. BODILY INJURY TO OTHERS

Under this Part, we will pay damages to people injured or killed by your auto in Massachusetts accidents. Damages are the amounts an injured person is legally entitled to collect for bodily injury through a court judgment or settlement. We will pay only if you or someone else using your auto with your consent is legally responsible for the accident. The most we will pay for injuries to any one person as a result of any one accident is $5,000. The most we will pay for injuries to two or more people as a result of any one accident is a total of $10,000. This is the most we will pay as the result of a single accident no matter how many autos or premiums are shown on the Coverage Selections page.

We will *not* pay:

1. For injuries to guest occupants of your auto.

2. For accidents outside of Massachusetts or in places in Massachusetts where the public has no right of access.

3. For injuries to any employees of the legally responsible person if they are entitled to Massachusetts workers' compensation benefits.

The law provides a special protection for anyone entitled to damages under this Part. We must pay their claims even if false statements were made when applying for this policy or your auto registration. We must also pay even if you or the legally responsible person fails to cooperate with us after the accident. We will, however, be entitled to reimbursement from the person who did not cooperate or who made any false statements.

If a claim is covered by us and also by another company authorized to sell auto insurance in Massachusetts, we will pay only our proportional share. If someone covered under this Part is using an auto he or she does not own at the time of the accident, the owner's auto insurance pays up to its limits before we pay. Then, we will pay up to the limits shown on your Coverage Selections page for any damages not covered by that insurance.

EXPANDING YOUR DEFINITION

The following strategies will help you expand your own definitions. Each strategy should be amplified by description and by synonyms or analogies whenever possible. Always begin an expanded definition with a formal sentence definition. In developing your expanded definition, use only the expansion strategies that serve your reader's needs.

Etymology

Often, the history of a word (origin, development, and changes in meaning) sheds light on the basic properties of the item it names. For example, arbitration is the legal process of settling a dispute by obtaining the binding (obligatory) judgment of a third party. This term is derived from the Latin *arbitrari,* meaning "to examine, give judgment." The word was first used in an informal sense in 1634: "To mediate in a friendly manner in a way of arbitration." And more explicitly in 1716: "To put their differences to the Arbitration of some of their Brethren." [1] Standard college dictionaries contain some of this information, but *The Oxford English Dictionary* and various encyclopedic dictionaries of science, technology, business, etc. are your best etymological sources.

Modern technical terms are often derived from two or more traditional terms. Thus, "transceiver" is derived from "transmitter" and "receiver," and defined as "a module composed of a radio receiver and transmitter."

---

[1] From *The Oxford English Dictionary* (New York, Oxford University Press, 1971). Used by permission.

History and Background

The definition of specialized terms like "radar," "bacteriophage," "laser," or "X ray" can often be clarified through a background discussion: discovery of the item, subsequent development, method of manufacture, changing applications, and possibilities for use in exploration, medicine, etc. Specialized encyclopedias are a good source of background information.

Example

A definition containing familiar examples is very helpful. Be sure to tailor your example to your reader's level of specialized knowledge. Thus a definition of "economic inflation," written for the general reader, could effectively use rising fuel prices as an example of higher costs per unit volume. Likewise, a definition of "clothing fashion" could be clarified by the example of changes in skirt lengths. For a more specialized reader, a definition of the mineral "borax," written for a student of ecology or chemistry, should mention its use as a cleaning agent and as a softener in detergents.

Graphic Illustration

A well-labeled diagram can clarify certain definitions. The figure should be introduced by an identifying sentence: "Figure 5-1 illustrates the construction of a spark plug." If your illustration is borrowed, credit your source at the bottom-left corner of the frame. Further explanation of the figure, if needed, should *follow* the illustration. Unless an illustration occupies one full page, or more, don't place it on a separate page. Include it within the text of your definition.

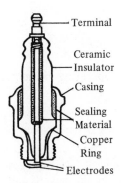

From *The McGraw-Hill Dictionary of Technical and Scientific Terms* (New York: The McGraw-Hill Book Company, 1974), p. 1388. Reprinted by permission.

FIGURE 5-1   A Spark Plug in Cross-section

Analysis of Parts

Many items or processes consist of several parts and are best defined through detailed explanation of each part.

> The standard frame of a pitched-roof wooden dwelling is composed of floor joists, wall studs, roof rafters, and collar ties.

> Psychoanalysis is an analytic and therapeutic technique consisting of four major parts: (1) free association, (2) dream interpretation, (3) analysis of repression and resistance, and (4) analysis of transference.

In discussing each part, of course, you would further define specialized terms like "floor joists" and "repression."

Comparison and Contrast

To compare is to identify similarities or likenesses; to contrast is to emphasize differences or dissimilarities. Whenever possible, compare the item with a more familiar one, or contrast it to various other models, sizes, etc.

Comparison	A cog railway, like a roller coaster, relies on a center cogwheel and a cogged center rail for transporting its cars up steep inclines.
Contrast	The X-55 fiberglass ski provides more edge control in icy conditions than its lighter but more durable aluminum counterpart, Model A-32.
Comparison and Contrast	Mediation, like arbitration, is a form of settling disputes; however, it differs from arbitration in that the decision of the mediator is not binding to the parties in the dispute.

Basic Operating Principle

Any mechanical item works according to a basic operating principle whose explanation should be part of your definition:

> The Model A-23 automobile jack operates on the principle of a simple lever. The lever is a machine consisting of a rigid bar pivoted on a fixed fulcrum and used for raising heavy objects.

> A clinical thermometer works on the principle of heat expansion: as the temperature of the bulb increases, the mercury inside expands and a thread rises into the hollow stem.

Even abstract items or processes can be explained this way:

Economic inflation functions according to the principle of supply and demand: if an item or service is in short supply, its price increases in proportion to its demand; this principle is evidenced by the effect of the recent United States grain sales to the Soviet Union.

Special Materials or Conditions Required

Some items and processes are highly sensitive or volatile. These may require special materials, conditions, or handling. A detailed definition should include this important information.

Fermentation is the chemical reaction that splits complex compounds into simpler substances (as when yeast converts sugar to carbon dioxide and alcohol). This process has several special requirements: (1) a causative agent such as yeast, (2) a controlled PH (acid-base balance), (3) airtight and sterilized containers, and (4) a narrow temperature range.

In order to obtain quality beer, the prospective home brewer needs to know all these requirements. More abstract subjects may also be defined in terms of their special conditions.

To be held guilty of libel, a person must have defamed someone's character through written or pictorial statements.

APPLYING THE STEPS

The following expanded definitions use several methods of amplification. As a study aid, the specific expansion strategies are identified in the right margin. Notice that each definition, like a good essay, is unified and coherent: each paragraph is developed around a central idea and logically connected to other paragraphs. The discussions are readable and easy to follow, with graphic illustrations incorporated into the text. Transitional words and phrases underscore the logical connection between related ideas. Each definition is written at a level of technicality that will connect with the intended audience.

EXPANDED DEFINITION
OF "SOLENOID"
(Written for Student Mechanics)

A solenoid is an electrically energized coil that forms an electromagnet capable of performing various mechanical functions. The term, "solenoid," is derived from the word, "sole," which in reference to

Formal sentence definition

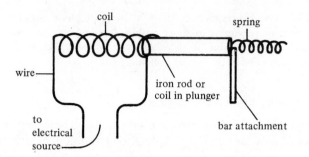

Graphic illustration

FIGURE 1   Lateral Diagram of a Plunger-Type Solenoid

electrical equipment means "a part of," or "contained   Etymology
inside, or with, other electrical equipment." The
Greek word *solenoides*, means "channel," or "shaped
like a pipe."

A simple, plunger-type solenoid consists of a coil   Description and
of wire attached to an electrical source, and an iron   analysis of parts
rod that passes in and out of the coil at right angles
to the spiral. A spring holds the bar outside the coil
when the current is deenergized, as shown in Figure 1.

When the coil receives electrical current, it becomes
a magnet and thus draws the iron bar inside, along   Special conditions
the length of its cylindrical center. With a lever at-   and principle of
tached to its end, the bar can transform electrical   operation
energy into mechanical force. The amount of mechanical force produced is determined by the product of the number of turns in the coil, the strength
of the exciting current, and the magnetic conductiv-
ity of the iron rod.

The plunger-type solenoid, shown in Figure 1,
is commonly used in the starter motor of an auto-   Example and analysis
mobile engine. It is 4½ inches long and 2 inches in   of parts
diameter, with a steel casing attached to the casing
of the starter motor. A linkage (pivoting lever) is
attached at one end to the iron rod of the solenoid,
and at the other end to the drive gear of the starter,
as shown in Figure 2. When the ignition key is
turned, current from the battery is supplied to the   Explanation of
solenoid coil and the iron rod is drawn inside the   illustration
coil, thereby shifting the attached linkage. The
linkage, in turn, engages the drive gear, activated

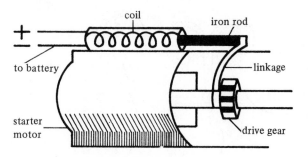

coil

iron rod

to battery

linkage

Graphic illustration

starter
motor

drive gear

FIGURE 2   Lateral Diagram of Solenoid and Starter-
Motor Assembly

by the starter motor, with the flywheel (the main rotating gear of the engine).

Because of the many uses of the solenoid, its size varies according to the amount of work it must do. Therefore, a small solenoid will have a small wire coil, hence a weak magnetic field. The larger the coil, the stronger the magnetic field; in this case, the rod in the solenoid is capable of doing harder work. It is easy to understand that an electronic lock for a standard door would require a much smaller solenoid than one for a large bank vault.

Comparison of sizes and applications

THE INTRAUTERINE DEVICE:
AN EXPANDED DEFINITION
(Written by a Nurse Practitioner for Patients)

The intrauterine device, or IUD, is a small plastic (occasionally metal) device that is placed semi-permanently inside the uterus by a trained person to prevent conception. The term is derived from the Latin *intra,* meaning "inside or within," and the Latin and French *uterus,* meaning "womb."

Formal sentence definition

Etymology

A primitive type of IUD has been used in camels for centuries. In the Middle East, camel herders prepared for crossing the desert by inserting smooth pebbles, the size of olive pits, into the uteri of their camels to prevent inopportune pregnancies.

History and background

Manufactured IUDs have been used in women for about one century. The first devices were metal prongs placed inside the uterus, with a stem and button-like disk outside the cervix. Heavy bleeding

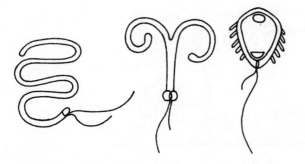

Graphic illustration

FIGURE 1   From Left to Right: Lippes' Loop, Saf-T-
Coil, and Dalkon Shield in Side View

and infection were common side effects; therefore,
the use of this device was soon discontinued.

In 1925 Dr. Ernst Graefenburg of Berlin theorized
that the portion of the IUD outside of the cervix
provided a ladder for bacteria, so he developed a
device contained entirely within the uterus. Graefenburg Rings, made of silkworm gut and silver wire,
were about the size of a nickel. But even with the
IUD contained entirely within the uterus, complications (primarily infection) developed. As a result,
the IUD was largely condemned by the medical
world.

In 1958 the first open-ended IUD was designed
at Mt. Sinai hospital in New York. This type had   Contrast
two major advantages: (1) The open-ended device
could be stretched into a straight line and loaded
into a soda straw-like inducer for insertion; this
process caused less pain because the cervix required
little dilation. After insertion, the device regained its
original spiral form. (2) Made of polyethylene (a
soft plastic), the device caused less inflammation
than metal.

The type of IUD most commonly used today is
called Lippes' Loop. It follows the construction of
the open-ended device. Other IUDs, the Saf-T-Coil
and the Dalkon Shield, are also used. All three are
shown in Figure 1. Lippes' Loop has the lowest ex-
pulsion rate of the three devices.   Comparison

The IUD is second to the birth-control pill in
effectiveness, being about 95 to 97 percent dependable in preventing conception. Women who have not   Contrast

had children are more likely to experience pain and difficulties with retaining the device; women who have given birth have fewer problems because of the size and elasticity of the uterus.

No one is yet sure exactly how the IUD works to prevent pregnancy. There are three current theories: — Principle of operation

1. The IUD touching the uterine wall at several points irritates the lining and keeps it from developing properly. Thus, the fertilized egg cannot find a good place to implant.

2. The IUD speeds up the muscular contractions — Contrast that move the egg down the fallopian tube to the uterus. The egg's normal journey of four to five days allows the uterine lining to develop; however, if the egg reaches the uterus too soon, the lining will not be ready for implantation.

3. The most recent theory holds that the uterine — Contrast wall responds to the foreign body by sending out macrophages (huge white blood cells) which try to destroy the IUD; failing that, they devour the egg or sperm or both.

This form of birth control has advantages and disadvantages. Its primary advantage is convenience. — Contrast Once the IUD has been inserted it only requires a periodic check of the length of the string protruding from the cervix. Checking the string allows the woman to determine if the IUD is still in the correct position. Also, there is no pill to remember to take every day and no fussing with other birth control devices. Primary disadvantages are an increase in menstrual flow and more frequent cramps.

The following conditions prevent a woman from using an IUD: endometriosis (inflamed uterine lin- — Special conditions ing), venereal disease, any vaginal or uterine infection, pelvic inflammatory disease, exceedingly small uterus, and an excessively heavy menstrual flow or cramping. One great danger of the IUD is that in rare cases it can penetrate the uterine wall and lodge in the abdominal cavity.

PLACING DEFINITIONS IN YOUR REPORT

If not carefully placed, definitions interrupt the flow of information. To avoid this threat to coherence, follow these suggestions for placement.

Parenthetical Definitions

If you have only a few informal definitions, place them in parentheses immediately after the terms. Several definitions per page will be disruptive and
should be placed elsewhere. Rewrite them as formal sentence definitions and
place them in a "Definitions" section of your report introduction, or place them
in alphabetical order in a glossary (as shown in Chapter 10).

Sentence Definitions

If your sentence definitions are few, place them in a "Definitions" section of
your report introduction. Otherwise, place them in a glossary. Any definitions
of terms in the report's title belong in your report introduction.

Expanded Definitions

Place expanded definitions in one of three locations:

1. If the definition is essential to the reader's understanding of the *entire*
report, place it in your report introduction. A report titled "The Effects of
Aerosol Spray on the Earth's Ozone Shield," for instance, would require expanded definitions of "aerosol" and "ozone" early in the report.

2. When the definition clarifies a major part of your discussion, place it in
the related section of your report. In a report titled "How Advertising Influ-
ences Consumer Habits," "operant conditioning" might form a major topic area
of the report; this term should then be defined early in the appropriate section.
Too many expanded definitions *within* a report, however, can be disruptive.

3. If the definition is an aid to understanding, but serves as a *secondary*
reference, it belongs in an appendix (see Chapter 10). An investigative report
on fire safety measures in a local public building, for example, might include
an expanded definition of "smoke detectors" in an appendix.

CHAPTER SUMMARY

Giving precise meanings is an important step in achieving clarity. Every field
has its own specialized language that has to be translated for general readers.
Also, many terms whose meanings seem clear — like "guarantee" or "disability"
— have different meanings in different contexts. Therefore, specify your meaning so that both you and your reader will understand what you are talking

about. Analyze your subject, purpose, and audience carefully in deciding when to use definitions and how long to make them. Sometimes a few words in parentheses will be enough; at other times you may need a full-sentence definition or an expanded definition, which could be several pages long. Regardless of length, write your definition in plain English, and be sure that it expresses the basic properties of the term objectively.

For a parenthetical definition (the simplest type), place a synonym or explanatory phrase, usually in parentheses, right after the word. For a sentence definition, indicate the term, the class in which the item it names belongs, and the features that distinguish the item from all others in its class. For an expanded definition, discuss your subject from as many of the following approaches as you can use to clarify its meaning:

1. etymology, or history, of the word
2. history and background of the item
3. familiar examples of the item or process
4. detailed diagrams with all parts labeled
5. detailed explanation of the various parts of the item
6. comparison of the item with similar ones or contrast with dissimilar ones to emphasize differences
7. the basic principle by which the item operates
8. special materials or conditions required

Place definitions where they do the most good. Depending on their length and number and their role in your report, you might place them in parentheses after the word, in a "Definitions" section of your introduction, in a glossary, at appropriate points in your discussion, or in an appendix.

REVISION CHECKLIST

Use this list to check the content, arrangement, and style of your definition.

Content

1. Have you chosen the type of definition (parenthetical, sentence, expanded) best suited to your subject, purpose, and reader's needs?
2. In defining a concrete and specific term (such as *transit* or *ophthalmoscope*), have you given all the details, and *only* those details, that will serve the reader's needs?

3. In defining an abstract and general term (such as *condominium* or *partnership*), have you specified its meaning in the context where you are using it?

4. Does the definition express the basic properties of the item (the unique features that make the item what it is)?

5. Is your definition objective (free of any implied judgments)?

6. For an expanded definition, have you used all applicable expansion techniques, and *only* those that will serve your reader?

7. Is the definition free of needless details?

8. Are all data sources documented?

Arrangement

1. Does your sentence definition follow the term-class-distinguishing-features structure?

2. Is your expanded definition a logical and coherent unit (like a good essay)?

3. Is there adequate transition between related ideas?

4. Have you placed your definition in the most appropriate location in your report?

Style

1. Is your definition written in plain English?
2. Will its level of technicality connect with the intended audience?
3. Is it free from needless words?
4. Are all sentences clear, concise, and fluent?
5. Is it written in correct English (see Appendix A)?

Now list those features of your definition that need improvement.

EXERCISES

1. In a three-paragraph essay discuss the differences among parenthetical, sentence, and expanded definitions, citing specific examples where each would be used for a certain purpose and audience.

2. In complete sentences identify the basic property of each of the following items:

desk	lamp	camel	music
bicycle	ski	bridge	mortgage
wood stove	subway	guitar string	depression (mental)
elevator	bed	clock	cancer

3. Adequate formal sentence definitions require precise classification and detailed differentiation. Tell whether you think each of the following definitions is adequate for a general reader. Rewrite those that seem inadequate. If necessary, consult dictionaries and specialized encyclopedias. Discuss your revisions in class.

    a. A bicycle is a vehicle with two wheels.
    b. A transistor is a device used in transistorized electronic equipment.
    c. Surfing is when one rides a wave to shore while standing on a board specifically designed for buoyancy and balance.
    d. Bubonic plague is caused by an organism known as *pasteurella pestis*.
    e. Mace is a chemical aerosol spray used by the police.
    f. A Geiger counter measures radioactivity.
    g. A cactus is a succulent.
    h. In law, an indictment is a criminal charge against a defendant.
    i. A prune is a kind of plum.
    j. Friction is a force between two poles.
    k. Luffing is what happens when one sails into the wind.
    l. A frame is an important part of a bicycle.
    m. Hypoglycemia is a medical term.
    n. An hourglass is a device used for measuring intervals of time.
    o. A computer is a machine that handles information with amazing speed.
    p. A Ferrari is the best car in the world.
    q. To meditate is to exercise mental faculties in thought.

4. Think of a situation in which someone failed to define a term adequately (as in giving instructions or an assignment) and caused you to misunderstand the message. Describe the situation and its consequences in two or three paragraphs.

5. *In class:* Without a dictionary, write down as many different sentence definitions for "stock" as you can think of. Compare your definitions with those of other class members and with those in a standard college dictionary. List five other terms with several meanings.

6. Both standard college dictionaries and specialized encyclopedias contain definitions. Standard dictionaries, however, define a word for the general reader, whereas specialized reference books provide definitions for the specialist. Choose an item in your field and copy the meaning as found (1) in a standard dictionary and (2) in a technical reference book. For the technical definition, label each expansion method carefully on a photocopy that

summons	economic inflation	pipette
generator	computer	calculator
dewpoint	golf club	t-square
clinical thermometer	contract	torque wrench
capitalism	hammer	wine vintage
marsh	frisbee	gourmet
economic recession	water table	editorial

you have made. Finally, rewrite the specialized definition so that it can be understood by a general reader.

7. Using reference books when necessary, write sentence definitions for the following terms or for selected terms from your major field. In order to express the basic properties of each item clearly, be sure to narrow your classification and to name its distinguishing features.

8. Select an item or concept from the list in question 7 or from an area of interest. Identify the particular audience and its particular needs. Begin with a sentence definition of the term. Then write an expanded definition for a first-year student in that particular field. Next, write the same definition for a layperson (client, patient, or other interested party). Leave a three-inch margin on the left side of your page to list your expansion strategies (use at least four in each version). Submit, with your two versions, a brief explanation of the changes you made in moving from the first to the second version.

9. The memo in Figure 5-2, written for laypersons, contains an expanded definition of "epilepsy," along with instructions for dealing with epileptic seizures in class. Your assignment here is twofold: (1) identify the expansion strategies used in the definition and (2) locate and copy an expanded definition of "epilepsy" written for technically informed readers. (Consult *The Merck Manual* or various medical textbooks, *not* general encyclopedias.) Document your source. Discuss the specific differences (in content, arrangement, and style) between the two versions. Submit your analysis and your copy of the second version to your instructor.

FIGURE 5-2   Memo Containing an Expanded Definition

POWNAL COLLEGE HEALTH OFFICE

October 20, 1981

TO: Faculty

FROM: Hester Pryor, Director of Health Services

SUBJECT: EPILEPSY AND THE ROLE OF FACULTY MEMBERS

Each year, a number of our students are stricken by seizures.
A seizure during a class period may not only disrupt the
entire class but also the future of the individual student
<u>unless the faculty member knows how to handle such incidents
wisely.</u>

Epilepsy: A Definition

The term <u>epilepsy</u> is derived from the Greek word for
"seizure." The concept of "being seized" (as though from
the outside of oneself) was passed down through the ages
until the nineteenth century, when Hughlings Jackson, the
great English neurologist, defined epileptic seizures as a
state produced by "a . . . sudden, violent, disorderly discharge of brain cells." Jackson's definition, implying the
discharge of excessive electrical (nervous) energy, has been
substantiated by brain wave studies made possible by the
development of the electroencephelograph.

Many seizures fall under the general classification of epilepsy, but for our purposes, only the three predominant
types of seizures will be discussed.

Petit Mal

The petit mal seizure is a brief interruption of consciousness characterized by the appearance of daydreaming.

2

The person generally has a blank look, and some small muscular movements may occur in the face. Usually the seizure
lasts only a few seconds. Such people will have no awareness
that they have had a seizure.

Grand Mal

From the point of view of the faculty member and the
other students, the grand mal seizure is the most alarming
type. The person may suddenly slump over or fall to the
floor. Muscles become rigid and then make convulsive movements. Bladder control may be lost. The person may have
difficulty breathing, and saliva will begin to collect and
run from the mouth. The attacks usually are brief, and after
resting, the person sometimes can continue with the class.

Psychomotor Seizures

Psychomotor seizures take many forms. In some cases, a
student will experience strange sensations (e.g., unpleasant
odors, or distorted perception). In others, the person will
seem to be making purposeful movements, but these will bear
no relation to the immediate situation. Some people will
make lip-smacking or chewing movements.

It is important to remember that with epileptics the attitudes of those around them are of vital importance. The
more normal a life they are able to lead, the better will be
their overall response to medical treatment.

What to Do During a Seizure

1. Do not restrain the person's movements any more
    than absolutely necessary to prevent self-injury.
    Loosen clothing. Keep the person away from hot or
    sharp objects. Do not force the mouth open, and

FIGURE 5-2 (*Continued*)

3

do not force anything between the teeth. If the
mouth is already open, however, you might place a
soft object (such as a folded handkerchief) between
the side teeth and turn the person on his or her
side to allow saliva to flow from the mouth.

2. Treat the occurrence matter-of-factly and explain
to the other students that there is no danger and
that the seizure will be over in a few minutes.

3. After the seizure stops and the person appears to
be relaxed, let her or him sleep or rest quietly in
the health office or dormitory.

4. It is usually not necessary to call a doctor unless
the attack lasts more than about ten minutes or is
followed by another major seizure. The health
office staff can answer any questions or help in
evaluation (extensions 8756 and 8757).

5. In describing the seizure to the health office try
to be as accurate as possible. An accurate description of the seizure is important to the physician
treating the patient.

FIGURE 5-2 (*Continued*)

6

Dividing and Organizing

DEFINITIONS

USING PARTITION AND CLASSIFICATION

GUIDELINES FOR PARTITION
Apply Partition to a Single Item
Make Your Partition Complete and
Exclusive
Make Your Division Consistent with
Your Purpose
Subdivide as Far as Necessary
Follow a Logical Sequence
Make All Parts of Equal Rank Parallel
Make Sure That Parts Do Not Overlap
Use Precise Units of Measurement
Choose the Clearest Format

GUIDELINES FOR CLASSIFICATION
Apply Classification to a Group of Items
Make Your Classification Complete,
Exclusive, and Inclusive
Limit Your Classification to Suit Your
Purpose
Choose Bases of Comparison Consistent
with Your Purpose
Express All Bases in Precise and
Objective Terms

Make All Categories of Equal Rank
Parallel
Make Sure That Categories Do Not
Overlap
Choose the Clearest Format

APPLYING THE TECHNIQUES OF
DIVISION

CHAPTER SUMMARY

REVISION CHECKLIST

EXERCISES

DEFINITIONS

Sometimes we divide a single thing into its parts to make sense out of it. At other times we divide an assortment of things into classes to sort them out. Partition and classification are the techniques we use to make these divisions. The two activities divide for different purposes: *partition* identifies the parts of a single item; *classification* creates categories for sorting similar items.

USING PARTITION AND CLASSIFICATION

Whether you choose to use partition or classification in a particular situation clearly depends on the nature of your subject. If you must describe a golf club, for example, you have little choice but to begin by partitioning it into its major parts: handle, shaft, and head. But if someone has unexpectedly given you 228 record albums, which you want to arrange in some way so that you can easily locate the particular record you want, partition will not help you. You will have to classify the records by dividing the pile into smaller categories. You might want to classify them as classical, jazz, rock, and miscellaneous. Or you might classify them according to your likes and dislikes.

The close examination of any complex problem almost inevitably requires the use of both partition and classification. If, for example, you are hired to plan a new supermarket, you must first partition the whole market into its functional parts to ensure the efficiency of each part.

- display and shopping area
- receiving and storage area
- meat refrigeration and preparation area

– checkout area
– small office area

Next, you will need to sort out your inventory by dividing the thousands of items into smaller groups or classes according to their similarities, for example:

– frozen foods
– dairy products
– meat, fish, and poultry
– pet foods
– fruits and vegetables
– paper products
– canned goods
– beverages
– baked goods
– cleaning products

In turn, you will divide each of these sections even further. You might divide "meat" into three smaller groups.

– beef
– pork
– lamb

Under these headings you will group the various cuts of meat (steaks, ribs, etc.) in each category. And you might carry the division further for certain meat products, like types of ground beef:

– regular
– lean
– extra lean
– diet lean

This kind of division continues until you have enough categories or classes to sort the hundreds of crates and cartons of inventory that sit in your receiving and storage area. On the one hand, you have divided your store into its parts. On the other hand, you have divided your inventory into classes in order to sort out the items. You have used both partition and classification.

Whether you are dividing by partition or classification, you must follow certain guidelines so that your division will make sense.

GUIDELINES FOR PARTITION

Apply Partition to a Single Item

Table 6-1 shows how a single item can be divided. Notice that the component nutrients add up to 100 percent; the total egg equals the sum of its parts.

TABLE 6-1   Approximate Composition of a Whole Goose Egg

Component	Percentage (by Weight)
Shell	14.0
Water	60.0
Protein	13.0
Fat	12.0
Ash	1.0

Make Your Partition Complete and Exclusive

Include *all parts* of the item. If you have omitted one or more parts for a good reason, say so in your title ("The Exterior Parts of a Typewriter"). On the other hand, be sure that each part belongs to the item. For instance, do not include "typewriter ribbon" in your partition of the exterior of a typewriter.

Make Your Division Consistent with Your Purpose

Most things can be divided in different ways for different purposes. You might divide an apple into the meat, skin, and core, but that division is useless to the nutritionist who wants to know the food value of an apple and is not interested in the fact that it has a core. You could divide a house in at least three different ways: (1) on the basis of the rooms it comprises, (2) on the basis of the materials used in construction, and (3) on the basis of the steps required to build it. The sum of the parts in each of these divisions equals 100 percent of the house. The basis you choose, however, will depend on your purpose. To interest a prospective buyer you might choose option 1; to provide cost or materials specifications you would probably choose option 2; to give instructions to the do-it-yourselfer you would choose option 3. Sometimes, in fact, you might need two or more partitions of the same item.

Let your title promise what the partition will deliver: "The Jones House, Partitioned According to Materials for Construction." Under this title, you would include every item, from nails to plumbing fixtures. If your purpose is less ambitious, limit your title: "Concrete Materials," "Electrical Materials," or the like.

Subdivide as Far as Necessary

Subdivide as much as necessary to show what makes up the item. This textbook, for instance, is divided into chapters. In turn, each chapter is subdivided into major topics such as "Guidelines for Partition." Major topics are again divided into minor topics such as "Apply Partition to a Single Item," and so on.

Follow a Logical Sequence

Most items have a particular logic of organization that determines the order for listing their parts. The division of a basic house, for example, most logically follows a *spatial sequence* — the foundation, the floor, the frame, the siding, and the roof — the sequence in which the parts are arranged. A progress report (a partition of your activity over a certain period) would follow a *chronological sequence* — from earliest to latest. A partition of your monthly budget might follow a descending scale of the *order of importance* of each expenditure. A problem can be analyzed by dividing it into its *causes and their effects*. Other specific orders of development are discussed in Chapters 3 and 7.

Make All Parts of Equal Rank Parallel

All parts at any one level of division (major parts, minor parts, subparts, etc.) are considered equal in rank. Therefore, list them in equal or parallel grammatical form.

> Faulty
>
> A deed contains the following seven items:
>
> 1. The buyer and seller must be identified.
> 2. A granting clause.
> 3. Consideration, not necessarily money, must be mentioned.
> 4. An explanation of the rights being transferred.
> 5. A full-length description of the property.
> 6. Proper execution, signature, seals, and delivery.
> 7. It must be properly recorded in the county where the property lies.

Here, items 1, 3, and 7 are not parallel to the others; they are expressed as complete sentences, whereas the others are expressed as phrases. Therefore, items 1, 3, and 7 should be revised to read as follows:

> 1. identification of buyer and seller
> 3. mention of consideration, not necessarily money
> 7. proper recording in the county where the property lies

Or, items 2, 4, 5, and 6 could be rewritten as complete sentences.

Make Sure That Parts Do Not Overlap

The logic of division requires that each item be exclusive of all others. Consider this partition of the executive branch of a corporation.

Faulty

chairman of the board	vice-president
board of directors	administrative officers
president	

These parts overlap because each of the first four positions may be listed under the heading "administrative officers."

Use Precise Units of Measurement

If the parts are in percentages, pounds, feet, or the like, say so, as in Table 6-2.

TABLE 6-2   The Parts of a Selected Multivitamin

Ingredient	*Quantity*
Vitamin A	15 International Units
Vitamin E	15 International Units
Vitamin C	60 mg
Folic Acid	0.4 mg
Vitamin B_1	1.5 mg
Vitamin B_2	1.7 mg
Niacin	20 mg
Vitamin B_{16}	2 mg
Vitamin B_{12}	25 micrograms
Vitamin D	400 International Units
Iron	18 mg

Choose the Clearest Format

Select the best format for your partition: prose discussion, list (as in an outline), table, or chart. Assume, for instance, that you need to partition the federal budget for 1974 in two ways: in terms of sources of income and types of expenditure. Here is how your prose version might read:

> The federal budget dollar for fiscal year 1974 can be divided into two broad categories: sources and expenditures. Income sources are broken down as follows: Individual income tax provided $0.42 of every dollar of federal income. Social insurance taxes and contributions provided $0.29, whereas corporation income taxes provided $0.14. The smaller income sources included excise taxes ($0.06), borrowing ($0.05), and miscellaneous sources ($0.04).
>
> The greatest federal expenditure was in human resources, which consumed $0.47 of every federal dollar. Next was national defense, requiring $0.30. Smaller expenditures were for physical resources ($0.10), interest ($0.07), and miscellaneous spending ($0.06).

TABLE 6-3   A Partition of the Federal Budget Dollar for Fiscal Year 1974

Where It Comes From	*Amount*
Individual income taxes	$0.42
Social insurance taxes and contributions	0.29
Corporations income taxes	0.14
Excise taxes	0.06
Borrowing	0.05
Other	0.04
Where It Will Go	*Amount*
Human resources	$0.47
National defense	0.30
Physical resources	0.10
Interest	0.07
Other	0.06

With an enumerative list like this, a prose discussion can be difficult to follow. Table 6-3 shows the same division in another format. Depending on the writer's purpose, any of the items in the table could be partitioned further. Notice that all items add up to 100 percent. Figure 6-1 shows pie-chart versions of the same data. For this partition, the pie chart seems to be most effective because it dramatizes the vast differences in sources of income and types of expenditure. See Chapter 11 for instructions on composing visual aids.

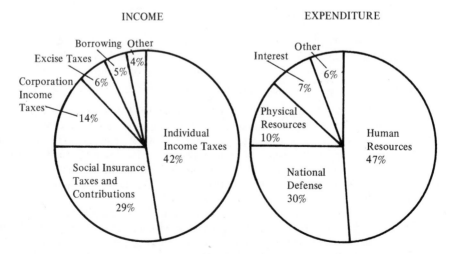

FIGURE 6-1   A Partition of the Federal Budget Dollar for Fiscal Year 1974

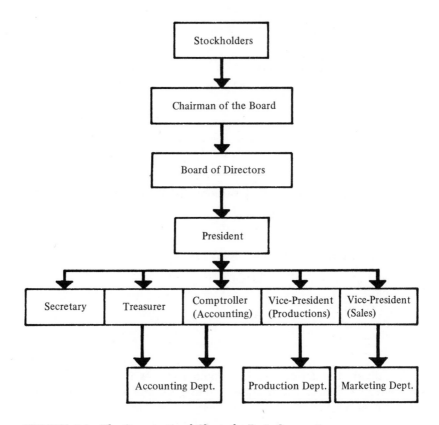

FIGURE 6-2   The Organizational Chart of a Basic Corporation

Different subjects for partition lend themselves to different formats. A subject whose parts are ranked in order of occurrence might best be partitioned as a flowchart while one whose parts are ranked in order of importance might call for an organizational chart, as in Figure 6-2.

GUIDELINES FOR CLASSIFICATION

Apply Classification to a Group of Items

Assume that you are writing a report on vegetarian diets and have decided to divide foods into three classes: fat sources, starch sources, and protein sources. You might classify the foods you've designated as protein sources as a table (see Table 6-4). Notice that the classification is limited by the term "selected."

TABLE 6-4   Selected Protein Sources

Brewer's yeast	Sunflower (seed)
Soybean (seed)	Whole egg
Groundnut (peanut)	Coconut
Cottonseed	Cow's milk (whole)
Sesame	Potato

Source: Adapted from Johnson and Peterson, *Encyclopedia of Food Technology,* p. 722. Used by permission of AVI Publishing Company.

Notice also that the whole egg partitioned in Table 6-1 is here one item in the classification system.

Make Your Classification Complete, Exclusive, and Inclusive

List *all items* that logically belong to a particular class.

> *Red Meats*
> – pork
> – lamb
> – beef

Also, divide the assortment into enough classes to contain every item you are sorting. A division of red meats into pork and beef, for example, would not be inclusive because lamb products (chops, legs, ribs, etc.) would have no place in which to be grouped. Finally, be sure that each item belongs in that particular class.

> *Red Meats*
> – pork
> – lamb
> – beef
> – frozen meats

"Frozen meats" are a function of temperature, not a type of meat. This system also overlaps because all three types of meats might be included in the category "frozen."

Limit Your Classification to Suit Your Purpose

Your classification becomes more specific and useful as you limit its focus. A résumé classification labeled "Work Experience," for instance, is more informative than one labeled "Experience." The classification "Vegetables" would require a list of every vegetable ever grown. Depending on your purpose, you might want to limit the classification to "Vegetables Sold by Our Food Co-op," "Canned Vegetables," "Green Vegetables," or another precise designation. Let your title promise what you will deliver.

Choose Bases of Comparison Consistent with Your Purpose

After limiting your classification you may need to select one or more bases by which to compare and contrast the items in your list. For example, in planning a high-protein diet, you would choose this basis: "Green Vegetables Classified on the Basis of Their Protein Content." Or you might choose several bases: protein content, caloric content, chlorophyll content, etc. A classification that includes one or more bases is called a formal classification. Table 6-5 gives the same material as Table 6-4, but presented as a formal classification. Because only one basis of comparison is used here, the items are arranged in a specific order of presentation — in this case, a descending order.

TABLE 6-5   Selected Protein Sources Classified in Descending Order
of Protein Content

Source	Protein (gm/100)
Brewer's yeast	38.8
Soybean (seed)	38.0
Groundnut (peanut)	25.6
Cottonseed	20.2
Sesame	18.1
Sunflower (seed)	12.6
Whole egg	12.4
Coconut	6.6
Cow's milk (whole)	3.5
Potato	2.0

Source: Adapted from Johnson and Peterson, *Encyclopedia of Food Technology*, p. 722. Used by permission of AVI Publishing Company.

Your purpose in classifying some types of food, however, may be to achieve a healthful diet; therefore, you would arrange your items according to several bases of nutrient content — protein, fat, iron, vitamins, etc. — all in a combined table such as Table 6-6. Here the bases are listed in the left vertical column so that the table can be contained within the width of one page.

TABLE 6-6   Nutrient Content of Red Meats, Poultry, and Fish

	Chicken	Nonfatty Fish	Herring	Beef	Lamb	Pork
		Per Ounce (28.35 gm) Raw Meat				
Protein (gm)	5.9	4.5	4.5	4.2	3.7	3.4
Fat (gm)	1.9	0.1	4.0	8.0	8.8	11.4
Calories (kcal)	41	19	54	89	94	116
Calcium (mg)	3	1	28	3	3	3
Iron (mg)	0.4	0.3	0.4	1.1	0.6	0.3
Vitamin D (mg)	—	—	6.38	—	—	—
Vitamin A (mg)	—	—	13	—	—	—
Vitamin B_1 (mg)	0.01	0.02	0.01	0.02	0.04	0.28
Vitamin B_2 (mg)	0.05	0.03	0.09	0.06	0.07	0.06
Niacin (mg)	1.7	0.8	1.0	1.4	1.4	1.4
Pantothenic acid (mg)	0.19	0.06	0.28	0.11	0.14	0.17
Vitamin B_6 (mg)	0.28	0.06	0.13	0.08	0.09	0.14
Biotin (mg)	2.83	2.83	x	0.85	0.85	1.1
Folic acid (mg)	0.85	14.1	x	2.83	0.85	0.85
Vitamin B_{12} (mg)	x	0.28	2.83	0.56	0.56	0.56
Vitamin E (mg)	0.06	x	x	0.17	0.23	0.19

Source: Johnson and Peterson, *Encyclopedia of Food Technology,* p. 626. Used by permission of AVI Publishing Company.
"—" indicates nutrient not present.
"x" indicates content not yet determined.

Express All Bases in Precise and Objective Terms

The bases you use must express specific similarities and differences; vague terms of assessment are useless. Modifiers such as "long," "nice," "heavy," "good," and "strong" are not specific enough. A classification titled "Good Vegetables" is meaningless unless "good" can be defined clearly and measured — with units of nutritional value expressed in grams, milligrams, international units, or the like.

Make All Categories of Equal Rank Parallel

Consider these class headings:

> *Frozen Foods*   *Dried Foods*   *Foods That Are Smoked*

This system is not parallel unless the last heading is revised to read "Smoked Foods."

Make Sure That Categories Do Not Overlap

Consider these headings:

Pork   *Beef*   *Ham*   *Lamb*

The headings overlap because ham is not an exclusive category; it is a pork product.

Choose the Clearest Format

Select the clearest format for your classification: either a prose discussion, a list (as in an outline), a table, or a graph. Remember that you have two purposes in classifying: (1) to help you organize your material and (2) to save your reader time and effort in interpreting data. Assume, for instance, that you are classifying types of red meat, poultry, and fish on the basis of caloric content. A prose version might read:

> Red meats, poultry, and fish have widely differing caloric contents. Among red meats, pork is the highest in calories, with 116 kilocalories per ounce of raw meat. Next in caloric content are the red meats, lamb and beef, containing 94 and 89 kilocalories per ounce. Herring follows red meats with 54 kilocalories per ounce. Slightly lower is chicken, with 41 kilocalories per ounce. Finally, nonfatty fish, with 19 kilocalories per ounce, provides the fewest calories of all six classes.

This prose version is tedious and more difficult to interpret than Table 6-6 or Figure 6-3, which shows a bar graph of the same data given in Table 6-6.

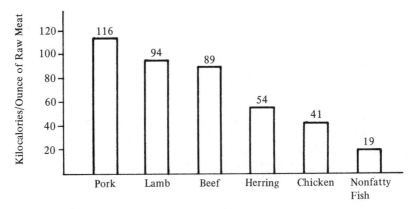

FIGURE 6-3   The Caloric Content of Red Meats, Poultry, and Fish

The significance of data can be lost to readers unless a precise explanation precedes the visual. It is therefore good practice to introduce all visuals, as shown below.

TABLE 6-7   Counties on Our State Classified According to Population and
Amount of Wholesale Trade, as of April 5, 1981

County	Population	Wholesale Trade in Dollars
Medford	1,499,386	3,841,108
Latah	821,262	5,134,250
Burns	721,814	732,816
Elrin	721,418	821,914
Worly	693,461	2,406,381
Nampa	502,016	724,572
Brighton	463,982	436,800
Stebbins	365,741	243,921
Brown	199,628	101,627
Dorin	156,524	35,996
Argot	111,428	41,824
Hollins	72,141	33,671
Baker	8,425	(−2,610)
Franklin	4,960	(−876)

Source: author.

Although Table 6-7 shows that some correlation exists between population level and wholesale trade among counties in our state, several inconsistencies can be noted. The most obvious inconsistency is found in comparing Medford and Latah counties in terms of population and wholesale trade. Both have a higher-than-average ratio of trade to population, but Latah, with little more than half the population of Medford, has one-third *more* wholesale trade. The reasons for this difference are as follows: (1) Gotham, our state's major trade center, is located in Latah County; and (2) Medford County encompasses the Route 138 industrial belt, which creates a large population that relies on Gotham as a trade center.

Of the two smallest counties, Baker and Franklin, it appears that Franklin is better prepared to handle business-cycle fluctuations and thereby minimize deficits.

Although population level has some effect on wholesale trade in respective counties, the major determinant seems to be the location of the county.

Sometimes, the bases in a classification cannot be expressed in simple units of measurement or by single words. Because the following classification contains detailed explanations of the various responsibilities held by dietitians, it is cast in a prose format (for those planning to major in nutrition).

DIETITIANS:
A CLASSIFICATION

A dietitian is a trained professional responsible for the nutritional care of individuals and groups. This person holds a bachelor's degree from an accredited college or university, with a major in food, nutrition, or institution management. After graduation, the dietitian completes an approved internship. To qualify for the title of Registered Dietitian, one must meet all requirements for membership in the American Dietetics Association, pass the registration examination, and continually satisfy requirements for further education.

Dietitians work in several capacities. The three major specialties in this field are in dietetics administration, therapeutic dietetics, and consultant dietetics.

The administrative dietitian is a member of the management team in an organization and is in charge of the food service systems. This person has direct authority and responsibility for the entire food service operation.

The therapeutic dietitian usually works in a hospital or clinic and plans modified menus for the patients' needs as determined by the physicians. He or she cooperates and participates in research and surveys on food and nutrition in the hospital, institution, or community.

Consultant dietitians work with a health care team to determine the nutritional requirements of patients. They provide guidance for the food service and dietetic personnel and may also evaluate their performance. In addition, they may develop budget proposals and recommend specific procedures for cost control. The consultant dietitian may be shared by two or more small hospitals or clinics.

In any of these roles, the dietitian applies the science and art of human nutrition to help people select the best foods in health and in disease.

APPLYING THE TECHNIQUES OF DIVISION

The bookkeeping form in Figure 6-4 typifies the organized records that are vital to the continued success of any business. Total monthly spending is partitioned into individual expenditures in the left column. In the right columns, expenses are sorted under eight general classes: ink, plates, films, and so on. Expenditures in each class are then totaled to provide an ongoing record of every penny spent.

CASH DISBURSEMENT JOURNAL

✓	CHECK NUMBER	CHECK ISSUED TO	DATE				CHECK AMOUNT		BANK BALANCE	
		BALANCE FORWARD ⟶							6 6/5	30
	0001	UNITED OFFICE SUPPLIES	3/2/76	OFFICE	SUP.		30	25	6 585	05
	0002	MAINE PAPER COMPANY	3/3/76	PAPER			2 6/5	10	3 969	95
	0003	JONES TRANSPORT COMPANY	3/3/76	FREIGHT			85	75	3 884	20
	0004	UNITED PARCEL SERVICE	3/4/76	FREIGHT			32	16	3 852	04
	0005	NEW ENGLAND TELEPHONE	3/5/76	UTILITIES			256	15	3 595	89
	0006	EASTERN GAS & ELECTRIC	3/5/76	UTILITIES			151	63	3 444	26
	0007	SMITH PRINTING COMPANY	3/7/76	PAYROLL			2 668	78	775	48
	0008	NORTHERN PRINTERS SUPPLY	3/7/76	PLATES			265	13	2 325 2 060	48 35
	0009	ACME SANITATION SERVICE	3/8/76	MAINTENANCE			43	50	2 016	85
	0010	JACKSON'S FILM PROCESSING	3/9/76	FILM			502	15	1 514	70
	0011	BOMARC REALTY	3/10/76	RENT			600	00	2 730 2 130	36 35
	0012	HAMSON OFFICE MACHINES	3/12/76	TYPEWRITER			559	75	1 570	60
	0013	HICKSVILLE WATER DEPT.	3/12/76	UTILITIES			6	23	1 564	37
	0014	SAM'S INDUSTRIAL LAUNDRY	3/14/76	MAINTENANCE			14	70	2 537 2 522	62 92
	0015	BLOTTO INK CORPORATION	3/14/76	INK			212	63	2 310	29
	0016	SCRUBBO CLEANING SERVICE	3/14/76	MAINTENANCE			60	00	2 250	29
	0017	DUMONT'S BOTTLED GAS	3/15/76	UTILITIES			48	00	2 209	29
	0018	ABCO INC.	3/17/76	OFFICE SUPPLIES			41	17	2 161	12
	0029	TRUE BLUE INK, INC.	3/28/76	INK			69	99	813	03
	0030	SCRUBBO CLEANING SERVICE	3/30/76	MAINTENANCE			60	00	753	03
							13 484	30	753	03

FIGURE 6-4   Bookkeeping Form

CHAPTER SUMMARY

When we divide a single thing into its parts we use partition. When we divide an assortment of things into specific categories we use classification. Partition is always applied to a single object. Its purpose is to systematically separate that whole object into its parts, pieces, or sections. Classification is always applied to an assortment of objects that have some similarities. Its purpose is to group these objects in a systematic way.

Whether you choose to apply partition or classification will depend on your subject and your purpose. In analyzing complex problems you will often have to use both techniques.

Follow these guidelines for dividing a single item into its parts:

– Apply partition to a single item (e.g., a golf club).
– Include all parts and be sure that each part belongs.

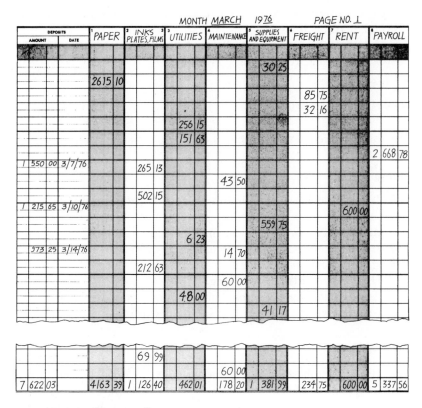

FIGURE 6-4   (*Continued*)

– Divide in a way consistent with your purpose.
– Subdivide as far as needed to show what makes up the item.
– Make sure your sequence of division follows the item's logic of organization (spatial, chronological, or the like).
– Make parts of equal rank parallel in grammatical form.
– Make sure that parts do not overlap.
– Choose the clearest format (prose, list, or visual aid).

Follow these guidelines for dividing an assortment of things into classes:

– Apply classification to a group of items.
– Include all items that belong to a particular class; be sure that each item belongs in that class; and divide the assortment into enough classes to contain every item you are sorting out.
– Limit your classification to a specific focus.

– Choose useful bases for comparison and contrast and express them in precise and objective terms.

– Make equal categories parallel and make sure categories do not overlap.

– Choose the clearest format.

REVISION CHECKLIST

Use this list to refine your system of partition or classification.

Content

1. Is the title clear and limiting? Does it promise exactly what you will deliver?

2. Are all data sources fully documented?

3. Is the system both complete and exclusive? Does it include all items that belong, but no extraneous ones?

4. Have you avoided overlapping?

5. Are all units of measurement given (grams, pounds, etc.)?

6. Are all items in your partition subdivided as far as necessary?

Arrangement

1. Have you chosen the clearest format for your purpose (table, chart, graph, prose discussion)?

2. Is the partition or classification integrated into the surrounding text (introduced and discussed)?

3. Does the partition follow a logical sequence (spatial, chronological, etc.)?

4. Are all items in the partition divided according to a basis that suits your purpose and audience?

5. Are all items in the classification arranged according to a basis that suits your purpose and audience?

6. If a chart, graph, or table is used, does it follow the criteria for visual aids discussed in Chapter 11?

Style

1. Are all items in parallel grammatical form?

2. Are the bases objective and precise?

3. Is the prose version written in correct English (see Appendix A)?

Now list those features of the partition or classification system that need improvement.

EXERCISES

1. In a short essay, discuss the types of partition and classification systems that you expect to use on the job. Be as specific as possible in identifying typical subjects, your intended audience, and uses to which your data will be put.

2. Is the following classification effective? Use the revision checklist as a guide for evaluation, and write out your suggestions for improvement.

1980 Beer Production
in the Top Six National Breweries

Langdon Brewing Co.	14,678,400
Kastel Inc.	21,739,200
Flagstaff Brewing Co.	7,635,800
King Inc.	10,478,300
Rothberg Brewing Co.	23,547,700
Case Brewing Co.	9,658,400

3. In a unified and coherent paragraph discuss the difference between classification and partition. List three items that can be classified and three that can be partitioned.

4. Jot down each of your activities on any given day. When your list is complete, group related activities under specific headings: ("Social Activities," "Schoolwork," etc.). Next, choose a basis of comparison for arranging items within a given class in a specific order: (descending order of time spent on each activity, order of desirability, etc.). Finally, choose one class of activities — "Schoolwork," for example — and present your formal classification as a bar graph, a table, and a prose discussion. Which form of presentation seems most "readable" here?

5. Classify the following items in terms of five different bases of comparison. Be sure to limit and define your classification with an explicit title: lettuce, celery, cabbage, asparagus, broccoli, spinach.

6. Assume that you are planning a week-long camping trip into a wilderness area. Make a list of all the items you will need. Next, group related items under specific headings in order to organize your inventory. Finally, take one item from the list (e.g., "tent") and partition it into its various parts (pegs, poles, canvas, guy lines, mosquito netting, etc.).

As an alternate assignment, assume that you are planning a one-week spring vacation in Bermuda, and perform the same tasks.

7. Some items listed in the following classification systems are not logically related. Identify the specific error in each group — faulty parallelism,

overlapping, inconsistent general meaning, or incompleteness — and correct it.

Classification of selected
alcoholic beverages
   beer
   wine
   whiskey
   bourbon
	Classification of
	winter sports
	   hockey
	   curling
	   handball
	   skiing
	   figure skating
	   sledding

Classification of automobiles
on the basis of body type
   sedans
   family cars
   roadsters
   coupes
   hardtops
   station wagons
	Classification of
	technical writing tasks
	   proposals
	   sending memos
	   specifications
	   work orders
	   progress reports
	   instructions
	   writing letters
	   reports which are formal

8. Find an example of a formal classification in a magazine, such as *Consumer Report,* or in a textbook. Is the classification effective? Using the criteria for effectiveness discussed in this chapter, compose a one-paragraph answer. Do the same for a partition.

9. Choose an item with a singular meaning (bicycle, digital calculator, clock, retail store, stereo system, etc.). Partition the item, for a specified audience and purpose, on as many levels as necessary.

(1st level)
Major parts of a bicycle
   frame
   wheels
   drivetrain
   attachments

(2nd level)

Parts of a frame	*Major parts of a wheel*	*Major parts* *of a drivetrain*	*Major attachments*
	hub spokes rim		

(3rd level)

Etc.	*Parts of a hub* etc.	Etc.	Etc.

Notice how this partition (composed for the repair-it-yourself bicycle owner) moves from division to subdivision, and so on. Hence, the partition is complete at each level. Complete the entire partition by filling in the required items, or construct one of your own, using a subject that you know well.

10. Keep a record of every penny you spend during one week, or review your checkbook record for the semester. Partition your spending, in chronological order, in the lefthand column of an accounting sheet like the one in Figure 6-4. Now, group your expenses under specific class headings, in terms of similarity of expense ("Rent and Household Expense," "Automobile Expense," "Food Expense," "Entertainment Expense," "School Expense," and any other categories needed to include specific expenditures). Next, add up your totals in each category, write a prose analysis of your findings, and make specific recommendations for overhauling your budget, if necessary. (Use Figure 6-4 as a model.) Present the same findings (1) in a bar graph and (2) in a pie chart.

11. Compose a classification of the jobs commonly available to graduates in your major. Write for an audience of entering freshmen, who will be reading your classification as part of the college's *Career Handbook*. Use the essay on page 123 as a model.

12. Assume that your company's vice-president has decided that it's time to purchase a new line of tools or equipment (chain saws, microscopes, drafting tables, cash registers, company cars, or the like). You've been asked to collect data for comparing various brands of the item and to submit your findings to the vice-president. Identify one or more bases that will best serve your reader's needs. Do the research and compose a classification system following the guidelines in this chapter. Use visuals where applicable.

7

Charting
Your Course:
The Outline

DEFINITION

THE PURPOSE OF OUTLINING

CHOOSING THE BEST TYPE OF
    OUTLINE
  The Informal Outline
  The Formal Topic Outline
    Roman Numeral–Letter–Arabic
     Numeral Notation
    Decimal Notation
  The Formal Sentence Outline

ELEMENTS OF AN EFFECTIVE FORMAL
    OUTLINE
  Full Coverage
  Successive Partitioning
  Logical Notation and Consistent Format
  Parallel Construction for Parallel Levels
  Clear and Informative Headings
  Parts in Logical Sequence
    Chronological Sequence
    Spatial Sequence
    Reasons For and Against
    Problem-Causes-Solution
    Cause and Effect
    Comparison-Contrast
    Simple to Complex
    Sequence of Priorities
  Items Relevant to Purpose

COMPOSING THE FORMAL OUTLINE
  Preliminary Steps
  Introduction
  Body
  Conclusion

THE REPORT DESIGN WORKSHEET

CHAPTER SUMMARY

REVISION CHECKLIST

EXERCISES

DEFINITION

Outlines come in many shapes and sizes: a few key words or phrases listed on a page is an outline; a grocery list is an outline; a daily reminder calendar is an outline. Each reminds the writer or reader of important things to be done. The size and complexity of your own outline will vary according to your specific writing task. Sometimes an informal list of words or ideas will be enough of a guide; at other times you will need a highly systematic arrangement of topics arranged in a formal numbering and lettering system. In any case, don't write the final draft of your report — long or short — until you have some sort of outline.

THE PURPOSE OF OUTLINING

Why does a writing assignment seem to fall easily into place for one writer but not for another? Because *the successful writer usually spends more time planning than writing*. To write effectively you need to take several distinct steps: (1) thinking about your main point until you see it clearly; (2) composing a thesis statement; (3) developing your idea; (4) dividing it into its parts; (5) arranging the parts into a discussion that continually relates to your original thesis; and (6) expressing your message in clear, correct English. Rushing through these steps all at once is like trying to build a house before the plans have been drawn up and the materials delivered.

Last-minute changes in planning and building a house are made more easily on the blueprints than by tearing down actual walls or moving a bedroom wing from one side to another. Likewise, you can modify your outline more easily than your written report. Like the blueprint, the outline is not a set of commandments but simply a tool for your convenience. In planning a long and complex research report, for example, you can sketch an informal outline long before you write the actual report. This list of ideas will provide a general direction for gathering evidence. As your investigation proceeds you can revise and refine your outline. Before writing the first draft of your report, you should compose a formal outline — the detailed, polished version that will guide you to your finished product.

CHOOSING THE BEST TYPE OF OUTLINE

The Informal Outline

An informal outline is a simple list that serves as a brief reminder. This list may be all you need for a short report. First, identify your approach by formulating a clear statement of purpose. Next, brainstorm your topic (as discussed in Appendix B) to generate related ideas. After selecting the most pertinent ideas (expressed as phrases), arrange them in the sequence that makes the most sense. Here is a sample statement of purpose, with its outline.

> The purpose of this report is to describe the evaluation of a lakefront building site to determine the appropriate waste-disposal system needed.

TOPICS
1. Location of the Site
2. Physical Description of the Site
3. Type of Water Supply
4. Instructions for Constructing the Gray-Water System (sink drain)
5. Instructions for Constructing the Privy (Outhouse) Pit
6. Instructions for Applying for a Gravel Waiver

In formulating this sequence, the writer observed the principle that anything to be discussed must first be described. With this kind of list the writer can now compose the report quickly and efficiently.

If you are planning a longer report, an informal outline is still a handy tool. In this case it is a tentative outline, or a working outline, because it is designed to keep you on track without excluding possibilities for additions, deletions, and other revisions as you plan and write your report.

Here is an informal outline for a report titled "An Analysis of the Advisability of Converting Our Office Building from Oil to Gas Heat." The writer begins by formulating her statement of purpose.

The purpose of this report is to inform the president of Abco Engineering Consultants, Inc. of the advisability of converting our office building from oil to gas heat.

As the writer brainstorms her subject she produces the following list of major topics for research and discussion.

- Estimation of Gas Heating Costs
- Removal of the Oil Burner and Tank
- Installation of a Gas Burner
- Description of Our Present Heating System
- Installation of a Gas Pipe from the Street to the Building

She then arranges these topics in the sequence that makes the most sense — in this case, a chronological sequence.

1. Description of Our Present Heating System
2. Removal of the Oil Burner and Tank
3. Installation of a Gas Pipe from the Street to the Building
4. Installation of a Gas Burner
5. Estimation of Gas Heating Costs

Notice that the topics are arranged according to the sequence of steps in the actual conversion process, beginning, of course, with a description of the present system. The writer now has a general plan for gathering data. She can expand this outline and make it more specific by adding subtopics from her brainstorming list, as shown below. When her data gathering is complete, the writer develops her informal outline into a formal outline *before* writing her final draft.

The Formal Topic Outline

The formal topic outline is a more detailed and systematic arrangement of topics and subtopics using a formal system of notation (numbers, letters, and other symbols marking logical divisions). You may choose between two common systems of notation: the roman numeral–letter–arabic numeral system or the decimal system.

Roman Numeral–Letter–Arabic Numeral Notation

Here are the five topics from the informal outline developed into a formal outline using roman numeral–letter–arabic numeral notation:

II. REPORT BODY (or COLLECTED DATA)
    A. Description of Our Present Heating System
        1. Physical condition

    2. Required yearly maintenance
    3. Fuel supply problems
       a. Overworked distributor
       b. Varying local supply
    4. Cost of operation
  B. Removal of the Oil Burner and Tank
    1. Data from the oil company
    2. Data from the salvage company
       a. Procedure
       b. Cost
    3. Possibility of private sale
  C. Installation of a Gas Pipe from the Street to the Building
    1. Procedure
    2. Cost of installation
    3. Cost of landscaping
  D. Installation of a Gas Burner
    1. Procedure
    2. Cost of plumber's labor and materials
  E. Estimation of Gas Heating Costs
    1. Rate determination
    2. Required yearly maintenance
    3. Cost data from neighboring facility
    4. Overall cost of operation
       a. Cost of conversion
       b. Cost of maintenance
       c. Cost of gas supply

The writer now adds the introduction and conclusion.

  I. INTRODUCTION
    A. Background
    B. Purpose of the Report
    C. Intended Audience
    D. Information Sources
    E. Limitations of the Report
    F. Scope
      1. Description of our present heating system
      2. Removal of the oil burner and tank
      3. Installation of a gas pipe from the street to the building
      4. Installation of a gas burner
      5. Estimation of gas heating costs

 II. BODY (as shown earlier)

III. CONCLUSION
    A. Summary of Findings
    B. Comprehensive Interpretation of Findings
    C. Recommendations

In a short report, the introduction and conclusion are still included, but usually shortened to one or two sentences apiece. The entire outline easily converts into a table of contents for your finished report, as shown in Chapter 10.

Decimal Notation

Here is a partial version of the same outline in decimal notation:

    2.0 Collected Data
        2.1 Description of Our Present Heating System
            2.1.1 Physical condition
            2.1.2 Required yearly maintenance
            2.1.3 Fuel supply problems
                2.1.3.1 Overworked distributor
                2.1.3.2 Varying local supply
            2.1.4 Cost of operation
        2.2 Removal of the Oil Burner and Tank
            2.2.1 Data from the oil company
            2.2.2 Data from the salvage company
                2.2.2.1 Procedure
                2.2.2.2 Cost
            2.2.3 Possibility of private sale

The decimal outline makes it easier to refer readers to various sections. Both systems, however, achieve the same organizing objective. Unless readers express a preference, use the system that you prefer.

The Formal Sentence Outline

The above outline is called a *topic outline* because each division is expressed as a topic phrase. Although this is the most popular type of outline, it may be expanded one step further before you write the report. This final version is called a *sentence outline*.

    II. COLLECTED DATA
        A. Our present heating needs are supplied by circulating hot air gen-
           erated by a Model A-12, electrically fired, Zippo oil burner fed by a
           275-gallon fuel tank. Both are twelve years old.
            1. Both burner and tank are in good working order and physical
               condition, as they have been carefully maintained.
            2. The oil burner and associated components require cleaning once
               yearly at a service charge of $18.00. The air filter, costing $2.25,
               is replaced three times yearly at a total cost of $6.75.
            3. On three occasions during the past two years our system has run
               out of fuel for periods ranging from twelve to twenty-four hours
               for one of two reasons:

      a. Our town has only one fuel oil distributor who lacks the manpower and machinery to provide immediate service to all customers during the peak heating season.

      b. The supply of fuel allocated to our local distributor has fluctuated rapidly and unpredictably in recent months; at times, his oil supplies are not adequate to satisfy local demand.

    4. Our firm's oil bill for 1976 was $533.53; this figure added to the yearly maintenance charges amounts to an overall cost of $558.28.

  B. Before the gas burner and fixtures could be installed, the oil burner and tank would have to be removed from the basement.

    1. etc.

ELEMENTS OF AN EFFECTIVE FORMAL OUTLINE

Full Coverage

Make your list of major topics broad enough to encompass your subject. For example, the outline for the heating conversion report would not be adequate without "Description of Our Present Heating System." Readers would lack a basis for comparing the two forms of heating.

Besides making your outline inclusive, make it specific enough so that you can discuss each topic in detail. Thus, the sample outline partitions "Estimation of Gas Heating Costs" into four subtopics. The fourth subtopic, "Overall cost of operation," is further partitioned into its three constituent sub-subtopics. This finite breakdown helps the reader understand each step in the cost determination.

Succesive Partitioning

Each division must have at least two parts. Also, place parts or subparts of equal rank at the same level:

Faulty

    C. Installation of a Gas Pipe from the Street to the Building
      1. Procedure
      2. Cost of installation
    D. Cost of Landscaping

These topics are not correctly partitioned: *D* is not equal to *C*, but only a subtopic of *C*, since the lawn has to be dug up and repaired as part of the installation procedure.

Each successive level is a division of the immediately preceding level. Therefore, the subdivisions must add up to the immediately preceding item at the next higher level. Thus, "Procedure," "Cost of installation," and "Cost of Landscaping" add up to "Installation of a Gas Pipe. . . ."

Logical Notation and Consistent Format

We have defined notation as the system of numbers, letters, and other symbols marking the logical divisions of your outline. Format, on the other hand, is the arrangement of your material on the page (the layout). Proper notation and format show the subordination of some parts of your topic to others. Because your formal outline contains both major and minor parts, be sure that all sections and subsections are ordered, capitalized, lettered, numbered, punctuated, and indented to show how each part relates to other parts, and to the overall discussion.

The general pattern of notation, then, is as follows:

I.
  A.
    1.
    2.*
  B.
    1.
    2.
      a.
      b.
        (1)†
        (2)
  C.
II. etc.

Here is the same general pattern of notation in decimal form:

1.0
  1.1
    1.1.1
    1.1.2
  1.2
    1.2.1
    1.2.2
      1.2.2.1
      1.2.2.2
        1.2.2.2.1
        1.2.2.2.2
  1.3 etc.

---

* Any division must yield at least two subparts. For example, you could not logically divide "Types of Strip Mining" into "1. Contour Mining," without other subparts. If you can't divide your major topic into at least two subtopics, change your original heading.

† Further subdivisions can be carried as far as needed, as long as the notation for each level of division is individualized and consistent.

Use indentation that is consistent from category to category and from level to level. The same applies for line spacing. If, for instance, you indent your first A notation five spaces from the margin, indent all other uppercase letter notations identically. If you leave a triple space between notations I and II, be sure to triple space between II and III. Likewise, if you single space your first set of arabic numerals — 1, 2, and 3 — single space all other sets of arabic numerals at this level of division.

Express topics at particular levels in consistent letter case: all BLOCK LETTERS; First Letter of Each Word in Caps (except articles, conjunctions, and prepositions); or First letter capitalized.

Parallel Construction for Parallel Levels

Make all items of equal importance parallel, or equal, in grammatical form. Then your outline will emphasize the logical connections among related ideas.

Faulty

E. Estimation of Gas Heating Costs
    1. Rate determination
    2. The system requires yearly maintenance
    3. Cost data were obtained from a neighboring facility
    4. Overall cost of operation

Each item at this level of division is presented as equal in importance to the other items that enter into the cost estimation, but 1 and 4 are phrases, whereas 2 and 3 are sentences.

Correct

E. Estimation of Gas Heating Costs
    1. Rate determination
    2. Required yearly maintenance
    3. Cost data from neighboring facility
    4. Overall cost of operation

Conversely, each item could be expressed as a complete sentence. See Appendix A for a full discussion of parallelism.

Clear and Informative Headings

As you compose your outline be sure that your headings contain *specific* information. Choose highly informative words. Under "Description of Our Present Heating System," for example, a heading titled "Fuel" is not as informative as one titled "Fuel supply problems."

Also, avoid repetitions that add no information.

Faulty

C. Environmental Effects of Strip Mining
   1. Effects on land
   2. Effects on erosion
   3. Effects on water
   4. Effects on flooding

Correct

C. Environmental Effects of Strip Mining
   1. Permanent land scarring
   2. Increased erosion
   3. Water pollution
   4. Increased flood hazards

In the correct version, each topic heading contains key phrases that summarize the message.

Parts in Logical Sequence

Which details does your reader need — and in what order? Which item comes first? Last? Does your subject have any special traits that might determine the sequence? Answer these questions as you plan your outline. Some possible sequences for ordering the body section follow.

Chronological Sequence

In a chronological sequence you follow the time sequence of your subject (for example, the sequence of steps in a set of instructions). Also, follow this sequence to explain the order in which the parts of a mechanism operate (for example, how the heart works to pump blood). Begin with the first step and end with the last.

Spatial Sequence

In a spatial sequence you follow the physical arrangement of parts (left to right, top to bottom, front to rear, etc.), as when you describe various rooms in a college building as possible sites for a radio station.

Reasons For and Against

In giving reasons for and against something, follow the sequence in which both sides of an issue are argued — first one side, then the other — as in an analysis of the value and danger of a proposed flu-vaccination program.

Problem-Causes-Solution

Follow the sequence of the problem-solving process from posing the problem, through diagnosis, to a solution. An analysis of the rising rate of business failures in your area would require this sequence.

Cause and Effect

In a cause-and-effect sequence you follow actions to their specific results. You would use this sequence in analyzing the therapeutic benefits of transcendental meditation.

Comparison-Contrast

In evaluating two items, you can discuss first their similarities and then their differences. An example would be the item-by-item comparison of two sites for locating a small business.

Simple to Complex

A complex subject is often best explained by beginning with its most familiar or simplest parts. An explanation of color television transmission, for example, follows the logic of the learning process by describing what we see on our TV screens *before* discussing the complex mechanism that creates the picture.

Sequence of Priorities

Sometimes, you will want to place items in sequence according to their relative importance, as in a proposal for increasing the school budget in your town.

These sequences are illustrated in the sample reports throughout this text, and especially in Chapter 3. Many reports will involve more than one sequence. The heating-conversion outline, for example, fuses the chronological sequence, the comparison-contrast sequence, and the cause-and-effect sequence. A paragraph generally follows *one* of these sequences, as shown on pages 35–39.

Items Relevant to Purpose

Be sure that all items in your outline add up to your statement of purpose. Omit irrelevant topics. A sixth topic titled "Temperature Forecast for Next Winter," for example, would be irrelevant to the stated purpose of the report outlined earlier.

COMPOSING THE FORMAL OUTLINE

Preliminary Steps

Complete the following steps *before* beginning work on your formal outline:

1. *Write out a full statement of purpose.* Decide specifically what you want to accomplish in this report and write out your intention in one or two sentences. You will then have a point of reference as you work.

2. *Brainstorm your subject.* Do some hard thinking about your subject to identify all related ideas and topic divisions. The more time you spend on this step, the more concrete details you will have to work with. Follow the suggestions in Appendix B for brainstorming.

3. *Construct your informal (or working) outline.* Organize the relevant major topics from your brainstorming list in the most logical sequence for discussing your subject according to your stated purpose.

4. *Collect your data.* Using your rough outline as a guide, collect all the detailed information you will need for your report. Your data may suggest ideas for additional topics or subtopics. If so, revise your working outline accordingly.

5. Compose a rough draft of the report. This material can be refined into a final plan (your formal outline). When you have answered the first essential question — *What is my intention, or goal?* — your formal outline will help you answer the second essential question — *How will I achieve my goal in a way that is clearest to my reader?* At this point, in addition to the major topics from your brainstorming list, you should have identified all subtopics.

The following model can be adapted to most reporting assignments directed toward reaching a decision. (For information-type reports the CONCLUSION becomes simply a Summary.)

GENERAL OUTLINE MODEL

I. INTRODUCTION
   A. Definition, Description, and History (and significance of the subject)
   B. Statement of Purpose
   C. Target Audience (including assumptions about the reader's prior knowledge)
   D. Information Sources (including research methods and materials)
   E. Working Definitions
   F. Limitations of the Report
   G. Scope of Coverage (major topics from your brainstorming list in the sequence in which you will discuss them)

II. BODY (the component parts of your subject, divided into their subparts, as necessary)
   A. First Major Topic
      1. First subtopic of A
      2. Second subtopic of A
         a. First subtopic of 2
         b. Second subtopic of 2
            etc. (subdivision carried as far as necessary to isolate the important points in your topic)
   B. Second Major Topic
      etc.

III. CONCLUSION (where everything is tied together)
   A. Summary of Information in II
   B. Comprehensive Interpretation of Information in II
   C. Recommendations and Proposals Based on Information in II

Suggestions for developing each section follow.

Introduction

Your introduction should contain the following parts:

1. *Definition, description, and history.* Define and describe your subject before discussing it, and outline its history, as needed. In short, introduce your reader to *what* you will cover in your body section.

2. *Statement of purpose.* In one or two sentences, state what you plan to achieve in your report. A statement of purpose is like your thesis statement in an essay. Why are you writing this report?

3. *Target audience.* Identify the audience. If you can, explain how the target reader will use your information. How much prior knowledge does your reader need to understand the report?

4. *Information sources.* If your report includes data from outside sources, identify them briefly here (you will identify them in detail in your footnotes). Outside sources might include interviews, questionnaires, library research, company brochures, government pamphlets, personal observation, etc.

5. *Working definitions.* Do you need to define any technical terms, such as "autoanalyzer," or general terms, such as "liability"? If you have a long list of terms (ten or more), save them for a glossary at the end.

6. *Limitations of the report.* State the reasons for any information that is incomplete. For instance, perhaps you were unable to locate a key book or interview a key person. Or perhaps your study can only be title "preliminary," instead of "definitive" (the final word on the subject), because facts that might throw new light on your subject have yet to be made public. Or perhaps your report discusses only *one side* of an issue, as in a study of the *negative* effects of strip mining in your county.

7. *Scope of coverage.* In your final subsection, preview the scope of your report by listing all major topics discussed in section II, the body.

Not all reports will require each of these subsections in the introduction. Sources, definitions, and limitations subsections might be optional.

Body

Your body is the heart of your report. In it you develop major topics and subtopics. Whether you are describing an item or a process, giving instructions, or analyzing an issue or problem, the facts in this section support and clarify your statement of purpose, your conclusions, and any recommendations. "Show me!" is the implied demand that any reader will make of your report. Your body section should satisfy that demand by giving your reader a step-by-step view of the process by which you move from your introduction to your conclusion. Any interpretations or recommendations will be only as credible as the evidence that supports them.

Whenever possible, give your body section an explicit title to reflect the specific purpose of your report. For example, if your report is written to present a physical description of an item, you might title your body section "Description and Function of Parts." The same section in a set of instructions might be titled "Instructions for Performance," or "Collected Data" in a report that analyzes a problem or answers a question. See the section titles of sample reports in this book.

Conclusion

Your conclusion contains no new findings. Instead it reinforces, interprets, or otherwise clarifies the body. The following subsections are most often used, but the subsections in your conclusion will vary with different types of reports. In a report describing a mechanism, for example, your conclusion might simply review the major parts of the mechanism discussed in the body, and briefly describe one complete operating cycle.

1. *Summary of information in the body.* When your discussion is several pages long, summarize it.

2. *Comprehensive interpretation of information in the body.* Tie your report together by giving an overall interpretation of your data and drawing conclusions based on facts.

3. *Recommendations and proposals based on information in the body.* Be sure to base recommendations or proposals directly on your interpretations and conclusions.

Although a good beginning, middle, and ending are indispensable, feel free to modify, expand, or delete any subsections as you see fit.

THE REPORT DESIGN WORKSHEET

Some writers may wish to use a planning sheet in mapping out a report. The report design worksheet shown in Figure 7-1 works along with an outline to help you zero in on your audience and purpose. Use it for any report or letter. Figure 7-2 shows a completed worksheet for the heating-conversion report outlined earlier.

REPORT DESIGN WORKSHEET

Preliminary Information

What is to be done? _____

Whom is it to be presented to, and when? _____

	Primary Reader(s)	Secondary Reader(s)
Audience Analysis		
Position and title:		
Relationship to author or organization:		
Technical expertise:		
Personal characteristics:		
Attitude toward author or organization:		
Attitude toward subject:		
Effect of report on readers or organization:		

Reader's Purpose

Why has reader requested it?

What does reader plan to do with it?

What should reader know beforehand to understand it as written?

What does reader already know?

What amount and kinds of detail will reader find significant?

What should reader know and/or be able to do after reading it?

Writer's Purpose

Why am I writing it?

What effect(s) do I wish to achieve?

FIGURE 7-1   Report Design Worksheet

Design Specifications

Sources of data:

Tone:

Point of view:

Needed visuals and supplements:

Appropriate format (letter, memo, etc.):

Rhetorical mode (description, definition, classification,
                etc.--or some combination):

Basic organization (problem-causes-solution, intro-instruc-
                tions-summary, etc.):

Main points in introduction:

Main points in body:

Main points in conclusion:

Other Considerations

FIGURE 7-1   (*Continued*)

REPORT DESIGN WORKSHEET

Preliminary Information
A report on the feasibility of converting
What is to be done?   _our home office from oil to gas heat_

Whom is it to be presented to, and when? _Charles Jones, company_
president; April 1

	Primary Reader(s)	Secondary Reader(s)
Audience Analysis		
Position and title:	_President, Abco Engineering Consultants_	_Company officers engineering staff_
Relationship to author or organization:	_Employer_	_supervisors, colleagues, junior members_
Technical expertise:	_nontechnical (for this subject)_	_nontechnical_
Personal characteristics:	_highly efficient; demands quality and economy_	_all serious-minded professionals_
Attitude toward author or organization:	_is considering me for promotion to assistant V.P._	_friendly and respectful; officers will vote on my promotion_
Attitude toward subject:	_highly interested because of last winter's inconvenience_	_interested_
Effect of report on readers or organization:	_will be read closely and acted upon_	_will be read and discussed at our next staff meeting_

Reader's Purpose

Why has reader requested it? _wants to make a practical decision_ _____

What does reader plan to do with it? _use the data to make the best choice_ _confer with the president about the choice_

What should reader know beforehand to understand it as written? _nothing special; history of problem is reviewed in report_ _same_

What does reader already know? _remembers last winter's problems_ _same_

What amount and kinds of detail will reader find significant? _brief description of conversion procedures and detailed cost analysis_ _same_

What should reader know and/or be able to do after reading it? _make an educated decision_ _advise the president about his decision_

Writer's Purpose

Why am I writing it? _to communicate my research findings clearly_

What effect(s) do I wish to achieve? _to have my readers conclude that conversion is not economically feasible; to persuade them to accept my recommendation of an alternative to conversion._

FIGURE 7-2   Completed Report Design Worksheet

Design Specifications

Sources of data: *gas company, our oil company representative; Tubo Plumbing Corp., Jumbo Salvage Co., Watt Electronics, Inc.*

Tone: *formal*

Point of view: *third-person*

Needed visuals and supplements: *title page, letter of transmittal, table of contents, informative abstract, data sheet appendix reviewing the procedure for cost analysis*

Appropriate format (letter, memo, etc.): *formal report format with full heading system*

Rhetorical mode (description, definition, classification, etc.—or some combination): *primary mode: analysis; secondary modes: description, process narration*

Basic organization (problem-causes-solution, intro-instructions-summary, etc.): *questions-answers-conclusions and recommendations*

Main points in introduction: *Background
Purpose
Intended Audience
Data Sources
Limitations Scope*

Main points in body: *Description of Present System
Removal of Oil Burner and Tank
Installation of Gas Pipe
Installation of Gas Burner
Estimation of Gas Heating Costs*

Main points in conclusion: *Summary of Findings
Interpretation of Findings
Recommendation*

Other Considerations *no frills; these readers are all engineers interested in hard facts.*

FIGURE 7-2   (*Continued*)

CHAPTER SUMMARY

An outline is a plan for anything you write. It partitions a subject into its parts and classifies these parts on the basis of their similarities. To stay in control, the successful writer usually spends more time planning than writing.

Choose the best type of outline, depending on your purpose and on the length and formality of your report:

1. *An informal outline.* A simple list of words or phrases, especially useful for a short report.

2. *A formal topic outline.* A highly systematic arrangement of topics using a formal system of notation (either roman numeral–letter–arabic numeral or decimal notation).

3. *A formal sentence outline.* A further development of the topic outline. Each major and minor topic phrase is developed as a complete sentence which, in turn, serves as a topic sentence for a paragraph in the report.

Make your formal outline broad enough to encompass your subject and specific enough so you can discuss each topic in detail. Make each division yield at least two subparts and place all subparts of equal rank at the same level. Use a logical system of notation and a consistent format, expressing all divisions in parallel form. In a formal topic outline use clear and explicit headings and choose the most logical sequence for arranging the parts. Be sure that all items are relevant to your statement of purpose.

Follow these steps in constructing your formal outline:

1. Before writing the actual outline, complete all preliminary steps.
    a. Formulate your statement of purpose.
    b. Brainstorm your subject for specific ideas and topic breakdowns.
    c. Write your informal outline.
    d. Collect your data.
    e. Write a rough draft.

2. In your introduction, map out your background discussion, which begins with a definition or description of your subject and its history. Then state your intention and describe your target audience. Next, identify outside sources of data and explain any limitations of your report. Place working definitions in the following section, and end your introduction with a list of the major topics discussed in the body.

3. In your body, divide major topics and subtopics in logical sequence, presenting all relevant evidence. Give your body section an explicit title that reflects its contents.

4. In your conclusion, summarize the main points in your body, give an overall interpretation of these points, and draw conclusions based on facts. Base any recommendations or proposals on your conclusions. Vary these subsections according to the nature of the report.

Add, delete, or modify any subsections in this three-section structure as needed.

REVISION CHECKLIST

Use this list to check the quality of your outline.

1. Is this the best type of outline for your purpose (informal, formal topic, formal sentence)?
2. Is your outline broad enough to encompass the full range of your topic?
3. Is it specific enough in its divisions so that all major and minor points are represented?
4. Does each division yield at least two subparts?
5. Are parts or subparts of equal rank placed at the same level?
6. Do subparts at any given level add up to the immediately preceding items at the next higher level?
7. Should some minor points be major points, or vice versa?
8. Is the system of notation logical (roman numerals for major areas; capital letters for major topics; arabic numerals for subtopics; lowercase letters for further division; "(1)" for even further division)?
9. Is your format consistent (uniform indentation, spacing, and letter case)?
10. Are items of parallel importance expressed in parallel grammatical form?
11. Are all headings clear and explicit?
12. Is the subject arranged in the most logical sequence?
13. Is every item in the best possible location?
14. Are all topics and subtopics directly relevant to the stated intention of the report?
15. Are all necessary topics and subtopics included?
16. Is the introduction-body-conclusion structure fully developed?
17. Does the total of all parts in the body add up to the statement of purpose?
18. Is the outline clear and easy to follow (does it make sense)?
19. Are all items consistently stated either as topic phrases or as complete sentences?

Now list those elements of your outline that need improvement.

EXERCISES

1. In one paragraph, explain the difference between an informal outline and a formal outline. What is the major function of each?

2. Locate a short article (two thousand words maximum) from a journal in your field and make a topic outline or sentence outline of the article. Use the sample outlines in this chapter as models. Does the article conform to the general outlining procedures discussed in this chapter? If not, how could the original article be improved? Discuss your conclusions in class or in a written evaluation of one or two paragraphs.

3. For each of the following report topics, indicate the most appropriate sequence for organizing the subject. (For example, a proposal for athletic fields at your college would best proceed in a spatial sequence.)

- a set of instructions for operating a power tool
- a campaign report describing your progress in gaining support for your favorite political candidate
- a report analyzing the weakest parts in a piece of industrial machinery
- a report analyzing the desirability of a proposed nuclear power plant in your area
- a detailed breakdown of your monthly budget to trim excess spending
- a report investigating the reasons for student apathy on your campus
- a report investigating the effects of the ban on DDT use in insect control
- a report on any highly technical subject, written for a general reader
- a report investigating the value and success of a no-grade policy at other colleges
- a proposal for a no-grade policy at your college

4. *In class:* Organize into groups of four or five. Choose *one* of the following topics and, *after* you have formulated a clear statement of purpose, brainstorm in order to divide it into its components. Extend your partition into as many parts and subparts as possible. Rearrange the parts in logical order and in a consistent format, as you would in developing the body section of an outline. When each group completes this outlining process, one representative can write the final draft on the board for class criticism and suggestions for revision.

- job opportunities in your career field
- a description of the ideal classroom
- how to organize an effective job search
- how the quality of your higher educational experience can be improved
- arguments for and against a formal grading system

5. Use the checklist at the end of this chapter to evaluate the formal topic outline on pages 133–134. Suggest any needed revisions. Do the same in class for the outlines produced by various groups in doing exercise 4.

6. Assume that you are preparing a report titled "The Negative Effects of Strip Mining on the Cumberland Plateau Region of Kentucky." After brainstorming your subject, you settle on the four following major topics for investigation and discussion:

– economic and social effects of strip mining
– description of the strip-mining process
– environmental effects of strip mining
– description of the Cumberland Plateau

Arrange these topics in the most effective sequence for presentation.

When your topics have been effectively arranged, assume that subsequent research and further brainstorming produces the following list of subtopics:

– method of strip mining used in the Cumberland Plateau region
– location of the region
– permanent land damage
– water pollution
– lack of educational progress
– geological formation of the design
– open-pit mining
– unemployment
– increased erosion
– auger mining
– natural resources of the region
– types of strip mining
– increased flood hazards
– depopulation
– contour mining

Arrange each subtopic (and perhaps some sub-subtopics) under their appropriate topic headings. Use an effective system of notation and a good format to create the body section of a formal outline.

Hint: Assume that your thesis sentence is the following: "Decades of strip mining (without reclamation) in the Cumberland Plateau have devastated this region's environment, economy, and social structure."

8

Describing Objects and Mechanisms

DEFINITION

PURPOSE OF DESCRIPTION

MAKING YOUR DESCRIPTION
    OBJECTIVE
  Subjective Description
  Objective Description
    Be Totally Familiar with the Item
    Record Observable Details Faithfully
    Use Precise and Factual Language

ELEMENTS OF AN EFFECTIVE
    DESCRIPTION
  Clear and Limiting Title
  Overall Appearance and Component
    Parts
  Function of Each Part
  Comparisons with More Familiar Items
  Introduction-Body-Conclusion Structure
  Graphic Aids
  Appropriate Details

ORGANIZING AND WRITING YOUR
    DESCRIPTION
  Clearest Descriptive Sequence
    Spatial Sequence
    Functional Sequence
    Chronological Sequence
    Combined Sequences

The Outline
  Introduction: General Description
  Body: Description and Function
    of Parts
  Conclusion: Summary and Operating
    Description

APPLYING THE STEPS

CHAPTER SUMMARY

REVISION CHECKLIST

EXERCISES

DEFINITION

To describe is to create a picture with words (and diagrams, as needed). Descriptions serve a specific purpose: to convey appropriate information about an item to someone who will use it, buy it, operate it, or assemble it, or to someone who has to know more about it for some good reason. Any item can be described in countless ways. Therefore, *how* you describe — your plan of attack — depends on your purpose and on the needs of your audience.

PURPOSE OF DESCRIPTION

Description is part of all writing. People in marketing and sales, for example, use descriptions to stimulate interest in products; banks require detailed descriptions of any business venture before approving a loan; architects and engineers describe and perfect their plans on paper before actual construction begins; medical personnel maintain periodic descriptions of a patient's condition to ensure effective treatment.

The questionnaire in Figure 8-1 is used by police to obtain a description of a suspect. When a witness provides the details, a police artist converts the word picture into a sketch. Clearly, such a description is more useful than "The suspect is tall and dark, with a medium build," or "The suspect is ugly and evil-looking."

Another illustration of description can be seen in job descriptions that outline duties, responsibilities, and requirements. Your own job description will spell out what the organization expects of you. If you become a manager, you in turn might write job descriptions for other positions. Figure 8-2 shows a job description for the director of a computer center at a community college. The details provide the guidelines for evaluating the employee's performance.

DESCRIPTION QUESTIONNAIRE

Case No: _____

Interviewer: _____   Witness: _____
Place of Interview: _____   Address: _____
Date: _____   Phone #: _____

Description of Suspect

Sex ____ Nationality _____ Age ____ Height _____ Weight _____
Build _____ Who does this person look like? _____
In what way? _____
Hair: Color _____ Long ____ Short ____ Bald ____ Curly ____
       Straight ____ Other ____
Face: Round ____ Oval ____ Square ____ Other ____
Race_____ Color Skin _____
Complexion: Light _____ Dark _____ Ruddy _____
Unusual Facial: Scars _____ Pockmarks _____ Dimples _____
                Other _____
Eyes: Color _____ Shape _____ Brows: Color _____ Bushy _____
                                       Thin _____ Average ____
Nose: Large _____ Small _____ Wide _____ Flat _____
      Pronounced _____ Nostril Shape _____
Mouth: Lip Shape _____ Large _____ Small _____ Wide ____
       Thin _____ Color _____
Teeth: Large _____ Pronounced _____ Crooked _____
       Missing _____ Stained _____
Speech: Manner of Talking _____
Words Spoken: _____
_____
Chin: Pronounced _____ Recessed _____ Wide _____ Narrow _____
      Dimple _____
Mustache: _____ Beard _____ Sideburns _____ Color _____
          Shape _____
Cheeks: Pronounced _____ Recessed _____ Flat _____ Color ____
Ears: Large ____ Small ____ Flattened ____ Protruding _____
Neck: Large _____ Thin _____ Long _____ Short _____
Shoulders: Wide ____ Narrow ____ Chest: Broad____ Flat_____
                                        Other _____
Hand: Which used _____ Shaking _____ Calm . _____
      Gloves Worn _____ Tattoos _____ Watch _____
      Unclean _____ Scars _____ Other _____

FIGURE 8-1   Part of a Descriptive Questionnaire Used by Police Investigators (The questionnaire continues with *fingers, fingernails, clothing,* etc.)

DIRECTOR OF THE COMPUTER CENTER

The Director of the Computer Center is charged with the administrative direction of the College's Computer Center.

The Director of the Computer Center is responsible to the President of the College for the proper operation of the Computer Center as a teaching resource of the College, and particularly for its data processing programs. He is also responsible for the execution of certain administrative requests for computer information. He:

> Advises the President regarding the policies and procedures needed for the effective utilization of computer facilities.

> Serves the College's Advisory Committee on Data Processing Program and serves as consultant for data processing programs.

> Advises the College Committee on Administrative Data Processing Procedures and the College Committee on use of Computing Facilities regarding programming time allocations, new equipment and other matters related to the instructional and administrative uses of the computer.

> Supervises the data processing of college computerized records.

> Oversees the development of all computer programs.

> Advises Divisional Chairmen, faculty and other professional staff regarding the possible application of the computer to their instructional or administrative duties.

> Submits budget requests to the Dean of Administration.

> Participates in local, state and national professional associations in the field of data processing.

> Serves on the Administrative Advisory Council.

FIGURE 8-2   A Typical Job Description

No matter what the subject, readers need the answers to some or all of these questions:

1. What is it?
2. What does it do?
3. What does it look like?
4. What is it made of?
5. How does it work?
6. How has it been put together?

These are the questions that description answers. The police questionnaire, for example, is calculated to answer "What does it (the person) look like?" The job description is calculated to answer the question "What is it?" which usually requires an answer to some of the other questions. In this case, we need a description of the parts that make up the whole job. The purpose of description, then, is to answer as many of these questions as are applicable.

MAKING YOUR DESCRIPTION OBJECTIVE

From the earliest grades we write descriptions that are mainly *subjective* or *objective* — that is, based either on opinion (a belief held without proof) or on fact (something whose existence can be shown). Subjective description emphasizes the perceiver's attitude toward the thing, whereas objective description emphasizes the thing itself.

Subjective Description

The details in a subjective description are dictated by feelings or opinions about an item. Essays describing "My Biggest Complaint," "An Unforgettable Person," or "A Beautiful Moment" are expressions of opinion; they are written from a personal point of view. Subjective description aims at expressing feelings, attitudes, moods, and emotions. You create an *impression* of your subject more than communicating factual information about it ("The weather was miserable" versus "All day we had freezing rain and gale-force winds").

Objective Description

Objective description is not influenced by emotions or likes and dislikes. If you have ever been in an auto accident, you probably completed a form similar to the one in Figure 8-3, which forces the writer to be objective. In it you are asked to describe the physical details of the area and to give a *factual* account

LOCATION	City or Town Where Accident Occurred		Nearest Mile Marker		Reserved for Registry	

Street Name and/or Route Number

at intersection with _____ N. S. E. W. Of nearest intersection, bridge, mile marker, railroad.

or _____ feet

Other Landmarks:

Which direction was each vehicle traveling?
Vehicle No. 1 N. S. E. W.   No. 2 N. S. E. W.

54
1 ☐ On ramp from route _____
2 ☐ Off ramp from route _____
3 ☐ At rotary

55
1 ☐ Area built up
2 ☐ Area not built up

SITUATION

What were vehicles doing prior to accident? Mark appropriate box.

Vehicle 58-60				Vehicle				Vehicle		
1	2			1	2			1	2	
1	Making right turn	8		Skidding	F			Parked		
2	Making left turn	9		Slowing or stopping	G			Stalled or disabled		
3	Making U turn	A		Crossing median strip	H			Stalled or disabled with flasher on		
4	Going straight ahead	B		Driverless moving vehicle	J			In process of parking		
5	Passing on right	C		Backing	K			Entering or exiting from alley or driveway		
6	Passing on left	D		Starting in traffic	L			Other		
7	Stop sign	E		Starting from parked position						

Where was pedestrian located at time of accident? Mark appropriate box.

x	61		x
1	At intersection	7	Getting on/off vehicle
2	Within 300 feet of intersection	8	Working on vehicle
3	More than 300 feet from intersection	9	Working in street
4	Walking in street with traffic	A	Playing in street
5	Walking in street against traffic	B	Not in street
6	Standing in street	C	Other

DIAGRAM

INDICATE ON THIS DIAGRAM WHAT HAPPENED
Use one of these outlines to sketch the scene of your accident, writing in street or highway names or numbers.

1. Number each vehicle and show direction of travel by arrow:
2. Use solid line to show path before accident ; dotted line after accident.
3. Show pedestrian by:
4. Show railroad by:
5. Show distance and direction to landmarks; identify landmarks by name or number.
6. Indicate north by arrow, as:

INDICATE NORTH BY ARROW

FIGURE 8-3 Portion of an Accident Report Form

of the accident. The objective details — if honestly recorded — should speak for themselves.

Objective description is not only concerned with concrete physical details, however. A psychologist's description of a patient's mental condition is objective, insofar as it is based on demonstrable fact.

With the exception of promotional writing, most of your descriptions on the job will be objective. Your employer is less interested in your feelings about the item than about its factual details. The six questions on page 156 require objective answers about an item's appearance or function — or both. Notice that the question "What is your opinion of the item?" is not on that list. If you *are* asked to give an opinion or an interpretation, do so only *after* giving all the objective details. Objective description records exactly what you see, which should be what any other person would see. Here are some guidelines for maintaining objectivity.

Be Totally Familiar with the Item

Objective description relies on precise factual details; therefore, study the item and learn every detail before you write.

Record Observable Details Faithfully

To identify observable details, ask yourself these questions: What characteristics could any observer recognize? What details would a camera record?

Subjective   His office has an *awful* view, *terrible* furniture, and a *depressing* atmosphere.

The italicized words only *tell;* they do not *show.*

Objective   His office has broken windows looking out on a brick wall, a rug with a six-inch-wide hole in the center, chairs with their bottoms falling out, missing floor boards, and a ceiling with pieces of plaster missing in three or four places.

Use Precise and Factual Language

Use high-information words. Name specific parts without calling them "things," "gadgets," "whatchamacallits," or "doohickeys." Avoid judgmental words ("good," "super," "impressive," "poor") unless your judgment is specifically requested and can be based on observable facts or statistics. The same holds true for words like "large," "small," "long," and "near"; substitute "exact measurements," "weights," "dimensions," and "ingredients."

Use words that specify location: "above," "below," "behind," "to the right," "adjacent," "interlocking," "abutting," and "overlapping." Use position words: "horizontal," "vertical," "lateral," "longitudinal," "in cross section," "parallel."

Indefinite	Precise
at high speed	eighty miles per hour
a small office	an eight-by-twelve-foot office
a heavy typewriter	a fifty-pound typewriter
a late-model car	a 1978 Ford Granada two-door sedan
a high salary	$50,000 per year
long hours	sixty hours weekly
an inside view	a cross-sectional, cutaway, or exploded view

a tall mountain	with a vertical rise of 250 feet from sea level
impressive gas mileage	forty miles per gallon, city; fifty, highway
a poorly made tool	a tool with brittle plastic fittings
right next to the foundation	adjacent to the right side
partially exposed	overlapping
a small red thing	a red activator button with a one-and-a-half-inch diameter

Don't confuse precise language with needlessly complicated technical terms, however. Don't say "phlebotomy specimen" instead of "blood" in describing a microscopic blood analysis. The clearest writing uses the simplest terms. Non-technical language is preferable to specialized terminology as long as the simpler terms do the job.

The following physical description of a stethoscope is written in indefinite language:

> The standard stethoscope is an *ordinary looking thing* which is *small* and *light* in weight. This *well-made gadget* consists of *several* parts that work together to perform an *important* function.

If you didn't know anything about a stethoscope, this description wouldn't help. In fact, it could be referring to just about any item with parts that do something. The italicized words tell the reader nothing whatsoever.

Here is a description with vital details about appearance and function.

> The stethoscope is roughly twenty-four inches long and weighs about five ounces. The instrument consists of a sensitive, sound-detecting and amplifying device whose flat surface is pressed against a bodily area. This device, in turn, is attached to rubber and metal tubing that transmits the body sound to a listening device, which is inserted in the ear.
>
> Seven interlocking pieces contribute to the stethoscope's Y-shaped appearance: (1) diaphragm contact piece, (2) lower tubing, (3) Y-shaped metal piece, (4) upper tubing, (5) U-shaped metal strip, (6) curved metal tubing, and (7) hollow ear plugs. These parts are assembled into a continuous unit.

ELEMENTS OF AN EFFECTIVE DESCRIPTION

Clear and Limiting Title

Limit your topic by promising exactly what you will deliver — no more and no less. A title like "A Physical Description of a Typical Ten-Speed Racing Bicycle" promises a description of the entire item, down to the smallest bolt or cotter pin. If your intention, however, is to describe the bicycle's braking mechanism only, be sure that your title so indicates: "A Physical Description of a Center-Pull Caliper Braking Mechanism."

Overall Appearance and Component Parts

Allow readers to visualize the item as a whole before you describe each part. See the second description of the stethoscope above.

Function of Each Part

Explain the role of each part. In the stethoscope description, the function of the diaphragm contact piece can be explained this way:

> The diaphragm contact piece is caused to vibrate by body sounds. This part is the heart of the stethoscope as it receives, amplifies, and transmits the auditory impulse.

Comparisons with More Familiar Items

The following comparison helps readers visualize the item:

> The diaphragm contact piece is a circular metal disk roughly the shape and size of a silver dollar.

Introduction-Body-Conclusion Structure

Follow the structure of all good communication: (1) introduce the subject, (2) discuss it, and (3) summarize your discussion.

Graphic Aids

Use drawings, diagrams, or photographs generously (see Chapter 11). Our overall description of the stethoscope is greatly clarified by using Figure 8-4.

Appropriate Details

Your choice of the kinds and amount of details is crucial. On one hand, you must provide enough details to give a clear picture. On the other hand, you must avoid burying readers in needless details. Identify your readers and their reasons for reading your description.

Assume that you are about to describe a certain brand and model of bicycle to an uninformed person. The picture you create will depend on the details you select. How will your reader use this description? What is his or her level of technical understanding? Is the reader a customer likely to be interested in

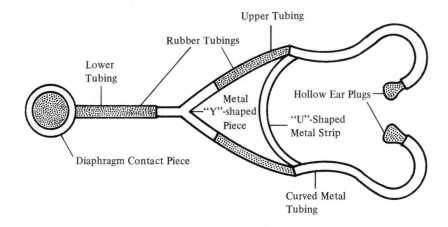

FIGURE 8-4   Stethoscope with Diaphragm Contact Piece

what the bike looks like — its flashy looks and racy style? Are you writing for a repair technician who needs to know the order in which the parts operate? Or is your reader a helper in your bicycle shop who needs to know how to assemble this particular kind of bicycle? Because anything could be described in dozens of ways, you need to identify your specific purpose and audience.

If you can't identify your reader, write for general readers and follow these guidelines: (1) details that are less technical are more widely understood and (2) too many details may be no better than too few.

Consider, for instance, a description of a particular brand of dishwasher that you would write for the average consumer. You can assume that your readers know what a dishwasher looks like. On the other hand, you do not want to bury them in details they cannot possibly use. Therefore, depending on your purpose, you might describe only the parts used in operating the appliance. Or, you might describe the interior chamber so that readers will know how many dishes the machine will clean. If your description were in a repair manual for service technicians, however, you would include many more technical details.

ORGANIZING AND WRITING YOUR DESCRIPTION

Before writing a word, answer these questions:

1. Why am I interested in this item?
2. Who is my reader?

3. How much does he or she already know?
4. Is my reader interested in the item's function or appearance — or both?
5. What details does this particular reader require, and in what order?

Your purpose in describing anything will be to tell readers what it looks like, how it has been put together, or how it works. Sometimes your purpose will be a combination of these. Once you have identified that purpose, you will need to devise a plan of attack. What descriptive sequence is best for helping the reader picture this item? Where do you begin? What comes next?

Clearest Descriptive Sequence

Like most items, yours should have its own logic of organization, based on: (1) the way it appears as a static object, (2) the way its parts operate in order, or (3) the way its parts are assembled. We can describe these relationships, respectively, through a spatial, functional, or chronological sequence.

Spatial Sequence

The spatial sequence is part of all physical descriptions. A spatial sequence answers these questions: What is it? What does it do? What does it look like? What parts and material is it made of? Use this sequence when you intend your reader to visualize the item as a static object or as a mechanism at rest (a house interior, a document, the Statue of Liberty, a plot of land, a tool, an electric razor at rest, and so on). Will your reader best understand the item from outside to inside, front to rear, left to right, top to bottom? (What logical path do the parts create?) A retractable pen, for example, would logically be viewed from outside to inside; a hammer, from top to bottom.

Sometimes you will want to emphasize certain parts of the item; then, you should try to design your description so those parts are either first or last in the sequence. You might describe a newly designed condominium high-rise in a top-to-bottom sequence if your intention is to emphasize the more dramatic penthouse suites. If, instead, you wished to emphasize the impressive lobby, lounge, pool, and playroom on the first floor, you might follow a bottom-to-top sequence. When striving for emphasis, base your order of description on the angle of vision that you wish to create for your reader.

Functional Sequence

Like the spatial sequence, the functional sequence answers the questions about identification, appearance, and parts and material; however, it also answers the question, How does it work? It is best used in describing a mechanism in action, such as a 35-millimeter camera, an electric razor, a smoke detector, or a stethoscope. The inherent logic of the item is reflected by the order in which its parts function. Therefore, the choice of direction should be simple: a mechanism usually has only one functional sequence.

In describing a solar home-heating system you would logically begin with the heat collectors in the roof, moving on through the pipes, pumping system, and tanks for the heated water, to the heating vents in the floors and walls — in short, from the source to the outlet. After describing the solar heating system according to the functional sequence of operating parts, you could describe each part according to a particular spatial sequence.

Chronological Sequence

In addition to answering the questions about identification, function, appearance, and parts and material, a chronological sequence answers the question, How has it been put together? The chronology is determined by the sequence in which the parts are assembled. Here again, there usually is only one possible direction for developing the description.

Use the chronological sequence for an item that is best understood in terms of its assembly (such as a piece of furniture, an umbrella tent, or a prehung window or door unit). Architects might find a spatial sequence best for describing a proposed beachhouse to clients; however, they would probably use a chronological sequence for describing the house to the contractor who will build it.

Combined Sequences

The description of a trout-activated feeder mechanism in Figure 8-5 employs all three sequences; first, a spatial sequence (top to bottom) for describing the overall mechanism at rest; next, a chronological sequence for explaining the order in which the parts are assembled; finally, a functional sequence for describing the order in which the parts operate.

AUDIENCE ANALYSIS

The audience for the description in Figure 8-5 (owners/operators of trout hatcheries) are technically informed. They already know about other types of feeder mechanisms. Therefore, they will need no background explanation about the function of feeder mechanisms in general. Nor will they need definitions of technical terms such as *surface feeders* and *gravity feed*. Instead, they are interested in the mechanism's *exact* dimensions and materials, the order in which the parts are assembled, and the way this mechanism works. Each section of the description can be greatly clarified by the insertion of visuals, including an enlarged drawing of the paper-clip attachment (the most crucial part of the mechanism at rest).

A DESCRIPTION OF AN

INEXPENSIVE TROUT-ACTIVATED

FEEDER MECHANISM

The Overall Mechanism

Our trout-activated feeder mechanism is designed so trout can obtain food at will. The purpose of this description is to explain the appearance, order of assembly, and order of operation of the feeder mechanism to trout hatchery owners and operators.

The feeder mechanism is made of a 1-pint plastic funnel, a 5-inch wooden strip, a 12-inch metal rod (1/4-inch diameter), a wooden disk, and a large paper clip. The mechanism is a cone-shaped device with a rod passing through its two openings. The large opening of the cone is on top. Centered on the rod under the smaller opening is a wooden disk of slightly larger diameter than the opening and fastened to the rod by a paper clip (see Figure 1).

Order of Assembly

The plastic funnel is large enough to hold more food than trout will need in one day. The wooden strip is

FIGURE 8-5   A Description Using Combined Sequences

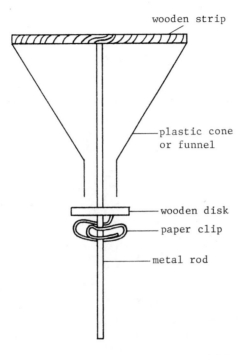

2

Figure 1. An Overall Lateral View of the Feeder Mechanism

placed across the upper, large funnel opening to hold the
metal rod that runs down through the cone. This rod passes
through the bottom, smaller opening until it touches the
surface of the water. The wooden disk is connected to the
rod just under the lower opening by a paper clip (as shown
in Figure 2). The paper clip is bent twice and forms a

FIGURE 8-5   (*Continued*)

3

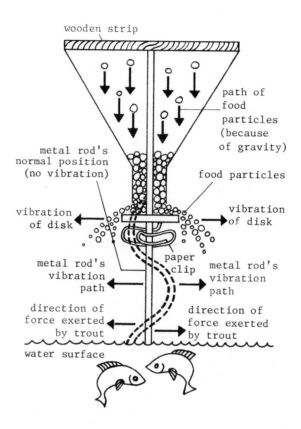

wooden strip

path of
food
particles
(because
of gravity)

metal rod's
normal position
(no vibration)

food particles

vibration
of disk

vibration
of disk

metal rod's
vibration
path

paper
clip

metal rod's
vibration
path

direction of
force exerted
by trout

direction of
force exerted
by trout

water surface

Figure 2. <u>A Lateral View of the Feeder Mechanism's
Assembly and Operation</u>

notch where the metal rod is inserted. One bend is made in

the inner section of the clip where there is a 180-degree

FIGURE 8-5   (*Continued*)

4

turn. The metal rod is inserted between the inner and outer

sections of the clip and the 180-degree head of the inner

section is bent around the rod at a 90-degree angle to the

plane of the clip. At the open end of the clip is a pointed

tip, which is bent 60 degrees straight upwards from the

clip's edge (as shown in Figure 3).

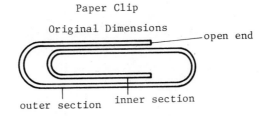

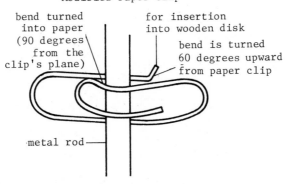

Figure 3.   A Side View of a Paper Clip and of Its
            Modifications for Attachment

FIGURE 8-5   (*Continued*)

5

<u>Mechanism in Operation</u>

As the trout surface they strike the tip of the metal
rod, which is touching the water. The vibration travels up
the rod and shakes the disk, which is covered with food from
the funnel's gravity feed. As the disk vibrates, it causes
food resting on it to fall to the water surface (see
Figure 2).

FIGURE 8-5   (*Continued*)

The Outline

Description of a complex mechanism almost inevitably calls for an outline. The following model can be adapted to virtually any description.

I. INTRODUCTION: GENERAL DESCRIPTION [1]
   A. Definition, Purpose, and Background of the Item
   B. Purpose of the Report and Specific Audience
   C. Overall Description of the Item (with graphic aid, if applicable)
   D. Principle of Operation (if applicable)
   E. List of Major Parts

II. DESCRIPTION AND FUNCTION OF PARTS
   A. Part One in Your Descriptive Sequence
      1. Definition
      2. Shape, Dimensions, Material (with specialized graphic aids)
      3. Subparts (if applicable)
      4. Function
      5. Relation to adjoining parts
      6. Manner of attachment (if applicable)
   B. Part Two in Your Descriptive Sequence
      etc.

III. SUMMARY AND OPERATING DESCRIPTION
   A. Summary (used only in a long, complex description)
   B. Interrelation of Parts
   C. One Complete Operating Cycle

Introduction: General Description

The introduction is the background section of your description, where you introduce the item. First, define the item, explain its purpose, and review its history. Next, explain the purpose of your description and identify your audience.

Definition and Purpose

The stethoscope is a listening device that amplifies and transmits body sounds to aid in detecting physical abnormalities. Its purpose is to assist doctors and nurse practitioners in diagnosing diseases.

History and Background

This instrument has evolved from the original wooden, funnel-shaped instrument invented by a French physician, R. T. Lennaec, in 1819. Because

---

[1] In most descriptions, the subdivisions in the introduction can be combined and need not appear as individual headings in the report.

of the modesty of his female patients, he found it necessary to develop a device, other than his ear, for auscultation (listening to body sounds).

Purpose of Report and Intended Audience
This report seeks to explain the structure and operating principle of the stethoscope to the beginning paramedical or nursing student.

Finally, give a brief, overall description of the item, discuss its principle of operation, and list its major parts. The overall description on page 159 follows a functional sequence.

Body: Description and Function of Parts

In the body you describe each major part. After arranging the parts in an overall sequence, follow the individual logic of each part. Begin your description of an individual part with a definition, followed by descriptions of the part's size, shape, and material. Also, discuss any important subparts. Finally, explain the part's function in the whole, its relation to adjoining parts, and its manner of attachment.

The intended readers of this description will use a stethoscope daily. Therefore, they will require intimate knowledge of its parts and their function.

Diaphragm Contact Piece
Definition, Size, Shape, and Material
The diaphragm contact piece is a circular metal and plastic disk, about the size of a silver dollar, which is caused to vibrate by body sounds.

Subparts
Three separate parts make up the piece: metal disk, plastic diaphragm, and metal frame, as shown in Figure 8-6.

The metal disk is made of stainless steel. Its inside surface is concave and designed with circular ridges that concentrate sound toward an opening in the center of the disk, then out through a hollow metal appendage. Lateral threads ring the outer circumference of the disk in order to accom-

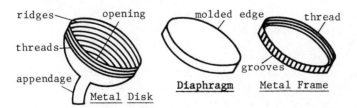

FIGURE 8-6   Frontal-Superior View of the Three Parts of a Diaphragm Contact Piece

modate the interlocking metal frame. A fitted diaphragm covers the four-inch circumference of the disk.

The diaphragm is a flat plastic disk, one millimeter thick, four inches in circumference, with a molded lip around the edge. It fits over the metal disk and vibrates sound toward the ridges. The diaphragm is held in place by a metal frame that screws onto the metal disk.

The stainless steel frame is four inches in circumference around its outer edge and three and one-half inches around its inner edge. A half-inch ridge between the inner and outer edge accommodates threads for screwing the frame onto the concave metal disk. The outside circumference of the frame is covered with perpendicular grooves — like those on the edge of a dime — which provide a gripping surface.

Function and Relation to Adjoining Parts

The diaphragm contact piece is the heart of the stethoscope as it receives, amplifies, and transmits the auditory impulse through the system of attached tubing.

Manner of Attachment

The diaphragm contact piece is attached to the lower tubing by an appendage on its upper end which fits inside the tubing.

Each part of the stethoscope, in turn, is described according to its own logic of organization.

Conclusion: Summary and Operating Description

In the conclusion, you review your description briefly (if it is complex) and explain how the parts combine to make the item function as a whole.

Summary

The stethoscope has been described according to the sequence of its functioning parts, in sufficient detail to acquaint the reader with its physical characteristics and operating principle.

Interrelation of Parts

The seven major parts of the stethoscope provide support for the instrument, flexibility of movement for the operator, and ease in auscultation.

One Complete Operating Cycle

In an operating cycle, the diaphragm contact piece is placed against the skin to pick up sound impulses from the body surface. These impulses cause the plastic diaphragm to vibrate. The amplified vibrations, in turn, are carried through a single tube to a dividing point. From here, the sound is carried through two separate but identical series of tubes to hollow ear plugs which transmit the amplified sounds to the listener's ears.

Add well-labeled visuals whenever they clarify your description. Use transitional sentences to provide logical bridges between sections. The sentence below leads from the diaphragm contact piece to the lower tubing:

> The diaphragm contact piece is attached to the lower tubing by an appendage on its upper end, which fits inside the tubing.

As long as you follow an introduction-body-conclusion structure, the outline and development of individual sections can be modified. Depending on your subject, purpose, and reader, you might delete or combine some areas.

APPLYING THE STEPS

The description of an automobile jack in Figure 8-7, written for a general reading audience, follows our outline model.

> AUDIENCE ANALYSIS
>
> Some readers of this description (written for an owner's manual) will have no mechanical background. Therefore, before they can follow instructions for *using* the jack safely, they will have to learn what it is, what it looks like, what its parts are, and how it works. They will *not* need precise dimensions (e.g., "The rectangular base is 8 inches long and 6½ inches wide, sloping upward 1½ inches from the front outer edge to form a secondary platform 1 inch high and 3 inches square"). The engineer who designed the jack might include such data in a description written for the manufacturer. For laypersons, however, only those dimensions are included that will help them recognize specific parts.

CHAPTER SUMMARY

In describing anything, identify it, explain its function, portray its appearance, and list its parts and materials. Also, explain how a mechanical item works and, if appropriate, how it has been put together.

Subjective description emphasizes your attitude toward the subject whereas objective description emphasizes the subject itself. To be objective, know your subject, stick to observable details, and use terms that communicate hard information.

An effective description begins with a clear and limiting title that promises exactly what you will deliver. First, the subject is described in terms of its overall physical characteristics; next, in terms of its individual parts and their respective functions. Whenever possible, a description uses comparisons with more familiar objects. The description has an introduction-body-conclusion

A DESCRIPTION OF THE

STANDARD FORD BUMPER JACK

INTRODUCTION -- GENERAL DESCRIPTION

Definition, Audience, and Purpose

    The standard Ford bumper jack is a portable device used
for raising the front or rear of a car by means of force
applied with a lever. It allows even a frail person to lift
one corner of a 2-ton automobile.

    This description is written for the general reader and
car owner who will sometimes have to raise a car to change a
flat tire.

Appearance, Function, and Major Parts

    The jack consists of a molded steel base supporting a
free-standing, perpendicular notched shaft. Attached to the
shaft are a leverage mechanism, a bumper catch, and a cylinder
for insertion of the jack handle (see Figure 1). Except for
the main shaft and leverage mechanism, the jack is made to be
dismantled and to fit neatly into the car's trunk.

    The jack operates on a leverage principle, with the
human hand traveling 18 inches and the car only three-eighths

FIGURE 8-7   A Descriptive Report

2

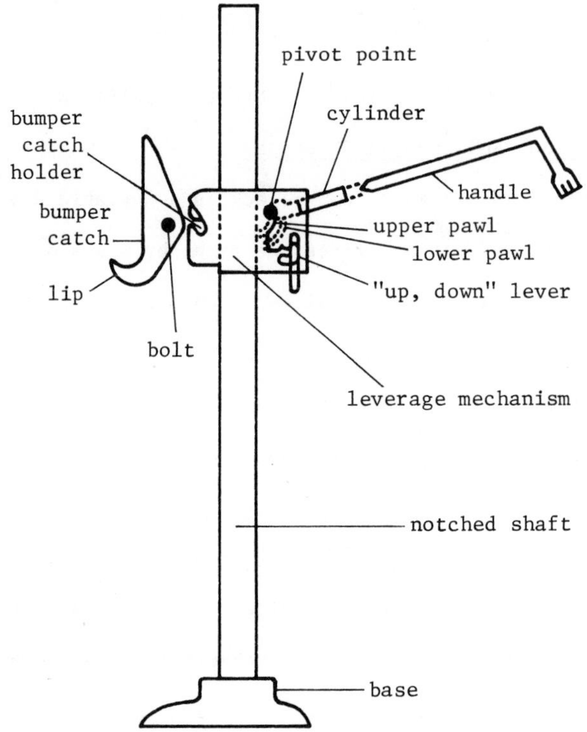

Figure 1. A Side View of the Standard Ford Bumper Jack

of an inch during a normal jacking stroke. Such a device requires many strokes to raise the car off the ground, but may
prove a lifesaver to a motorist on some deserted road.

FIGURE 8-7   (*Continued*)

3

Five main parts make up the jack: the base, the notched
shaft, the leverage mechanism, the bumper catch, and the
handle.

DESCRIPTION OF PARTS AND THEIR FUNCTION

The Base

The rectangular base is a molded steel plate that provides support and a point of insertion for the shaft (see
Figure 1). The base slopes upward to form a platform. In
this platform is a 1-inch depression that provides a stabilizing well for the shaft. Stability is increased by a
1-inch cuff around the well. As the base rests on its flat
surface, the bottom end of the shaft is inserted into its
stabilizing well.

The Shaft

The notched shaft is a steel bar that provides a vertical
track for the leverage mechanism. This triangular-shaped
piece is 32 inches long. The notches, which enable the
mechanism to maintain its position on the shaft, face the
operator.

The shaft gives vertical support to the raised automobile,

FIGURE 8-7 (*Continued*)

4

and attached to it is the leverage mechanism, which rests on individual notches.

The Leverage Mechanism

The leverage mechanism provides the mechanical advantage needed for the operator to raise the car. It is made to slide up and down the notched shaft. The main body of this molded steel mechanism contains two units: one for transferring the leverage and one for holding the bumper catch.

The leverage unit has four major parts: the cylinder, which connects the handle and a pivot point; a lower pawl (a device that fits into the notches to allow forward and prevent backward motion), which is connected directly to the cylinder; an upper pawl, which is connected at the pivot point; and an "up-down" lever, which applies or releases pressure on the upper pawl by means of a spring. (See Figure 1.) Moving the cylinder up and down with the handle causes the alternate release of the pawls, and thus movement up or down the shaft -- depending on the setting of the "up-down" lever. The movement is transferred by the metal body of the unit to the bumper-catch holder.

FIGURE 8-7   (*Continued*)

5

The holder consists of a downsloping groove, partially blocked by a wire spring. (See Figure 1.) The spring is mounted in such a way as to keep the bumper catch in place during operation.

The Bumper Catch

The bumper catch is a steel device that provides the attachment between the leverage mechanism and the automobile bumper. This 9-inch molded plate is bent to conform to the shape of the bumper. Its outer half inch is bent up into a lip (see Figure 1), which hooks behind the bumper to hold the catch in place. The two sides of the plate are bent back 90 degrees to leave a 2-inch bumper contact surface, and a bolt is riveted between them. The bolt slips into the groove in the leverage mechanism and provides the point of attachment between the leverage unit and the car.

The Jack Handle

The jack handle is a steel bar that serves both as a lever and a lug-bolt remover. This round bar is 22 inches long, 5/8 inch in diameter, and is bent 135 degrees 5 inches from its outer end. Its outer end is a wrench made to fit

FIGURE 8-7   (*Continued*)

6

the wheel's lug bolts. Its inner end is flattened for removing the wheel covers and for insertion into the cylinder
on the leverage mechanism that connects to the pivot point.

SUMMARY AND OPERATING DESCRIPTION

Interrelation of Parts

The jack's five main parts (base, notched shaft, leverage mechanism, bumper catch, and handle) make up an efficient,
lightweight device for raising a car to change a tire.

One Complete Operating Cycle

The jack is quickly assembled by inserting the bottom of
the notched shaft into the stabilizing well in the base, the
bumper catch into the groove on the leverage mechanism, and
the flat end of the jack handle into the cylinder.

The bumper catch is then attached to the bumper, with
the lever set in the "up" position. As the operator exerts
an up-down pumping motion on the jack handle, the leverage
mechanism gradually climbs the vertical, notched shaft until
the wheel is raised above the ground. Conversely, with the
lever in the "down" position, the same pumping motion causes
the leverage mechanism to descend the shaft.

FIGURE 8-7   (*Continued*)

structure, with graphic aids used generously. The type and amount of detail should be appropriate to the subject and the reader's needs.

Follow these steps in planning and writing your description:

1. After defining your purpose, identify the descriptive sequence or sequences that follow the logic of your subject.

   a. Use a spatial sequence for describing static objects or mechanisms at rest.

   b. Use a functional sequence for describing mechanisms in action.

   c. Use a chronological sequence for describing the order in which an item — either static or mechanical — has been assembled.

2. Make a detailed outline and develop your report from it.

   a. In your introduction, first define the item and explain its purpose and background. Next, discuss the aim of your description and specify your audience. Proceed to an overall description of the item and an explanation of its operating principle. Finally, list the major parts described in the body section.

   b. In your body section, describe and explain the function of each major part, in order, giving its definition, shape, dimensions, material, subparts, function, relation to adjoining parts, and manner of attachment.

   c. In your conclusion section, summarize the main points in the body, discuss the interrelation of parts, and describe one complete operating cycle. Within this three-section structure, delete or modify internal parts to suit your subject, purpose, and the reader's needs.

REVISION CHECKLIST

Use this list to refine the content, arrangement, and style of your description.

Content

1. Does the description have a clear and limiting title that promises exactly what has been delivered?

2. Have you described the item's overall features, as well as each of its parts?

3. Have you defined each part before discussing it?

4. Have you explained the function of each part?

5. Have you used comparisons with more familiar items whenever possible?

6. Have you used visuals whenever they provide clarification?

7. Will the intended reader(s) be able to visualize the item from your details?

8. Are there any unnecessary or confusing details, given your stated audience?

9. Is your description keyed to your stated purpose (will it do what it is supposed to do for your audience)?

10. Does the description answer all the questions on page 156 that are applicable for this audience?

Arrangement

1. Does your description follow an introduction-body-conclusion structure?

2. Does your description follow your outline faithfully?

3. Does your description follow the clearest possible sequence or combination of sequences (spatial, functional, chronological)?

4. Are your headings appropriate and adequate?

5. Are there enough transitions between related ideas?

Style

1. Is your description objective, with all relevant, observable details recorded in precise, factual language?

2. Will the level of technicality connect with the intended audience?

3. Is your description written in plain English?

4. Is each sentence clear, concise, and fluent?

5. Is the description written in correct English (see Appendix A)?

Now list those elements of your description that need improvement.

EXERCISES

1. In two or three paragraphs, discuss the kinds of descriptive writing assignments you expect to have in your field. Be as specific as possible in identifying the subjects, the situations, and the readers for whom you will write. After giving examples, illustrate with a detailed scenario that answers the questions about purpose on pages 161–162.

2. Rewrite these subjective statements, making them objective by describing observable details:

    a. This classroom is (attractive, unattractive).

    b. The weather today is (beautiful, awful, mediocre).

    c. (He, she) is the best-dressed person in this class.

    d. My textbook is in (good, poor) condition.

    e. This room is (too large, too small, just the right size) for our class.

3. Write a three-paragraph subjective description titled "My Best Friend" in which you express your feelings and attitude. Next, use the sample descriptive questionnaire in Figure 8-1 as a guide for writing a physical description of the same person so readers could recognize him or her in a crowd. Which of these descriptions is easiest to write, and why?

4. After consulting library sources, faculty members, or workers in your field, write a one-page job description of the position you hope to hold in ten years. Or write a job description for the job that you now hold. Using the sample description in Figure 8-2 as a model, include a description of function, duties, responsibilities, and qualifications.

5. Select three of your most valuable material possessions (try to choose simple objects that can easily be described). Write a one-paragraph physical description of each in enough detail that a police officer could recognize the items if they were lost or stolen.

6. The following descriptive statements are indefinite or subjective. Choose one statement from the list and expand it into an objective description in one paragraph:

   a. Come to my place for dinner and I will cook a meal you will never forget.
   b. During my vacation, I traveled everywhere.
   c. The job opportunities in my major are numerous.
   d. I plan to be highly successful in my career.
   e. Attending a small college has (advantages, disadvantages).
   f. The room in which this course is held is (cheerful, depressing).

7. Television commercials are, supposedly, a source of factual information about various products. But do we actually learn anything from descriptions that include modifiers like "super," "new," "improved," "jumbo," "nutritious," "elegant," and "exciting"? Often, the purpose of this indefinite language is to conceal the product's similarity to other products (such as other brands of aspirin).

In class, analyze recordings or scripts of commercials. What do you really learn from them? Try to isolate objective descriptions from subjective or downright indefinite ones.

8. Following is a list of items for description. How would each description best be organized? Map out your approach. When you have decided on your purpose and audience needs, choose the descriptive sequence that best describes the overall item. Next, identify the best sequence for describing each part of the item.

An overall description of a ski binding as a mechanism at rest, intended for the recreational skier, might best follow a spatial sequence that emphasized its appearance and function. In contrast, an overall description of the ski binding as a working mechanism, intended for the ski-shop technician who will repair and adjust bindings, might best follow an operational sequence.

In both cases, individual parts would be described according to their own logic of spatial organization. Choose a plan of attack that works best for you and your reader.

a handsaw	an accident
a pogo stick	a flower
a sailboat	a courtroom
a textbook	the members of your family
a retractable ball-point pen	a cigarette lighter
a steeplechase track	a bunsen burner
a pocketknife	a microscope
a pair of binoculars	a rock group

9. a. Choose an item from the following list, from exercise 8, or from your major. Using the general outline in this chapter as a model, outline a descriptive report. Write the report for a specific use by a specified audience. (Attach your written audience analysis to your report, if your instructor so requires.) Write one draft of the report, modify your outline as needed, and write your second draft. Proofread, following the revision checklist at the end of this chapter, and exchange reports with a classmate for further suggestions about revision.

a soda-acid fire extinguisher	a 100-amp electrical panel
a breathalyzer	a computer
a sphygmomanometer	a Skinner box
a ditto machine	a simple radio
a typewriter	a club or fraternity/sorority
a transit	a distilling apparatus
a saber saw	a bodily organ
a drafting table	a Wilson cloud chamber
a pet	a voodoo ritual
a blowtorch	an accomplishment

Remember that you are simply describing the item, its parts, and function; do not provide directions for its assembly or operation.

b. As an optional assignment, describe a place you know well. You are trying to convey a visual image, not a mood; therefore, your description should be objective, discussing only the observable details.

10. The bumper-jack description in this chapter is aimed toward a general reading audience. Evaluate its effectiveness by using the revision checklist as a guide. In one or two paragraphs, discuss your evaluation and include suggestions for revision.

11. *In class:* Compose an outline of the bumper-jack description, subdividing as necessary for subparts. Also, identify the various descriptive sequences.

9

Explaining a Process

DEFINITION

THE PURPOSE OF PROCESS
    EXPLANATION

TYPES OF PROCESS EXPLANATION
    Instructions
    Narrative
    Analysis

ELEMENTS OF EFFECTIVE
    INSTRUCTIONS
    Clear, Limiting, and Inclusive Title
    Logically Ordered Steps
    Appropriate Level of Technicality
        Background Information
        Details
    Visual Aids
    Warnings, Cautions, and Notes
    Appropriate Terms, Phrasing, and
        Paragraph Structure
        Active Voice and Imperative Mood
        Transitions to Mark Time and
            Sequence
        Parallel Phrasing
        Short Sentences and Paragraphs
    Introduction-Body-Conclusion Structure

ORGANIZING AND WRITING A SET OF
    INSTRUCTIONS
    Introduction
    Body
    Conclusion

APPLYING THE STEPS

ELEMENTS OF AN EFFECTIVE PROCESS
    NARRATIVE
    Phrasing in a Process Narrative
    Appropriate Level of Technicality
    Introduction-Body-Conclusion Structure

ORGANIZING AND WRITING A
    PROCESS NARRATIVE

APPLYING THE STEPS

ELEMENTS OF AN EFFECTIVE PROCESS
    ANALYSIS
    Phrasing in a Process Analysis
    Appropriate Level of Technicality
    Introduction-Body-Conclusion Structure

ORGANIZING AND WRITING A PROCESS
    ANALYSIS

APPLYING THE STEPS

CHAPTER SUMMARY

REVISION CHECKLIST

EXERCISES

DEFINITION

A process is a series of actions or changes leading to a product or result. A process explanation, then, is an account of how these actions or changes occur.

THE PURPOSE OF
PROCESS EXPLANATION

A process explanation is designed to answer the reader's question "How?" — posed in one of three ways:

1. How do I do (or make) something?
2. How did you do something?
3. How does something happen?

When you explain a process, you discuss causes and their effects. On the job, for example, you might instruct a colleague or customer on how to do something: how to analyze a soil sample; how to program a computer; how to assemble a precut log cabin.

Besides explaining how to do something, you may have to give two other kinds of process explanations. You might give accounts of various procedures to an employer or client: how you repaired a piece of equipment; how you inspected a building site; how you conducted an experiment. Or, you might have to explain to a new employee how certain things happen: how the promotion system in your company works; or how the budget for various departments is determined. Each type is discussed in the following section.

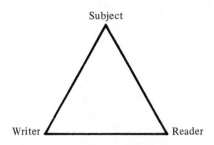

FIGURE 9-1   A Rhetorical Triangle

TYPES OF PROCESS EXPLANATION

Depending on your purpose and your readers' needs, you can design your explanation to answer one of the "how" questions, respectively, with *how to do it; how I did it*, or *how it happens*. As an illustration of how purpose and audience dictate your approach, consider Figure 9-1. Written communication occurs when these elements interact, but each type of process explanation emphasizes a different element. A set of instructions is reader-oriented; a process narrative is writer-oriented; and a process analysis is subject-oriented.

Instructions

Instructions tell readers how to do (or make) something. Almost anyone with a responsible job writes instructions, and almost everyone has to read some sort of instructions. The new employee needs instructions for operating the company's office machines; the employee going on vacation writes instructions for the person filling in. Owners of new cars read the operator's manual for servicing and operating instructions. The person who buys a stereo-component set reads instructions for connecting the turntable, amplifier, and speakers. Thus, instructions *emphasize the reader's role*, explaining each step in enough detail so that readers can complete the procedure safely and efficiently.

Narrative

Related to an explanation of *how to do something* is an explanation of *how you did something*, along with a description of your results, as, for example, in a lab report. Police and fire personnel write accounts of investigations and inspections. Medical laboratory technicians write accounts of tests and results. Construction supervisors write daily accounts of work completed on a project. This type of explanation *emphasizes your role* so that readers will understand the steps you followed and the results you achieved.

Analysis

An explanation of how things work or how something happens divides the process into its parts and principles. Colleagues and clients need to know how stock and bond prices are governed, how fermentation occurs, how your bank reviews a mortgage application, how your town decided on its zoning laws, how radio waves are transmitted, how low-glass fiber conducts electricity, and so on. Process analysis emphasizes *the process itself* instead of the reader or writer. The process is dissected so that readers can follow it as "observers."

ELEMENTS OF EFFECTIVE INSTRUCTIONS

Your purpose in writing instructions is to answer all possible questions about procedure.[1] After all, your reader may be dangling from a steel girder or surrounded by hot wires, with nothing but words on the page for directions. In some cases foggy instructions can be disastrous. Make your instructions readable by including the following elements.

Clear, Limiting, and Inclusive Title

Make your title promise exactly what your instructions will deliver — no more and no less. Consider this title: "Instructions for Cleaning the Carriage, Key Faces, and Exterior of a Typewriter." It is effective because it tells readers what to expect: instructions for performing a specific procedure on selected parts of an item. A title like "The Typewriter" would give readers no idea of what to expect: the report might be a history of the typewriter, typing instructions, or a description of each part. On the other hand, a title like "Instructions for Cleaning a Typewriter" would be misleading if the instructions were for cleaning selected parts only.

Logically Ordered Steps

Good instructions not only divide the process into steps; they also guide readers through the steps in *order,* so they won't go wrong. For example,

> You can't join two wires to make an electrical connection before you have removed the insulation. To remove the insulation, you will need. . . .

The logic of a process requires that you explain the steps in chronological order, as they are performed. Thus, instructions for building a house properly begin with the foundation, and move, in order, to the floor, walls, and roof.

---

[1] To write good instructions you must know the specific procedure down to the smallest detail. Unless you have performed the task successfully, don't try to write instructions for it.

Appropriate Level of Technicality

Unless you know that readers have a technical background and technical skills, write for general readers and be sure to do two things. First, give them enough background to understand *why* the instructions should be followed. Second, give sufficiently detailed explanations so that they can understand the instructions.

Background Information

Begin by explaining the reason for the procedure:

> Some people think they don't have the time or skill to change their car's motor oil and engine filter. They prefer to have the changes done at a garage, service station, or car dealership. At today's prices, however, you may spend fifteen dollars or more for a service you can perform yourself in thirty minutes. These instructions will help you change your oil and filter so that you can save up to eight dollars.

Also, state your assumptions about your reader's level of technical understanding:

> For building a wind harp, the reader is expected to have a general familiarity with hand tools and woodworking techniques in order to avoid troublesome mistakes. Also, if power tools are to be used, and used safely, a good working knowledge of their operation is essential.

Before writing the step-by-step instructions, define any special terms:

> Jogging is a kind of running — a slow trot consuming not more than seven minutes per mile.

When your reader understands *what* and *why,* you are ready to explain *how.*

Details

Most of us know the frustration of buying a disassembled item, opening the box, reading the instructions for assembly, and finding ourselves lost because the instructions are unclear. Even some do-it-yourself books and gourmet cookbooks are notorious for their lack of detail and overly technical instructions. For instance, how does one "treat a rack of lamb gently with truffle essence"? Vague instructions are based on the writer's mistaken assumption that readers have a strong technical background. Consider this set of fog-bound instructions for treating an electrical-shock victim:

1. Check vital signs.
2. Establish an airway.
3. Administer external cardiac massage if needed.

4. Ventilate, if cyanosed.
5. Treat for shock.

These instructions might be clear to the medical expert, but they mean little to the average reader. Not only are the details inadequate, but terms such as "vital signs," "cyanosed," and "ventilate" are too technical for the layperson. If written for employees in a high-voltage industrial area they would be useless. The instructions need to be rewritten with clear illustrations and detailed explanations as in a Red Cross First Aid manual.

It's easy to assume that others know more than they do about a procedure; this is especially true when you can perform the task almost automatically. Think about the days when a relative or friend was teaching you to drive a car; or perhaps you have tried to teach someone else. When you write instructions, remember that the reader knows less than you do. A colleague will know at least a little less; a layperson will know a good deal less — maybe nothing — about this procedure. Analyze your reader's needs carefully.

On the one hand, you need enough details for readers to understand and perform the task successfully. On the other hand, you should omit general information that the average person can be expected to know in advance (for example, the difference between a hand tool and a power tool). Excessive details add clutter and may interrupt the sequence of steps. The following instructions contain just enough details for the general reader:

BREAKING INTO A JOG

After completing your warming-up exercises, set a brisk pace walking. Exaggerate the distance between steps, making bountiful strides and swinging your arms freely and loosely. After roughly one hundred yards of this pace, you should feel lively and ready to jog.

Immediately break into a slow trot: let your torso lean forward and let one foot fall in front of the other (one foot leaving the ground while the other is on the pavement) at the slowest pace possible, just above a walk. *Do not bolt out like a sprinter!* The worst thing is to start fast and injure yourself.

Relax your body while jogging. Keep your shoulders straight and your head up, and enjoy the scenery — after all, it is one of the joys of jogging. Keep your arms low and slightly bent at your sides. Move your legs freely from the hips in an action that is easy, not forced. Make your feet perform a heel-to-toe action: land on the heel; rock forward; take off from the toe.

These instructions are clear and uncluttered because the writer has correctly assessed the general reader's needs. Terms like "bountiful stride," "torso," and "sprinter" should be clear to the general reader, without definitions or illustrations. In contrast, "a slow trot" is given a working definition because some people could have differing interpretations of this term. When in doubt about the need for certain details, you are safer to overexplain than to underexplain.

Visual Aids

Whenever you can use a visual aid to illustrate a step, do so (as discussed in Chapter 11). Incorporate it within your discussion of that step, as shown in this procedure for raising the roof section of an umbrella tent:

> Lift the ridge pole by hand until the pin of the support post can be inserted into the hole that is two inches behind the cap of the ridge pole (see Figure 9-2). Push the ridge pole up until the support post is perpendicular to the ground.

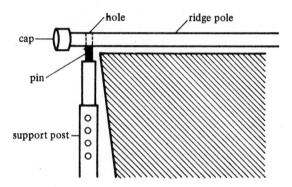

FIGURE 9-2   A Lateral View of the Support-Post Insertion into the Ridge Pole

Warnings, Cautions, and Notes

The only items that properly interrupt the order of steps are warnings, cautions, and notes about a particular step. Refer to these in your introduction, advising readers of their locations within the text. Place the warnings and cautions themselves, *clearly marked,* immediately before the respective steps.

> *Caution:* No one over thirty should be jogging without a doctor's approval. A person of any age who is overweight or who has not recently exercised should also have a physical checkup. Jogging, for most people, is healthful, but for some it can be deadly.

Overuse of warnings and cautions, however, might cause readers to ignore a crucial warning.

Appropriate Terms, Phrasing, and Paragraph Structure

Of all technical communication, instructions have the most demanding requirements for clarity and phrasing. Here, the words on the page lead to *immediate action.* Most readers are impatient and might not read the entire set of instructions before plunging into the first step. Poorly phrased and misleading instructions could have frustrating — or even tragic — results. Imagine astronauts receiving unclear instructions during launching or reentry. Closer to home, the results of pilot confusion over garbled landing or takeoff instructions are all too clear.

Like descriptions, instructions name parts, use location and position words, and state exact measurements, weights, and dimensions. In addition, there are four elements of phrasing and paragraph development that require unusually close attention for well-written instructions. They are: (1) consistent voice and mood — normally the active voice and the imperative mood; (2) time- and sequence-marking words like "first," "next," "after drying," etc.; (3) parallel phrasing; and (4) short sentences and paragraphs.

Active Voice and Imperative Mood

The active voice ("He opened the door") speaks more directly than the passive ("The door was opened by him"). Likewise, the imperative mood ("Open the door") lends more authority to your instructions than the indicative mood ("You open the door").

> Weak   The rudder arm should be moved toward the sail to cause the boat to "come about."

> Stronger   *Move* the rudder arm toward the sail to cause the boat to "come about."

> Weak   The valve that regulates the flow of air is turned all the way to the right.

> Stronger   *Turn* the valve that regulates the flow of air all the way to the right.

The imperative makes instructions more definite because the action verb — the crucial word that tells what the next action will be — comes first. Instead of burying your verb in the middle of a sentence, begin with an action verb ("raise," "connect," "wash," "insert," "open") to give readers an immediate signal.

Transitions to Mark Time and Sequence

Transitional words are like bridges between related ideas (see Appendix A). Some transitional words ("in addition," "next," "meanwhile," "finally," "on Tuesday morning," "in ten minutes," "the next day," "before," "the following afternoon") are designed to mark time and sequence. They help readers understand the task as a step-by-step process, as shown in these instructions for preparing the ground before pitching a tent:

PREPARING THE GROUND

Begin by clearing and smoothing the area that will be under the tent. This step will prevent damage to the tent floor and eliminate the discomfort of sleeping on uneven ground. *First,* remove all large stones, branches, or other debris within a level 10 × 13-foot area. Use your camping shovel to remove half-buried rocks that cannot easily be moved by hand. *Next,* fill in any large holes with soil or leaves. *Finally,* make several light surface passes with the shovel or a large, leafy branch to smooth the area.

Parallel Phrasing

Like any other items in a series, steps in a set of instructions should be expressed in identical grammatical form, that is, in parallel construction (see Appendix A). Such parallelism is important in all writing but particularly in instructions, because the repetition of the same grammatical forms can further emphasize the step-by-step organization of the instructions:

Incorrect

The major steps in sprouting an avocado pit are as follows:

1. Peeling the pit.
2. Insert toothpicks to hold the pit in position with the flat end down.
3. The water level should be maintained so that it covers two-thirds of the pit.
4. Keep the pit in a warm, dark place until the roots are developed.

Notice that step (1) is expressed as an "ing" phrase; steps 2 and 4 are expressed as complete sentences in the imperative mood; and step 3 is expressed as a complete sentence in the indicative mood. These should be rewritten so that each step is expressed as an "ing" phrase. (The instructions for performing each step will, of course, be written as imperative sentences.)

Correct

The major steps in sprouting an avocado pit are as follows:

1. Peeling the pit
2. Inserting toothpicks to hold the pit in position with the flat end down
3. Maintaining the water level so that it covers two-thirds of the pit
4. Keeping the pit in a warm, dark place until the roots are developed

Parallel construction makes the steps more readable and lends continuity to the instructions.

Short Sentences and Paragraphs

As a rule, each paragraph should explain one major step, and individual sentences within the paragraph should explain individual minor steps. In this way you clearly separate each activity. The previous instructions titled "Preparing the Ground" are a good example of effective paragraph and sentence division.

Do not shorten sentences by omitting articles ("a," "an," "the"). Instead of writing "Cover opening with half-inch plywood sheet," write "Cover the opening with a half-inch plywood sheet."

Introduction-Body-Conclusion Structure

Introduce readers to the procedure, explain the steps, and review your explanation. This structure is discussed in the next section.

ORGANIZING AND WRITING A SET OF INSTRUCTIONS

The introduction-body-conclusion structure can be adapted to any instructions. Here is a general outline:

   I. INTRODUCTION
     A. Definition (or Background) and Purpose of the Procedure
     B. Intended Audience
     C. Knowledge and Skills Needed
     D. Brief Overall Description of the Procedure
     E. Principle of Operation
     F. Materials, Equipment (in order of use), and Special Conditions
     G. Working Definitions (*Note:* Because definitions are vital to effective performance, always place them in your introduction.)
     H. Warnings, Cautions, and Notes
     I. List of Major Steps

  II. INSTRUCTIONS FOR PERFORMANCE
     A. First Major Step
       1. Definition and purpose
       2. Materials, equipment, and special conditions needed for this step
       3. Substeps (if applicable)
         a.
         b.
         c.
     B. Second Major Step
       etc.

III. CONCLUSION
  A. Summary of Major Steps
  B. Interrelation of Steps

This outline is only tentative because some elements might be modified, deleted, or combined. Revise it as needed for your subject, purpose, and audience.

Introduction

In your introduction, give readers the background (when, where, why). To avoid later interruptions, define the procedure, identify your intended audience, and state your assumptions about the knowledge and skills needed to follow your instructions. Also, describe the process briefly and the item's principle of operation (especially if these are repair instructions). Next, identify required materials, tools, and special conditions. Then define specialized terms and alert readers to warnings, cautions, and notes. Finally, list the major steps.

Below is an introduction from instructions for *making something*. Written for interested laypersons as well as accomplished gardeners, the instructions will be printed in an organic gardening magazine.

HOW TO SET UP A HYDROPONIC FLOODING SYSTEM

Definition, Purpose, and Audience

Hydroponics is the process of growing plants without soil. This process is particularly useful in environments where soil and space are limited. The instructions below give the knowledgeable gardener an alternative to soil gardening and also provide directions for anyone interested in gardening.

Description of the Procedure

An automatic hydroponic flooding system can be assembled and made to function easily. The assembled equipment and materials for the system are shown in Figure 1.

Operating Principle

Hydroponic gardening operates on the principle that plants will grow quickly and abundantly without soil, as long as their roots are oxygenated, kept moist at all times, and supplied with the elements essential for healthy plant growth.

Materials and Equipment

1. foundation (3 x 2 foot cart for rollaway, with two shelves)
2. plant container (waterproof, nontoxic)
3. inert aggregate (growth medium other than soil)
4. 2 feet of rubber tubing (1-inch diameter)

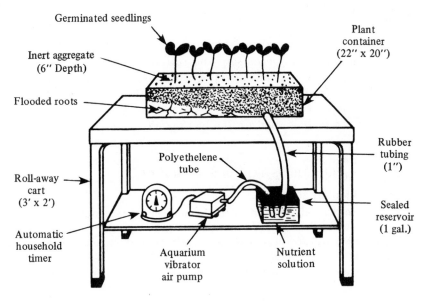

FIGURE 1   An Automatic Hydroponic Flooding System

5. sealed reservoir (one-half to one gallon capacity)
6. nutrient solution (mixture of water and 13 elements)
7. aquarium vibrator air pump (with polyethylene tubing)
8. automatic household timer

Working Definitions

Inert aggregate: a growth medium that will support plant roots, keep them moist between feedings, and allow aeration (e.g., gravel, chipped bricks, cinders, sand, marble chips, perlite, vermiculite). Perlite and vermiculite are soft media often used for below-ground plants such as carrots, turnips, and beets.

Nutrient solution: a solution of water and 13 elements essential for plant growth (iron, potassium, calcium, magnesium, nitrogen, phosphorus, sulfur, manganese, boron, zinc, copper, molybdenum, and chlorine). Commercial mixtures such as Hyponex or Rapid-Grow can be mixed with water and used.

Major Steps

The major steps for setting up an automatic hydroponic system are as follows: (1) placing the cart, (2) arranging the top shelf, (3) arranging the lower shelf, (4) mixing the nutrient solution, (5) sowing the plant seeds, and (6) designating the hour of feeding.

Body

In your body section (labeled *Instructions for Performance*), give the actual instructions, explaining each step and substep in order. Insert warnings, cautions, and notes as needed. Begin your discussion of each step by stating its definition and/or purpose. Readers who understand the reasons for a step can do a better job in carrying it out.

Here is the body section for the hydroponic system instructions:

INSTRUCTIONS FOR PERFORMANCE

1. Placing the Cart

Use the rollaway cart as the foundation for the assembly. Place the cart in an area where sunlight or artificial light (high-intensity, fluorescent, or grow light) will reflect on the upper shelf. With the foundation in place, you can begin to assemble the other parts. (Use Figure 9-3 as a reference for materials, equipment, and assembly.)

2. Arranging the Top Shelf

On the top shelf, place a plant container that is waterproof and nontoxic to plants (preferably concrete, earthenware, ceramic, or plastic).

Bore a 1-inch drainage/feeding hole in the plant container for accommodating the rubber tubing. Insert one end of the rubber tube a half inch into the container and seal the fitting with hot wax.

Fill the container with an inert aggregate. *Note:* Depending on the type of plant to be grown, use a hard or soft growth medium. (See *Working Definitions.*)

3. Arranging the Lower Shelf

On the lower shelf, place a sealed reservoir to hold the nutrient solution. Next to the reservoir, place an aquarium vibrator air pump and an automatic household timer.

To assemble the lower unit, drill two holes through the cover of the sealed reservoir: one to accommodate the polyethylene tube from the aquarium pump, and the other, the one-inch rubber tubing from the plaint container.

Next, take the polyethylene tube attached to the aquarium pump and insert it through the hole in the cover of the sealed reservoir so that it reaches bottom. Do likewise with the free end of the 1-inch rubber tubing from the plant container. Seal both tubes with wax at their point of entry into the reservoir cover.

4. Mixing the Nutrient Solution

Fill the sealed reservoir with a nutrient solution. (See *Working Definitions.*) A number of good commercial products (Eco-Grow, Hyponex, Rapid-Grow) work well in this system. After filling the reservoir with nutrient solution, replace the cover and seal its edges with wax.

5. *Sowing the Plant Seeds*

Sow preferred seeds in the inert aggregate according to instructions on the seed package. Planting seeds two inches apart in consecutive rows is a most effective method for hydroponic systems.

6. *Designating the Hour of Feeding*

Connect the plug from the aquarium pump to the automatic timer and set the timer to the desired hour for feeding. Your hydroponic system is now in operation.

Conclusion

In your conclusion, allow readers to review their performance by summarizing the major steps and discussing how the steps interrelate to bring about the desired result.

Summary

Setting up a hydroponic system is simple. After you obtain and assemble the materials and equipment (foundation, plant container, inert aggregate, rubber tubing, sealed reservoir, nutrient solution, aquarium pump, and automatic timer), growing plants becomes effortless.

Interrelation of Steps

At the designated time, the aquarium pump will activate air through the polyethylene tube into the nutrient solution. In turn, the solution will ascend the rubber tubing into the plant container, flooding the inert aggregate and plant roots. When the plant roots become saturated, the nutrient solution will drain back down the rubber tubing into the sealed reservoir, ready to be recycled at the next designated time.

Throughout your instructions, use headings generously and appropriately to serve as sign posts and to segment your information into digestible portions. Add well-labeled visuals whenever they clarify your explanations. Use transitional sentences to provide logical bridges between sections.

APPLYING THE STEPS

The instructions for *doing something* (felling a tree) in Figure 9-3 are patterned after our general outline, shown earlier. They will be printed in one of a series of brochures for forestry students who are about to begin summer jobs with the Idaho Forestry Service.

AUDIENCE ANALYSIS

These instructions are aimed at partially informed readers who know how to use chainsaws, axes, and wedges but who are approaching this dangerous procedure for the first time. Therefore, no visuals of cutting equipment (chainsaws, etc.) are included because the audience already knows what these items look like. Basic information (such as what happens when a tree binds a chainsaw) is omitted because the audience already has this knowledge also. Likewise, these readers need no definition of general forestry terms such as *culling* and *thinning*, but they *do* need definitions of those terms that relate specifically to tree felling (*barberchair, undercut, holding wood,* etc.).

To ensure clarity, the final three steps are illustrated with visuals. The conclusion, for these readers, is kept short and to the point — a simple summary of major steps with an emphasis on safety.

ELEMENTS OF AN EFFECTIVE PROCESS NARRATIVE

In a process narrative you explain how you did something. With a few exceptions, the elements of an effective narrative are almost identical to those of effective instructions.

Review pages 187–193 for a discussion of titles, arrangement, transitions, and visuals and pay especially close attention to phrasing, level of technicality, and structure.

Phrasing in a Process Narrative

Because you are not giving instructions, but rather describing what you did, use the indicative instead of the imperative mood. Also, use the passive voice from time to time ("The experiment was completed" rather than "I completed the experiment") so that you place the emphasis on the process. Otherwise, repeated use of the first person "I" sounds egocentric.

Appropriate Level of Technicality

Identify your readers precisely. As an illustration of technicality adapted to a specific audience, assume that you're an engineer who has just found a way of reducing stress on load-carrying girders in new buildings; your report will explain this procedure and its results to your colleagues and supervisors. Because your readers are informed and expert, you can exclude detailed background discussions. In this case, a brief review of the underlying principles would provide enough introduction to your account of the procedure itself.

INSTRUCTIONS FOR FELLING A TREE

INTRODUCTION

Definition and Purpose

Felling is the act of cutting down trees. Forestry
Service personnel may be called upon to fell trees for any of
several purposes: to cull or thin a forested plot; to eliminate the hazard of standing, dead trees near powerlines; to
clear an area for road or building construction; to furnish
wood for heating; etc.

Intended Audience and Required Skills

These instructions are written for personnel who need to
remove moderate-sized trees and who already know how to use
a chainsaw, axe, and wedge safely.

General Description

Felling always follows the same procedure. Of the total
time devoted to felling, more is given to planning the operation and preparing the surrounding area. Next, two cuts are
made with a chainsaw, severing the tree from its stump.
Finally, depending on the direction of the cuts and the
weather and terrain, the tree falls into a predetermined

FIGURE 9-3   A Set of Instructions

2

opening on the ground, where it can be cut into smaller units
and removed.

Cautions

Although these instructions cover the basic process,
felling can be a dangerous operation in certain cases.
Trees, felling equipment, and terrain can vary greatly.
Therefore, have a skilled tree feller demonstrate the process
before you try it.

The main consideration is safety. Chainsaws, axes, and
trees are all potentially dangerous. A number of professional fellers are killed each year because of errors in
judgment or misuse of tools.

Materials and Equipment

A 3 to 5 horsepower chainsaw with a 20-inch blade, a
single-blade splitting axe, and one or more 12-inch steel
wedges are the basic tools. Keep fuel and minor repair
tools available for maintaining the chainsaw.

If conditions prohibit the use of the chainsaw, the
axe may serve as a substitute, but it slows the operation.
Otherwise, use the axe to drive wedges or to clear brush for
an escape path.

FIGURE 9-3   (*Continued*)

3

List of Major Steps

   The major steps in felling a tree are (1) choosing the
lay, (2) making provisions for an escape path, (3) making the
undercut, and (4) making the backcut.

   NOTE: For the safest and most successful operation,
give the most effort to planning and the most concentration
to cutting.

INSTRUCTIONS FOR PERFORMANCE

1. Choosing the Lay

   To "choose the lay" is to decide where you want the
tree to fall. If the tree stands on level ground in an open
field, it makes little difference which way you direct the
fall, but since these ideal conditions are rare, you will be
limited by special conditions.

   Obstacles on the ground, surrounding trees, ground
topography, and the tree's condition should all be considered. Decide on your escape path and the location of
your cuts by considering surrounding houses, electrical
wires, and trees. Since most trees lean downhill, try to
direct the fall downhill. Use wedges to deflect the fall

FIGURE 9-3   (*Continued*)

4

if the tree is leaning away from the desired felling direction.

CAUTION: <u>If the tree is dead and leaning substantially, do not try to fell it without professional help.</u> Many dead, leaning trees have a tendency to "barberchair" (split up the trunk).

Use the following steps for choosing the lay:

a. Make sure the tree is still living.

b. Determine the direction and the amount of lean.

c. Find an appropriate opening in the direction of lean, or as close to the direction of lean as possible.

<u>2. Providing for an Escape Path</u>

After determining where to fell the tree, clear a path through which you can exit if necessary. Because trees are unpredictable, the best way to prevent injury is to build an escape path by following the steps below:

a. Locate the path in the opposite direction from which you plan to direct the fall. (See Figure 1.)

b. With the axe, clear a path roughly two feet wide and twenty feet long.

FIGURE 9-3 (*Continued*)

5

c. Place the axe where you will not stumble on it, but

close enough to the tree for later use.

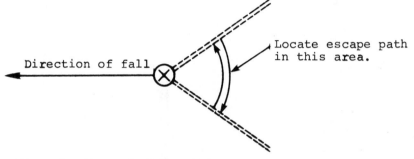

Direction of fall

Locate escape path
in this area.

Figure 1. Escape Path Location

3. Making the Undercut

Making the undercut is the process of cutting a slab of

wood from the tree trunk on the side where you want the tree

to fall. The undercut both directs and controls the fall.

Use the following steps for making the undercut:

a. Start the chainsaw.

b. Holding the saw with the blade parallel to the

ground, make the first cut 2 to 3 feet above the

FIGURE 9-3   (*Continued*)

6

ground, cutting horizontally into the tree to one-

third of the diameter. (See Figure 2.)

c. Make a downward sloping cut, starting 4 to 6 inches

above the first so that the cuts intersect at one-

third of the diameter. (See Figure 2.)

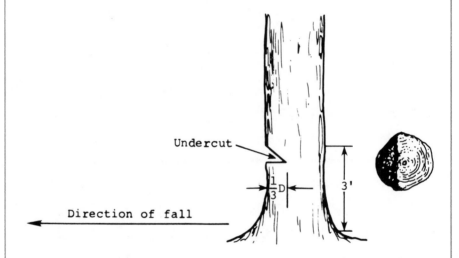

Figure 2. <u>Making the Undercut</u>

FIGURE 9-3   (*Continued*)

7

4. Making the Backcut

After the undercut is completed, the backcut is made to sever the stem from the stump. Use the following steps for making the undercut:

a. Start the cut 1 to 3 inches above the undercut, on the opposite side of the trunk. (See Figure 3.)

b. Do not cut completely through to the undercut, but leave a narrow strip of "holding wood" (Figure 3) to act as a hinge.

CAUTION: Observe tree movement closely while cutting. If the tree starts to bind the chainsaw, hammer a wedge into the backcut with the blunt side of the axe head. Then continue cutting.

c. As the tree begins to fall, withdraw the chainsaw and turn it off. Step back a foot or two and watch the tree movement closely.

WARNING: If the tree comes your way, use your escape path.

FIGURE 9-3 *(Continued)*

8

Backcut

Undercut

Holding wood

Direction of fall

Figure 3. Making the Backcut

SUMMARY

Felling can be a complex and dangerous operation.
Choosing the lay, providing for an escape path, making the
undercut, and making the backcut are steps for the most
basic procedure, but trees, terrain, and other circumstances
can vary greatly. For the safest operation, seek professional assistance whenever you foresee complications.

FIGURE 9-3 (*Continued*)

A college lab report is another type of process narrative written for the informed or expert reader (your instructor). See the audience analysis on page 208 that accompanies a sample lab report.

Introduction-Body-Conclusion Structure

Use an introduction-body-conclusion structure; the body section should describe the procedure in chronological order. The three sections, in order, tell the reader what you did, and why, when, and where you did it; how you did it; and what results you observed or achieved. Often the emphasis in a process narrative is on the results (as shown in the sample lab report in this chapter).

ORGANIZING AND WRITING A PROCESS NARRATIVE

The introduction-body-conclusion structure can be adapted to the writing of any process narrative. Here is how the outline is organized:

    I. INTRODUCTION
        A. Purpose of the Procedure
        B. Intended Audience
        C. Brief Description of the Procedure
        D. Principle of Operation
        E. Materials, Equipment (in order of use), and Special Conditions
        F. List of Major Steps

   II. STEPS IN THE PROCEDURE
        A. First Major Step
           1. Purpose (include a definition when writing for the general reader)
           2. Materials, equipment, and special conditions
           3. Substeps (if applicable)
              a.
              b.
              c.
        B. Second Major Step
          etc.

  III. CONCLUSION
        A. Summary of Major Steps
        B. Description of Results

Depending on your subject and audience, you might wish to modify, delete, or combine some of the elements in the sample outline. Revise it as necessary to suit your purpose.

APPLYING THE STEPS

The process narrative in Figure 9-4 is patterned after our sample outline. Because it was written by a student for her instructor and classmates, the report reflects clear assumptions about the readers' levels of technical understanding (expert and informed). This writer chose to add a subsection titled "Significance of the Experiment" to her conclusion; in it she describes her perception of the lab exercise as a learning experience.

AUDIENCE ANALYSIS

The informed and expert readers of this report will need a minimum of background explanation. A brief discussion of theory is included, however, to demonstrate the *writer's* understanding of the procedure. Items commonly used by chemists (crucible, bunsen burner, analytical scale, etc.) are not illustrated or defined for this audience. A version written for laypersons would include theoretical explanations, definitions of specialized terms, equipment, and materials, as well as visuals to illustrate certain equipment and techniques. The less readers know, the more explanation they must be given.

ELEMENTS OF AN EFFECTIVE PROCESS ANALYSIS

In a process analysis you explain *how* something happens by breaking down the process into its individual steps. The elements of an effective process analysis are similar to those of instructions and the process narrative.

Review pages 187–193 for a discussion of titles, arrangement, transition, and visuals and observe the following additional suggestions for phrasing, level of technicality, and structure.

Phrasing in a Process Analysis

Write in the third person to keep the emphasis on the process, not on the writer or reader. Use the indicative mood throughout and the passive voice selectively.

Appropriate Level of Technicality

Adjust your level of technicality to your specific audience. As an illustration, assume that you need to write an analysis explaining the process of desalination (conversion of salt water to fresh water). Consider your audience. Are they potential investors, political figures, or other laypersons with no apparent technical background; new members of your organization who are informed but

REPORT OF A LABORATORY EXERCISE IN DETERMINING

THE PERCENT COMPOSITION OF A COMPOUND

INTRODUCTION

Purpose of the Procedure

On January 28, 1981, Leslie Foye performed a laboratory experiment in determining the percent composition of elements in a compound of barium chloride hydrate. The specific purpose of the procedure was to determine the relative amounts of barium chloride salts ($BaCl_2$) and water ($2H_2O$) in the compound. This report describes that procedure and its results.

Intended Audience

The report is written for readers who have at least a basic knowledge of chemistry principles and procedure.

Principle of Operation

All compounds are homogeneous substances whose chemical compositions are determined by the definite proportions of elements they contain. These elements can be separated through appropriate chemical techniques. In the case of barium chloride hydrate, the water can be separated from the salt through heating and evaporation.

FIGURE 9-4   A Process Narrative

2

Brief Description of the Procedure

A specific amount of barium chloride hydrate was
weighed, heated, and reweighed. The difference in the two
weighings equaled the amount of water in the compound.
With these data, the percentage of water (hydrate) and the
percentage of barium chloride salt in the compound were calculated. Complete recording of all data can be found on the
data sheet later in this report.

Materials, Equipment, and Special Conditions

The following materials and equipment were used:

1. roughly three grams of a barium chloride hydrate
   sample

2. one 1½-inch diameter crucible and cover

3. one bunsen burner

4. an analytical scale

All steps were performed under supervision in a chemistry
laboratory where appropriate safety precautions were taken.

List of Major Steps

This procedure comprised three major steps: (1) determining the weight of the compound, (2) heating and reweighing
the compound, and (3) calculating the percentages of salt and
water in the compound.

FIGURE 9-4   (*Continued*)

3

STEPS IN THE PROCEDURE

Determining the Weight of the Compound

The purpose of the first step was to determine the precise weight of the compound.

First, the crucible and its cover were heated over the bunsen burner until red hot to burn off any particles that may have interfered with their mass weight. After being allowed to cool to room temperature the crucible and cover were weighed on the analytical scale: weight of crucible and cover = 18.4128 grams.

Next, a sample of barium chloride hydrate weighing approximately three grams was added to the preweighed crucible. The covered crucible and contents were then weighed on the analytical scale: weight of crucible, cover, and sample = 21.4373 grams.

The weight of the barium chloride hydrate sample was then calculated by subtracting the weight of the crucible and cover from the weight of the crucible with contents covered (as shown on the data sheet): weight of the barium chloride hydrate sample = 3.0245 grams.

Heating and Reweighing the Compound

Heating and reweighing were performed to determine the

FIGURE 9-4   (*Continued*)

4

amount of water in the compound by causing the water to
evaporate and calculating the difference in weight.

First, the covered crucible containing the compound
was gradually heated over the bunsen burner for ten minutes.
After being allowed to cool to room temperature, the covered
crucible and contents were weighed on the analytical scale:
<u>weight of crucible, cover, and sample after primary heating</u> =
<u>20.9913 grams</u>.

Next, the covered crucible and contents were heated for
five minutes, allowed to cool to room temperature, and
weighed on the analytical scale: <u>weight of crucible, cover,</u>
<u>and sample after secondary heating = 20.9912 grams</u>.

<u>Calculating Percentages of Water and Salt</u>

The weight of the water in the compound was determined
before the percentage of water could be calculated. The
weight of the water was determined by subtracting the weight
of the sample after the second heating from the weight of
the sample before heating (as shown on the data sheet):
<u>weight of water = 00.4461 grams</u>.

The percentage of water in the sample was calculated by
dividing the weight of the water by the weight of the
compound before heating and multiplying the quotient by 100

FIGURE 9-4 (*Continued*)

5

(as shown on the data sheet): <u>percentage of water in the</u>
<u>compound = 14.76 percent</u>.

Finally, the percentage of barium chloride salt in the
compound was determined by subtracting the percentage of
water from 100 percent (as shown on the data sheet):
<u>percentage of barium chloride salt = 85.24 percent</u>.

CONCLUSION

<u>Summary of Major Steps</u>

In this experiment, the weight of a sample of barium
chloride hydrate was determined; the sample was then heated
and reweighed twice to ensure that all water had been
evaporated; from these data, the percentages of water and
barium chloride in the compound were calculated.

<u>Significance of the Experiment</u>

The experiment provided practical insights into the
physical chemistry of a compound and its formation, along
with practice in the use of selected laboratory equipment.

<u>Description of Calculations and Results</u>

All data, calculations, and results of this experiment
are listed in the data sheet that follows.

FIGURE 9-4 (*Continued*)

6

Data Sheet for Calculating Percent Composition of Barium Chloride Hydrate

DATA:

Weight of crucible, cover and sample . . . . . 21.4373 grams

Weight of crucible and cover . . . . 18.4128 grams

Weight of sample before heating . . . . 3.0245 grams

Weight after primary heating . . . . 20.9913 grams

Weight after secondary heating . . . . 20.9912 grams

FINAL CALCULATIONS:

Weight of water (equals loss of weight from heating)

weight of crucible, cover, and sample	21.4373 grams
weight after secondary heating	− 20.9912
	00.4461 grams of water

Percentage of water

$$\frac{\text{weight of water}}{\text{weight of compound before heating}} \times 100 = \text{percentage of water}$$

$$\frac{0.04661 \text{ grams}}{3.0245 \text{ grams}} \times 100 = 14.76\% \text{ water}$$

Percentage of $BaCl_2$

100.00 percent
− 14.76 (percentage of water)
 85.24% barium chloride

FIGURE 9-4   (*Continued*)

not expert; technicians who will design the desalination apparatus; company executives who have a broad technical background but who have done no complex technical work for years; colleagues with your level of expertise who are unfamiliar with this process? The specific needs of each class of reader will differ. If you can't identify your readers or if your audience is diverse, write at the lowest level of technicality; give full background explanations, define all specialized terms, and translate technical data into plain English.

The sample analysis of color television transmission later in this chapter, intended for a general reading audience, is written at the lowest level of technicality. The same explanation for readers with a working knowledge of electronics would include schematic diagrams, symbols, wave theory, and equations.

Introduction-Body-Conclusion Structure

Use an introduction-body-conclusion structure with the body following the chronological order of the process. The introduction tells the reader what it is and why, when, and where it happens. The body tells how it happens by analyzing each step in sequence. The conclusion summarizes the steps and briefly describes one complete cycle of the process.

ORGANIZING AND WRITING A PROCESS ANALYSIS

The introduction-body-conclusion structure can be adapted to the writing of any process analysis. Here is a general outline:

    I. INTRODUCTION
        A. Definition, Background, and Purpose of the Process
        B. Intended Audience
        C. Knowledge Needed to Understand the Process as Described
        D. Brief Description of the Process
        E. Principle of Operation
        F. Special Conditions Needed for the Process to Occur
        G. Definitions of Special Terms
        H. List of Major Steps

    II. STEPS IN THE PROCESS
        A. First Major Step
           1. Definition and purpose
           2. Special conditions (needed for the specific step)
           3. Substeps (if applicable)
              a.
              b.

B. Second Major Step
   etc.
III. CONCLUSION
   A. Summary of Major Steps
   B. One Complete Process Cycle

Feel free to modify, delete, or combine the elements in this outline to suit the structure of the process you will explain.

APPLYING THE STEPS

The report in Figure 9-5 is patterned after the sample outline. It was written by a student to explain the process of color television transmission and reception to a general reader. Notice how the writer translates a highly technical subject into terms that any reader should be able to understand. At the same time, he wisely omits information that the general reader can be expected to know in advance (e.g., the definition of words such as "antenna" or a diagram of the exterior of a TV set).

CHAPTER SUMMARY

There are three basic types of process explanations: instructions (how to do it), narrative (how you did it), and analysis (how it happens).

When writing instructions, begin with a clear, inclusive, and limiting title that promises exactly what you will deliver. Explain each step of the procedure, in order of its performance, at a level of technicality appropriate for your audience. Use graphic aids generously and clearly indicate all warnings, cautions, and notes. Write in the active voice and imperative mood, using enough transitions to mark time and sequence, and parallel phrasing to emphasize the step-by-step explanation. Explain each major step in a single paragraph and each minor step in a single sentence so that your instructions will be easy to follow.

Follow these steps in planning and organizing your instructions:

1. In your introduction, prepare your reader for the task. First, define the process, explain its purpose, and specify your audience — stating your assumptions about the knowledge and skills readers need to understand your instructions. Next, describe the process briefly and mention all needed materials, equipment, and special conditions, along with the location of upcoming warnings, cautions, and notes. Finally, list the major steps to be explained in your body.

HOW A

COLOR TELEVISION IMAGE

IS TRANSMITTED

AND RECEIVED

INTRODUCTION

Background

Anyone who watches TV is familiar with the images (accompanied by sound) that are generated on the screen. Whether its source is a tape, film, or live event, the image is transmitted by a sophisticated electronic process. This report seeks to explain that process.

Brief Description and Principle of Operation

The color TV camera, located at the studio or source, takes a continuous picture of a scene. Instead of recording the scene on film, the camera converts the image (and sound)

FIGURE 9-5   A Process Analysis

2

into a radio signal, which is then sent out into space.
The TV receiver picks up this signal out of the atmosphere (through a TV antenna) and reconverts it for display
as pictures and sound. What the viewer sees on the screen
is an instantaneous representation of the scene that is
before the TV camera at that moment.

List of Major Steps

The major steps discussed in this process explanation
are (1) the role played by the TV station and (2) the role
played by the color TV receiver.

STEPS IN THE PROCESS

The Role Played by the TV Station

The purpose of the TV station is to capture the visual
image of a scene, to convert the image into electrical energy,
to coordinate the signals needed for transmitting the image
intact, and to transmit the composite signal to receiving
antennae in the area. Figure 1 diagrams this process.

The camera shown in Figure 1 picks up the light reflected
off the objects in the scene and transforms it into electrical
energy. Specifically, the image is captured by the camera

FIGURE 9-5   (*Continued*)

3

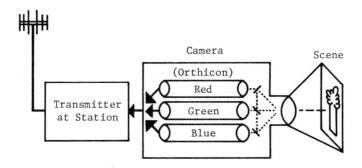

Figure 1. <u>A Block Diagram of the Role Played
by the TV Station</u>

through a scanning process whereby a beam of light moves
over the surface of the scene to reproduce an image. This
image is then sent to the image orthicon tubes, which divide
the image into 525 horizontal lines, positioned one under
the other. Each bright and dark area of individual lines is
converted by the image orthicon tubes into small signal
voltages.

As shown in Figure 1, each of the image orthicon tubes
is sensitive to red, green, and blue, respectively. Thus,
three signals are generated: a red signal, a green signal,
and a blue signal. Because few scenes are simply red, green,
and blue the signals vary in intensity: for example, weak

FIGURE 9-5 (*Continued*)

4

blue, strong green, and moderate red. As when paints are
mixed, depending on the <u>amounts</u> of each color, a new color
is produced. Thus, each small signal is of the correct
strength to later combine and yield the appropriate color.

These signal voltages are now joined to the sound (if
any) and assembled into a signal group called the <u>color</u>
<u>video</u> <u>signal</u>. In order for the combined signal to arrive
at the receiver simultaneously, the TV camera at the studio
and the TV receiver at home must be synchronized. For this
purpose, the camera sends out synchronizing signals along
with the color video signal. Thus, as each of the 525
horizontal lines mentioned earlier is being scanned, the TV
set is showing the identical line at the same moment.
Finally, the camera sends out a color-burst signal along
with the other signals in order to tell your TV set that a
program is in color.

When the composite signal is fully assembled it is sent
to the TV transmitter. Here it is mixed with a powerful
voltage and thrust into space as electrical energy.

FIGURE 9-5 (*Continued*)

5

The Role Played by the Color TV Receiver

The purpose of the color TV receiver is to perform essentially the same steps performed by the TV camera in <u>reverse</u> order. Figure 2 diagrams this process.

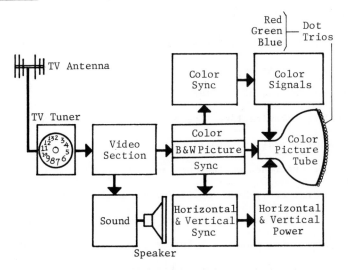

Figure 2. <u>A Block Diagram of the Role Played by the Color TV Receiver</u>

The TV tuner (controlled by the knob you turn to select a station) takes the composite signal from the antenna and sends it to the video section, where it is separated from its sound components. In turn, the color-burst signal instructs the TV set to activate its color section and to add the color information to the black-and-white picture.

FIGURE 9-5   (*Continued*)

6

Without the color-burst signal the picture would appear in

black and white.[1]

Simultaneously, the color-synchronizing signals and the

horizontal and vertical synchronizing signals synchronize

the set with the studio camera. If the studio camera is

scanning a red scene, the set should display a red scene.

If the camera is scanning line 357 (of the 525 lines), the set

will display the 357th line at that instant because of hori-

zontal synchronization. The picture is kept from rolling

through vertical synchronization.

The one function that the set performs by itself is to

provide extremely high voltage to the picture tube. It

requires high-intensity power to present clear, bright

pictures. This power is supplied by the section marked

"Horizontal and Vertical Power" in Figure 2.

The video, color-burst, and synchronizing signals are

channeled to the picture tube along with the power supply in

---

[1] As shown in Figure 2, there are actually two signals
going to the picture tube: a color signal and a black-
and-white signal. If the viewer turns the color control
knob completely counterclockwise (assuming the set is
not automatic-color controlled), he or she will see only
a black-and-white picture.

FIGURE 9-5 (*Continued*)

7

order to generate the actual picture. Just as the color

camera has three image orthicons, the picture tube has

three sections called <u>guns,</u> one gun for each basic color --

red, green, and blue. The three color signals, in their

respective strengths, are applied to each respective gun.

   The face of the picture tube is printed with colored

dots (or lines). Each gun will hit the dots that correspond

to its own color signal: the red gun will hit the red dots,

the green gun will hit the green dots, etc. The dots are

printed on the face of the tube in trios (a red dot, a blue

dot, and a green dot). Each gun hits the appropriate dot in

each trio with a high-intensity electron beam. The degree

of intensity is controlled by the strength of the signal

emitted from the image orthicon tubes in the camera. This

varying intensity results in the correct color mixture for

the scene being scanned and, in turn, being projected as an

image on the TV screen. In short, the light energy that was

converted into electrical energy by the image orthicon tubes

is now converted back to its original form by the picture

tube to produce the actual image on the screen.

FIGURE 9-5 (*Continued*)

8

CONCLUSION

Summary of Major Steps

    The color TV station captures the image, converts it
into an electrical signal, combines it with other signals,
and transmits it to the TV receiver. The receiver, in turn,
performs the reverse order of these steps to produce the
actual image on the screen.

One Complete Process Cycle

    In a full cycle this process occurs as follows:

1. The station's function

    a. The studio camera scans the scene to produce a light
       reflection of the image.

    b. The camera then converts this light energy into color
       video signals and adds synchronizing and color-burst
       signals to form the composite video signal.

    c. This composite video signal is impressed on a high-
       voltage carrier and hurtled into space by the trans-
       mitter.

2. The receiver's function

    a. The TV antenna picks up the composite signal.

    b. The receiver properly disassembles the signal into

FIGURE 9-5   (*Continued*)

9

picture (video signal) and sound components, color-

burst signal, and synchronizing signals.

c. Assisted by high-voltage input, the picture tube con-

verts the electrical signal back to its original form

as light energy in order to display the appropriately

colored image on the screen.

FIGURE 9-5   (*Continued*)

2. In your body, explain the instructions for completing the task, describing each major step and substep in chronological order. State the definition and purpose of each step, and insert warnings, cautions, and notes as required.

3. In your conclusion, summarize the major steps in your body and explain how the steps interrelate in completing the task.

In a process narrative follow the same guidelines used for instructions, with a few exceptions. Write in the indicative instead of the imperative mood and use the passive voice occasionally to avoid sounding egocentric. Follow an introduction-body-conclusion structure, but emphasize the results you observed or achieved.

Follow these steps in planning and organizing your narrative:

1. In your introduction, tell your reader what you did and when, where, and why you did it.

2. In your body, explain how you did it — step by step.

3. In your conclusion, summarize what you did, and describe your results.

In a process analysis, follow the same guidelines used for the process narrative. Write in the indicative mood and use the passive voice selectively. Again, follow an introduction-body-conclusion structure.

1. In your introduction, tell your reader what it is, and why, when, and where it happens.

2. In your body, explain how it happens — step by step.

3. In your conclusion, summarize the steps and briefly describe one complete process cycle.

Modify or delete subsections in this three-section structure to suit your purpose.

REVISION CHECKLIST

Use this list to refine the content, arrangement, and style of your process explanation.

Content

1. Is the process explanation keyed to your stated purpose (instructions, narrative, analysis)?

2. Does your report have a clear, inclusive, and limiting title that promises exactly what you have delivered?

3. Have you given readers enough background to understand and be interested in the process?

4. Have you defined each step and discussed its purpose before explaining it (except in a narrative for informed or expert readers)?

5. Have you included visuals whenever they can clarify your explanation?

6. In a process narrative, does your conclusion emphasize results?

Arrangement

1. Does your report have a fully developed introduction, body, and conclusion?

2. Does your report follow your outline faithfully?

3. Does your explanation follow the exact order of the steps in the process or procedure?

4. Have you used a single paragraph for each major step and a single sentence for each minor step?

5. Have you used adequate transitions as time and sequence markers?

6. Are your headings appropriate and adequate?

Style

1. Does the level of technicality connect with your intended audience?

2. Are items of equal importance expressed in parallel grammatical form?

3. For instructions, have you used the imperative mood and active voice?

4. For a process narrative or analysis, have you used the indicative mood throughout and the passive voice selectively (pages 41–43)?

5. Is each sentence clear, concise, and fluent?

6. Is the report written in correct English (see Appendix A)?

Now list those elements of your process explanation that need improvement.

EXERCISES

1. Use the revision checklist to evaluate the sample reports in this chapter. Compare your evaluation with those of your classmates in a classroom-workshop session.

2. In a short essay, discuss the similarities and differences among the three types of process explanation. Give examples of each, paying particular attention to intended audience and varying levels of technicality.

3. In a short essay, discuss the types of process explanations that you would write on the job. As specifically as you can, identify typical subjects, your expected audience, and its level of technical understanding.

4. Identify a situation from your own experience where you received faulty, incomplete, or unclear instructions, either written or spoken. Describe the situation, the specific deficiency, and its consequences.

5. In your textbooks or in library sources, locate a one- or two-page process explanation written for the general reader (or a beginning student). Using the revision checklist, evaluate the effectiveness of its organization and the appropriateness of its level of technicality. Include suggestions for improving its effectiveness. Submit your essay and a copy of the process explanation to your instructor.

6. *In class:* Select a simple but specialized process from your major field that you understand well (how gum disease develops, how a savings bank makes a profit, how the heart pumps blood, how an earthquake occurs, how steel is made, how various petroleum fuels are made, how electricity is generated, how a corporation is formed, how a verdict is appealed to a higher court, how a bankruptcy claim is filed, how a Xerox machine produces copies, how nicotine affects the body, how paper is made, etc.). Write an essay explaining the process. Exchange your essay with a classmate in another major. Study your classmate's explanation for fifteen minutes and then write an explanation of that process in your own words, referring back to your classmate's paper as needed. Now, evaluate your classmate's version of your original explanation for accuracy of content. Does it show that your explanation was understood? If not, why not? Discuss your conclusions in a short essay and submit all samples to your instructor.

7. *In class:* Draw a map of the route from your classroom to your dorm, apartment, or home — whichever is closest. Be sure to include identifying landmarks. When your map is completed, write a set of instructions for a classmate who will try to duplicate your map from the information given in your written instructions. Be sure that your classmate does not see your map! Exchange your instructions and try to duplicate your classmate's map. Compare your results with the original map. Discuss the conclusions you have drawn from this exercise.

8. Select a recent event in which you participated (a lab experiment, a meeting, etc.). Write a process narrative, using one paragraph for each major step of your activities. Include enough details for a general reader to understand your role in this event.

9. Make the following instructions more readable by rewriting them in the appropriate voice, mood, phrasing, and sentence or paragraph division. Insert transitions wherever necessary.

WHAT TO DO BEFORE JACKING UP YOUR CAR

Whenever the misfortune of a flat tire occurs, some basic procedures should be followed before the car is jacked up. If possible, your car should be positioned on as firm and level a surface as is available. The engine has to be turned off, the parking brake set, and the automatic transmission shift lever placed in "park," or the manual transmission lever in "reverse." The wheel diagonally opposite the one to be removed should have a piece of wood placed beneath it to prevent the wheel from rolling. The spare wheel, jack bar, jack stand, hook, and lug wrench should be removed from the luggage compartment.

10. Locate a technical manual in your field or a set of instructions written for a general reader. Evaluate the sample for effectiveness. Are the instructions clear and easy to follow? Are the technical terms defined? Are graphic aids used when necessary, and are they integrated into the directions? In a paragraph, discuss the positive and negative elements of the instructions, based on the revision checklist.

11. Choose a topic from the following list or from your major field or an area of interest. Using the general outline in this chapter as a model, outline a set of instructions for a procedure that requires at least three major steps. Write the instructions for a general reader and be sure to include a title page; a clear title; major topic headings; transitional sentences between major topics; cautions, notes, and warnings; and strict chronological organization. Write one draft, modify your outline as needed, and write a second draft. Proofread in accordance with the revision checklist, and exchange your instructions with another class member, preferably in your field, for further suggestions for revision.

achieving a golden tan
making a water pipe
growing tomatoes
filleting a fish
building a bird house
beginning a food co-op
changing a car's oil and filter

changing spark plugs
seeing Europe on $10.00 a day
making beer or wine at home
safely losing ten pounds in four weeks
taking a job interview
jump-starting a car battery

10

Designing an Effective Format and Supplements

DEFINITIONS
    Report Length and Complexity: Informal
        or Formal
    Format
    Supplements

PURPOSE OF AN EFFECTIVE FORMAT

PURPOSE OF REPORT SUPPLEMENTS

CREATING AN EFFECTIVE FORMAT
    High-Quality Paper
    Neat Typing
    Uniform Margins, Spacing, and
        Indentation
    Consistently Numbered Pages
    Introduction-Body-Conclusion Structure
        Introduction
        Body
        Conclusion
    Section Length
    Effective Headings
        Make Your Headings Informative
        Make Your Headings Comprehensive
        Make Your Headings Parallel
        Lay Out Headings by Rank

COMPOSING REPORT SUPPLEMENTS
    Cover
    Title Page
        Title
        Placement of Title Page Items
    Letter of Transmittal
        Introduction
        Body
        Conclusion
    Table of Contents
    Table of Illustrations
    Informative Abstract
    Glossary
    Endnote Pages
    Bibliography Pages
    Appendix

CHAPTER SUMMARY

EXERCISES

DEFINITIONS

Report Length and Complexity: Informal or Formal

Depending on your specific job and reporting responsibilities you may be writing brief (informal) reports, longer and more complex (formal) reports, or both.

Informal reports vary in length and arrangement. Often you might write brief memos of less than a page, as well as progress reports and proposals of several pages. These are usually written for readers within your organization. From time to time you might write formal reports, which are usually at least ten pages long and contain information about projects or studies too complex to be covered in an informal report. Accordingly, formal reports contain parts not found in informal reports. Formal reports are often written for readers outside as well as inside your organization and may form part of a permanent record. Both types follow certain format conventions that make their information more accessible and give them a professional appearance.

There is no rule for deciding whether any given data should be presented in an informal or a formal report. For instance, a short proposal to improve safety procedures in your laboratory might be a memo, whereas a proposal for merging with another company would probably be a formal report. Likewise, a set of instructions for a colleague who will fill in as department supervisor during your absence might be informally written in a few pages, whereas instructions for operating new equipment in your inhalation therapy unit would most likely be formally written and permanently filed as a manual. In short, the specific reporting task will determine the formality of the report. Sometimes, in fact, a certain report may not fit neatly into either the informal or formal category, but may fall somewhere in between. The point is to make your report only as detailed and as formal as it has to be to get the job done.

Whether you write an informal or formal report, you will need to create a professional format. In a formal report you will have to include some or all supplements discussed in this chapter.

Format

Format is the mechanical arrangement of words on the page: indentation, margins, spacing, typeface, headings, page numbering, and division of report sections. Format determines the physical appearance of your report.

Supplements

Supplements make a report more accessible to readers. The title page, letter of transmittal, table of contents, and abstract give summary information about the content of the report. The glossary and appendixes either provide supporting data or help readers follow certain technical sections. They may refer to these supplements or skip them according to their needs. Footnotes and bibliography identify sources of data. All supplements are written *after* you complete your report proper.

PURPOSE OF AN EFFECTIVE FORMAT

Your writing should be impressive in appearance and readability as well as in content. *What your report looks like* and *how it is arranged* are just as important as what it has to say. A good format helps you look good.

No matter how accurate and vital your information, a ragtag presentation will surely alienate readers. How seriously, for instance, would you take the information in a textbook that looked like Figure 10-1? How would you judge its author, publisher, and credibility? Imagine trying to follow such instructions! Figure 10-2 shows the same information after a format overhaul. As readers, we take good format for granted; that is, we hardly notice format *unless* it is offensive.

Your format is the wrapping on your information package. Just as there are many techniques and styles for wrapping a package, there are many effective formats. In fact, many companies have their own requirements. Here you will study one style, which you may later modify according to your needs.

<u>Giving the IM Injection</u>

First of all, the preparation for giving the injection
must be ~~taken care of~~ carried out. This includes: selecting
the correct medication, preparing the needle, and drawing
the medication. In selecting the medication, it must be
triple cheecked to ensure that the right medication and
dosage is being given. This is done by checking the
order against the medication card, against the label on the
drug container. Haveing the needle ready to go ~~is important~~
~~because it~~ in order to prevent fumbling with the needle and
medication bottle when drawing up the medication. ~~It is~~
~~important to~~ Make sure that the needle is tight to the ~~syri~~
syringe and that it is the right size. Freeing the plunger
so that it will draw back and push forward easily is a good
idea as it prevents fighting with it when it may be in an
awkward position, like in the patient's leg. Drawing up
the medication has several points that are important in
avoiding contamination of the needle, or medication, and in
ensuring that the right dosage is being given.

For ease of expl anation I am assuming that the medication is in liquid form in a container with a rubber seal
that is supposed to be the correct dose.

FIGURE 10-1   An Ineffective Format

GIVING THE INTRAMUSCULAR INJECTION

Selecting the Correct Medication and Dosage

CAUTION: Triple-check the physician's order against the medication card and the label on the medication container, to ensure that you administer the correct medication in precise dosage.

After selecting the correct medication and dosage, prepare your needle and syringe.

Preparing Your Needle and Syringe

1. Choose a twenty-six (26)-gauge needle and affix it tightly to the neck of your syringe.

2. Free the syringe plunger so that it will draw back and push forward easily, to avoid later difficulties when the needle is in the muscle.

With your needle and syringe prepared, you are ready to draw up the medication.

Drawing up the Medication

CAUTION: Use aseptic technique to avoid contamination of the needle and/or medication; recheck the correct dosage on the container label.

*   *   *   *   *

FIGURE 10-2   An Effective Format

PURPOSE OF REPORT SUPPLEMENTS

The central message of a formal report is in its introduction-body-conclusion sections. Supplements to the report help readers grasp the material by providing "a place for everything." Supplements can be time savers: a busy person may read only your letter of transmittal, table of contents, and abstract for key information. If you have ever written a college research paper you probably used the same approach — skimming prefaces, tables of contents, and introductions to many library books to locate relevant information. Typical formal reports in fact are likely to be read by many people for a wide variety of purposes. Technically qualified persons may be studying the actual body of the report and the appendixes for supporting data such as maps, graphs, or charts. Executives and managers may read only the letter of transmittal and the abstract. If they read any of the report proper, they are likely to read only the recommendations. Most supplements, then, are designed to accommodate readers with various purposes.

CREATING AN EFFECTIVE FORMAT

High-Quality Paper

Type your final draft on 8½-by-11-inch plain, white paper. Use heavy (20 lb. or higher) bond paper with a high fiber content (25 percent minimum). Erasures with typewriter correction tape are the neatest. Use onionskin for carbon copies only.

Neat Typing

The quality of your typing job should match that of your paper. Keep erasures to a minimum and retype all smudged pages. Use a fresh typewriter ribbon and keep your typewriter keys clean.

Uniform Margins, Spacing, and Indentation

Leave the following margins on each page: top margin, 1¼ inches; bottom, 1 inch; left 1½ inches; right 1 inch (the larger left margin leaves space for binding your report).

Make your report easy to scan by double spacing within and between paragraphs. Set off a long quote (four or more lines) by separating it from your discussion with double spaces and single spacing the quote itself. Indent the entire quotation five spaces from both right and left margins. When indenting a long quote, use no quotation marks.

Indent the first line of all paragraphs five spaces from the left margin. Be sure that no illustrations extend beyond the inside limits of your margins. As a rule, avoid hyphenating a word at the end of a line; if you must hyphenate, check your dictionary for the correct syllabic breakdown of the word.

Consistently Numbered Pages

Count your title page as page i, without numbering it, and number all pages up to and including your table of contents and abstract with small roman numerals. Use arabic numerals (1, 2, 3, etc.) for subsequent pages, numbering the first page of your report proper as page 1. Place all page numbers in the upper right corner, two spaces below the top edge of the paper and five spaces to the left of the right edge.

Introduction-Body-Conclusion Structure

Organize your report as you would organize any well-structured communication: simply expand the basic three-part structure into larger units of information.

Introduction

Preview the discussion in your body section by giving all necessary background information. This section in your report functions like a topic sentence in a paragraph. Here, the "topic sentence," expanded many times over, promises what you will deliver in your body section. Before discussing any subject, introduce it by describing and defining it.

Body

The body section is the heart of your report. Here, you divide your subject into its parts and their subparts. You also deliver what you promised in your introduction by presenting your collected evidence and findings.

Conclusion

The final section contains no new information. In a conclusion you tie everything together. First, you review the evidence and findings discussed in your body section. Next, if appropriate, you draw conclusions based on the above material. Finally, you make recommendations based on your conclusions, if they are needed.

Section Length

The length of each section depends on your subject. A set of instructions (as shown on pages 194–197) usually begins with a detailed introduction listing materials, equipment, cautions, and so on. The body, in turn, enumerates each step and substep. These sections are followed by a brief conclusion, because the key information has been given earlier.

On the other hand, a problem-solving report (as shown on pages 489–495) may often have a brief introduction outlining the problem. The body, however, may be quite long, explaining the possible and probable causes of the problem. Because the conclusion contains a summary of findings, an overall interpretation of the evidence, and clearly specified recommendations, it will most likely be developed in detail. Only when your investigation uncovers one specific answer or one definite cause will your body section be relatively short.

Examples of varying section length, according to subject, are found in the sample reports in this text.

Effective Headings

Like textbooks, many reports are not read in a linear order, from cover to cover, as a novel would be read. Just as you often refer back or jump ahead to specific sections in a textbook, so do readers of a report. Without headings, the text is confusing — and often useless, because the readers can't find what they need without wasting time.

Headings help you as well as your readers. Like points on a road map they keep you on course and provide transitions from one point to another. They signal your readers that a part of your discussion has ended and another is about to begin. They make a report less intimidating by partitioning large and diverse areas of data into smaller, more readable parts. Finally, they are good attention-getting devices, because they help readers focus on the parts of your report they find most important. Guidelines for using headings are discussed below.

Make Your Headings Informative

Phrase all headings so that they are short but informative and serve to advance thought. Preview your topic. Compare, for instance, these versions of a heading in a set of typewriter-cleaning instructions:

Uninformative   **The Carriage**

Informative   **Brushing the Dirt from Beneath the Carriage**

The second version creates an explicit context for readers, showing them exactly what to expect.

Express your headings as phrases ("Frequency of Lubrication") rather than as awkward pieces of sentences ("Needs Frequent Lubrication").

Make Your Headings Comprehensive

Headings must be comprehensive in two ways: individual headings must be inclusive and enough headings must be provided to cover all categories. As a frame or fence encloses a specific visual or geographic space, your heading "encloses" or contains an information category. When framing a picture you would logically frame the entire visual area, without leaving out, say, the upper right corner. Follow the same logic of inclusiveness with headings. If your discussion is about the relationship of apples to oranges, be sure that your heading is not simply "Apples."

Also, provide enough headings to contain each discrete information category. If apples, oranges, and elephants are three *separate* discussion items, provide a heading for each. In other words, don't combine your discussion of oranges *and* elephants under "Oranges." Take major and minor headings from your outline, as we do throughout this book.

Make Your Headings Parallel

Because in any report all major topics are equally important, express them in the same grammatical form to emphasize that equality.

Nonparallel Headings
1. Brushing the Dirt from Beneath the Carriage
2. Brushing the Dirt from the Key Faces and Surrounding Areas
3. Clean the Carriage Cylinder with Rubbing Alcohol
4. Cleaning the Paper-Bail Rolls with Rubbing Alcohol
5. Rubbing Alcohol Is Used to Clean the Key Faces
6. Wiping the Dirt from the Typewriter Exterior
7. It Is Crucial That All Dirt Be Wiped from the Work Area

Headings 1, 2, 4, and 6 are participial phrases, whereas headings 3, 5, and 7 are sentences. Moreover, each nonparallel heading is in a different mood: heading 3, giving a command, in the imperative mood; heading 5, making a declarative statement, in the indicative mood; heading 7, making a recommendation, in the subjunctive mood. This mood shift obscures the logical relationship between individual steps and confuses readers. To make these steps parallel, rephrase them as follows:

Parallel Headings
3. Cleaning the Carriage Cylinder with Rubbing Alcohol
5. Cleaning the Key Faces with Rubbing Alcohol
7. Wiping All Dirt from the Surrounding Area

Parallelism is discussed further in Appendix A.

Lay Out Headings by Rank

Within its three general areas, a report will contain major topics; often, major topics will contain subtopics; in turn, subtopics may contain sub-subtopics, depending on the amount of detail the report requires. (Use the logical divisions within your outline as a model for your heading arrangement.) The logic of your divisions will be clear if your headings reflect the rank of each item. The headings in Figure 10-3 vary in typeface, indentation, and position, according to rank. Notice that the sentence immediately following your heading stands independent of the heading. Don't begin your sentence with a pronoun such as "this" or "it" to refer back to the heading.

In keeping with the logic of division, make sure that each section (or subsection, etc.) contains at least two headings. Show major and minor headings in the same relationship in your outline or table of contents.

COMPOSING REPORT SUPPLEMENTS

Report supplements can be classified in two groups:

1. *supplements that precede your report* (front matter): cover, title page, letter of transmittal, table of contents (and illustrations), and abstract

2. *supplements that follow your report* (end matter): glossary, footnotes, bibliography, appendix(es)

Cover

Use a sturdy, plain, light cardboard cover with good page fasteners. With the cover on, the pages still should be flat as they are turned. In general, use a cover only for longer, formal reports, not for those only a few pages long.

MAJOR AREA HEADING

Write major headings in full capitals, centered horizontally on the page, two spaces above your following text (and three spaces below your preceding text, for subsequent area headings). Do not underline or italicize.

Major Topic Heading

Begin each word (except for articles and prepositions) in your major topic heading with a capital letter. Make the heading abut the margin, two spaces below the preceding text and two spaces above the following text. Underline or italicize this heading.

Subtopic Heading

Begin each word in your subtopic heading with a capital letter. Indent it five spaces from your left margin and place it two spaces below the preceding text and two spaces above the following text. Underline or italicize this heading.

Sub-Subtopic Heading. Begin each word in your sub-subtopic heading with a capital letter. Indent it five spaces from the left margin and place it two spaces below your preceding text and on the same line as the first sentence of your following text (separated by a period). Underline or italicize this heading.

FIGURE 10-3   A Sample Heading System

Center the report title and your name four to five inches from the upper edge:

AN ANALYSIS OF THE EFFECTIVENESS OF THE FRESHMAN REMEDIAL

PROGRAM AT CALVIN COLLEGE

by

Francis Freeman

Title Page

The title page signals readers by providing certain vital information: report title, author's name, name of person or organization to whom the report is addressed, and date of submission.

Title

Your title promises what your report will deliver by stating the report's purpose and content. The following title is effective because it is clear, accurate, comprehensive, specific, concise, and appropriately phrased:

An Effective Title   **AN ANALYSIS OF THE EFFECTIVENESS**
OF THE FRESHMAN REMEDIAL
PROGRAM AT CALVIN COLLEGE

Notice how slight word changes can distort this title's signal.

An Unclear Title   **THE FRESHMAN REMEDIAL PROGRAM**
AT CALVIN COLLEGE

This version does not state clearly the purpose of the report. Its signal is confusing. Is the report *describing* the program, *proposing* the establishment of such a program, *giving instructions* for setting it up, or *discussing its history?*

An Inaccurate Title   **THE EFFECTIVENESS OF THE FRESHMAN**
REMEDIAL PROGRAM AT CALVIN COLLEGE

This version does not state accurately the purpose of the report. In fact, it presents a distorted signal by suggesting that the program's effectiveness is already proven, instead of being an issue in question. Insert key words in your title ("analysis," "instructions," "proposal," "feasibility," "description," "progress," "proposal," etc.) which state accurately your purpose.

A Noncomprehensive Title

AN ANALYSIS OF THE EFFECTIVENESS OF THE REMEDIAL WRITING PROGRAM AT CALVIN COLLEGE

Here is a title that fails to name all that the report will cover: namely, an analysis of *all* remedial programs, including mathematics, reading, and writing.

A Nonspecific Title

AN ANALYSIS OF THE EFFECTIVENESS OF FRESHMAN PROGRAMS AT CALVIN COLLEGE

This version has the reverse deficiency of the one above: it promises an analysis of all freshman programs, without focusing on its proper subject — the freshman remedial program. The signal is too broad and imprecise.

A Long-Winded Title

AN EXHAUSTIVE ANALYSIS OF THE OVERALL EFFECTIVENESS OF THE SPECIAL PROGRAM IN REMEDIATION FOR FRESHMEN WHICH WAS RECENTLY IMPLEMENTED AT CALVIN COLLEGE

Words like "exhaustive," "overall," "special," "recently," and "implemented" provide no useful information; they merely obscure the intended signal.

An Inappropriately Phrased Title

THE FRESHMAN REMEDIAL PROGRAM AT CALVIN COLLEGE: WINNER OR LOSER?

Phrase your title to stimulate reader interest, but not like a commercial or a sideshow gimmick. Word choice should reflect the significance of your effort. A "catchy" title can make a serious report seem trivial. To be sure that your title promises what your report delivers, write its final version *after* completing your report.

Placement of Title Page Items

Do not number your title page, but count it as page i of your prefatory pages. Center the title horizontally on the page, three to four inches below the upper edge, using all capital letters. If the title is longer than six to eight words, center it on two or more single-spaced lines, as in Figure 10-4. Place the items in

AN ANALYSIS OF THE EFFECTIVENESS
OF
THE FRESHMAN REMEDIATION PROGRAM AT CALVIN COLLEGE

for

Professor John Johnson
Technical Writing Instructor
Calvin College
Kalamazoo, Minnesota

by

Sarah Jane Robertson
Student in English 266

December 25, 1981

FIGURE 10-4   A Title Page for a Formal Report

the following spacing, order, and typescript:

1. seven spaces below the title and horizontally centered, the word "for" in small letters
2. two spaces below, your reader's name followed by his or her position, organization, and address, horizontally centered with the first letter of each word capitalized
3. seven spaces below, the word "by" in small letters
4. two spaces below, your name, position, and organization on each of three separate single-spaced lines with the first letter of all words capitalized
5. ten spaces below, the date of report submission written out in full

You may work out your own system, as long as your page is balanced.

Letter of Transmittal

Your letter of transmittal comes immediately after your title page and is bound as part of your report. Include a letter of transmittal with any formal report or proposal addressed to a specific reader. Your letter adds a note of courtesy besides giving you a place for adding personal remarks or opinions. Figure 8-5 shows a sample letter of transmittal.

Whereas your informative abstract (discussed later in this chapter) summarizes all major findings, conclusions, and recommendations in the report, your letter of transmittal calls attention to items that might be of special interest to a specific reader. Depending on your reporting situation, your letter might:

– Acknowledge those who helped with the report.
– Refer readers to sections of special interest: unexpected findings, key charts or diagrams, major conclusions, special recommendations, and the like.
– Discuss the scope and limitations of your study, along with any special problems encountered in gathering data.
– Discuss the need and approaches for follow-up investigations.
– Describe any personal observations, for example, off-the-record comments.
– Briefly explain how and why your report data are useful and offer suggestions for practical uses of the information.

In many cases the letter of transmittal can be tailored to a particular reader. Thus if a report is being sent to a number of people who are variously qualified and bear various relationships to the writer, the accompanying letters of transmittal may vary. The letter itself follows an introduction-body-conclusion structure.

Introduction

Open with a cordial statement referring to the date and content of the reader's original request for the report. Briefly explain your reasons for submitting your letter and report.

Maintain a confident and positive tone throughout the letter. Indicate your pride and satisfaction in the work you have done. Avoid apologetic statements, such as "I hope this information is adequate," or "I hope that this report meets your expectations."

Body

In the body of your letter, include items from the prior list of possibilities (acknowledgments, special problems, limitations, unexpected findings, special conclusions, recommendations, and personal observations). Although your abstract summarizes major findings, conclusions, and recommendations, the body of your letter provides an overview of the *entire project*, from the moment you receive the original request to the moment you submit the report.

Conclusion

Tie earlier parts together with a statement of your willingness to answer any questions or discuss the findings. End your letter on a positive note, with something like "I believe that the data in this report are accurate, that they have been analyzed rigorously and objectively, and that the recommendations presented are sound."

Figure 10-5 shows a sample letter of transmittal. Individual items and sections are labeled in the right margin. (Your own letter of course would contain no such labels in final draft.)

Table of Contents

Your table of contents serves as a road map for your readers and an inventory checklist for you. Because it is based on your outline, this supplement is easy to compose. Simply assign appropriate page numbers to headings in your outline.

Here are some guidelines for composing a table of contents:

1. List preliminary items (cover letter, abstract) in your table of contents, numbering the pages with small roman numerals. (The title page, although

                                        43 Ocean Avenue
                                        West Harwich, Massachusetts 02046
                                        December 1, 1981
Professor Grand Savant
Technical Writing Instructor
Wishbone College
Plymouth, Massachusetts 03456

Dear Professor Savant:

Enclosed is the report you requested on October 3, 1981,
analyzing the alleged advantages of the Wankel rotary com-
bustion engine over the reciprocating, piston-driven V-8.
Comparison of the two suggests that the Wankel might be the
engine of the future.

Although long-term testing will take years, the Wankel shows
economic and manufacturing advantages, as well as unique
operational qualities. Simple design, economy, and high performance make it an attractive alternative to the conventional
piston-driven engine, as shown in the comparative specifica-
tion chart in Appendix A. (I am grateful to Professor John
Jones, of the Engineering Department, for his help with this
chart.) Absence of data prevented me from including the
diesel engine in the above comparison.

Because it is light, easy to build, and durable, the Wankel
should find many practical applications in automobiles, snow-
mobiles, boats, lawnmowers, generators, and tractors. In
fact, you might wish to delay your own planned purchase of a
new inboard engine until next spring, when rotary boat
engines will be available.

As a result of my analysis, I'm convinced that the Wankel
engine holds great promise for energy-efficient use. If
you have any questions after reading the enclosed report,
please call me at 430-356-9856.

                                        Sincerely,

                                        Ronald Varg

                                        Ronald Varg

FIGURE 10-5   A Letter of Transmittal for a Formal Report

not listed, is counted as page i.) Also, list glossary, appendix, notes, and bibliography sections; number these pages with arabic numerals, continuing the page sequence of your report proper, where page 1 is the first page of your report text.

2. Include no headings in the table of contents not listed as headings or subheadings in the report; your report text may, however, contain certain subheadings not listed in the table of contents.

3. Phrase the headings in your table of contents exactly as you phrase those in the report text.

4. List headings at various levels in varying typescript, capitalization, and indentation to reflect their rank.

5. Use double-spaced horizontal dots (. . . .) to connect the individual heading to its page number.

The table of contents in Figure 10-6 is adapted from the outline for "An Analysis of the Advisability of Converting Our Office Building from Oil to Gas Heat." The formal outline was shown in Chapter 7.

Table of Illustrations

Following your table of contents is a table of illustrations, if needed. If your report contains more than four or five illustrations, place this table on a separate page. Figure 10-7 shows a table of illustrations from a report titled "The Negative Effects of Strip Mining on Kentucky's Cumberland Plateau."

Informative Abstract

For some readers your abstract (as discussed in Chapter 4) is the most important item in your report. Because it summarizes your work, the abstract is always written after your report proper is completed. Sometimes a busy person will read only your abstract, expecting to find the key points reviewed.

Writing the abstract gives you the chance to measure your own understanding of the material. Because you are cutting your report's text to roughly 10 percent of its original length — or much less for an abstract of a long report — you must establish logical connections among related items. If you can't effectively summarize your report, you probably haven't fully mastered your material. In this case, more homework and revision are probably necessary.

Use the following guidelines for writing your abstract:

1. Make your abstract unified, coherent, and able to stand alone in meaning and emphasis — a mini-report.

2. Make your abstract intelligible to the general reader. Remember that the people who read it will vary widely in technical competence, perhaps much

iii

TABLE OF CONTENTS

LETTER OF TRANSMITTAL ii

INFORMATIVE ABSTRACT v

INTRODUCTION 1

   Background 1

   Purpose 1

   Intended Audience 1

   Data Sources 1

   Limitations 2

   Scope 2

COLLECTED DATA 2

   Description of Our Present Heating System 2
     Physical Condition 2
     Required Yearly Maintenance 3
     Fuel Supply Problems 4
     Cost of Operation 4

   Removal of the Oil Burner and Tank 5
     Data from the Oil Company 5
     Data from the Salvage Company 5
     Possibility of Private Sale 6

FIGURE 10-6   A Table of Contents for a Formal Report

iv

TABLE OF CONTENTS (continued)

Installation of a Gas Pipe from the Street to the
Building 6
    Procedure 6
    Cost of Installation 7
    Cost of Landscaping 7

Installation of a Gas Burner 8
    Procedure 8
    Cost of Plumber's Labor and Materials 9

Estimation of Gas Heating Costs 10
    Rate Determination 10
    Required Yearly Maintenance 10
    Cost Data from Neighboring Facility 10
    Overall Cost of Operation 11

CONCLUSION 11

Summary of Findings 11

Comprehensive Interpretation of Findings 12

Recommendations 12

GLOSSARY 13

APPENDIX 14

FIGURE 10-6 (*Continued*)

TABLE OF ILLUSTRATIONS

Figure 1. A Contour Map of the Appalachian Region 3

Figure 2. A Contour Map of the Cumberland Plateau 4

Figure 3. Graph of a Ten-Year Contrast of Population
           Figures in the Cumberland Plateau Region . . . 13

Figure 4. Table of a Ten-Year Comparison of Population
           Figures among the Counties of the Cumberland
           Plateau 14

Figure 5. A Map of the Range of the Tennessee Valley
           Authority 20

FIGURE 10-7   A Table of Illustrations for a Formal Report

more so than the people who read the report itself. Therefore, translate technical data for nontechnical readers (managers, clients, and other laypersons).

3. Add no new information. Simply summarize the report.

4. Follow the chronology of your report. If it discusses apples, oranges, and elephants, in this order, follow the same order in your abstract.

5. Emphasize only major points. Omit prefaces, supporting details, computations, and lengthy arguments. Often, your table of contents or your report headings provide a good master plan for your abstract.

The informative abstract in Figure 10-8 accompanies the heating conversion report outlined in Chapter 7.

Glossary

A glossary is an alphabetical listing of specialized terms and definitions, immediately following your report. Many reports with technical subjects and specialized terminology contain glossaries, especially when written for both technical and nontechnical readers. A glossary allows you to make key definitions available to nontechnical readers without interrupting technical readers. If fewer than five terms need to be defined, place them instead in the introduction section of your report, listing them as working definitions (as discussed in Chapter 5) — or you might use footnote definitions. If you use a separate glossary, inform readers of its location: "(see the glossary at the end of this report)."

Follow these guidelines in composing your glossary:

1. Define all terms unfamiliar to a general reader.

2. Define all terms that may have a special meaning in your report (e.g., "In this report, a small business is defined as . . .").

3. Define all terms by giving their class and distinguishing features (as discussed in Chapter 5), unless some terms need more expanded definitions.

4. List your glossary and its first page number in your table of contents.

5. List all terms in alphabetical order. Underline each term and use a colon to separate it from its single-spaced definition.

6. Do not define terms whose meanings are commonly known. In doubtful cases, however, overdefining is better than underdefining.

7. Upon first use, place an asterisk in the text by each term defined in the glossary.

Figure 10-9 shows part of a glossary for a comparative analysis of two techniques of natural childbirth, written by a nurse practitioner for expectant mothers and student nurses. The term *natural childbirth* receives a more expanded definition because it is the subject of the report.

INFORMATIVE ABSTRACT: AN ANALYSIS OF THE ADVISABILITY

OF CONVERTING OUR HOME OFFICE FROM OIL TO GAS HEAT

The rising cost and declining availability of heating

oil has led our company to consider the advisability of con-

verting our home office from oil to gas heat. Our present

heating needs are supplied by circulating hot air generated

by an electrically fired Zippo oil burner, which is fed by a

275-gallon oil tank. Both burner and tank are twelve years

old and in good condition. Because of short supply and an

overworked local fuel distributor, however, our system has

run out of fuel three times in the past two years. In 1976

our total heating costs were $558.28.

Conversion to gas heat would first require removal of

the oil burner and tank from the basement. Junko Salvage

Company will remove these items at no cost, provided they

keep both burner and tank. Next, a gas line would need to be

installed from the street to the building. Although the gas

company will provide free installation, our landscaping costs,

after installation, would be $85.00. Added to this figure is

the plumber's labor and materials charge of $570.00 for

FIGURE 10-8   An Informative Abstract

installing the gas burner. These conversion costs amount to $65.50 yearly over ten years. With a projected gas-supply and maintenance cost of $621.75 yearly, the overall yearly cost of gas heating would be roughly $129.00 higher than that of oil heating. Because the advantage of constant fuel supply does not offset the expense and inconvenience of conversion, we should retain our present system and consider installing an auxiliary 500-gallon underground oil tank. This tank would ensure adequate supply during peak heating months.

FIGURE 10-8 (*Continued*)

GLOSSARY

Analgesic: a medication given to relieve pain during the
     first stage of labor.

Anesthetic: a substance administered to cause loss of
     consciousness or insensitivity to pain in a region of
     the body.

Cervix: the neck-shaped anatomical structure which forms
     the mouth of the uterus.

Childbirth Education: the process of instruction providing
     parents with information about childbirth and preparing
     them for active participation in the delivery.

Dilation: the act of cervical expansion occurring during
     the first stage of labor.

Episiotomy: an incision of the outer vaginal tissue, made by
     the obstetrician just before the delivery, to enlarge
     the vaginal opening.

First Stage of Labor: the stage in which the cervix dilates
     and the baby remains in the uterus.

Induction: the process of stimulating labor by puncturing
     the membranes around the baby or by giving an oxytoxic
     drug (uterine contractant), or both.

Natural Childbirth: (also called Prepared Childbirth, Par-
     ticipating Childbirth, Educated Childbirth, and Coopera
     tive Childbirth) a process of giving birth in which
     parents actively participate and which is based on an
     understanding of the body, muscular relaxation,
     breathing techniques, and emotional support. It is not
     a primitive process where modern knowledge plays no
     part, or a rigid system forbidding obstetrical inter-
     ference regardless of circumstances. Medication may or
     may not be used, according to individual need. The
     natural childbirth experience is presently defined
     by the woman's preparation and knowledge of how to
     cooperate actively in her delivery.

FIGURE 10-9   A Glossary Page for a Formal Report

Endnote Pages

The endnote pages list each of your references in the same numerical order as they are cited in your report proper. See Chapter 13 for a discussion of documentation.

Bibliography Pages

The bibliography lists, in alphabetical order, each reference you consulted — whether or not you quoted, paraphrased, or otherwise referred to them in your report. See Chapter 13 for a discussion of bibliographies.

Appendix

An appendix comes at the very end of your report. Information in the appendix illustrates further an item already discussed, without cluttering your report text. Typical items in an appendix include:

- details of an experiment
- statistical or other measurements
- maps
- complex formulas
- long quotes (one or more pages)
- photographs
- texts of laws, regulations, etc.
- related correspondence (letters of inquiry, etc.)
- interview questions and responses
- sample questionnaires and tabulated responses
- sample tests and tabulated results
- some visual aids occupying more than one full page

Thus, an appendix serves as a catchall for items that are important but difficult to integrate within your text.

Do not rely too heavily on appendixes by including needless information or by not properly integrating appropriate data into your report proper. Readers should not have to turn to appendixes every few seconds to understand a particular point. The following guidelines should help you to use appendixes effectively:

1. Include only material that is relevant but difficult to fit into your report.
2. Use a separate appendix for each major item.
3. Title each appendix clearly: "Appendix A: A Sample Questionnaire."

4. Do not use too many appendixes. If you have four or five appendixes in a ten-page report you have probably not taken the time to organize the report material effectively.

5. Limit your appendix to two or three pages, unless greater length is absolutely necessary.

6. Mention your appendix early in your introduction and refer readers to it at appropriate points in the report: "(See Appendix A)."

As a rule of thumb, use an appendix for any item that relates to the purpose of your report, but that would harm the unity and coherence of your discussion. Remember, however, that readers should be able to understand your report without having to turn to the appendix. Distill the essential facts from your appendix and place them in your report text.

Improper Reference	The whale population declined drastically between 1976 and 1977 (see Appendix B for details).
Proper Reference	The whale population declined *by 16%* from 1976 to 1977 (see Appendix B for a statistical breakdown).

Figure 10-10 shows the first page of an appendix to a report on modern whaling techniques.

CHAPTER SUMMARY

Whether your report is informal or formal you will need to follow certain conventions to give it a professional appearance and make its information more accessible. Format and supplements determine your report's appearance, the arrangement of its parts, and the accessibility of your material. Both are important elements of a professional presentation. To achieve an effective format follow these guidelines:

1. Use high-quality paper.
2. Type your page neatly.
3. Use uniform margins, spacing, and indentation.
4. Number your pages consistently.
5. Use an introduction-body-conclusion structure.
6. Use headings effectively and extensively.

A formal report is prefaced by supplements: cover, title page, letter of transmittal, table of contents, and informative abstract. Include the following supplements — glossary, footnotes, bibliography, and appendixes — only as needed. Compose your supplements *after* your report's text is written.

APPENDIX A

A CLASSIFICATION OF WHALE SPECIES ACCORDING TO SIZE, DIET, RANGE, GESTATION PERIOD, AND LEVEL OF ENDANGERMENT BY MODERN WHALING

Species	Size(ft.)	Diet	Range	Gestation Period	Level of Endangerment
Sperm	M 47 F 37 baby 14	Giant squid, cuttlefish.	Cosmopolitan. F & baby remain in warm water. M have harems. Bachelors in polar seas in summer. Travels singly or in schools.	11–16 months. One calf every three years.	Threatened.
Grey	M 43 F 46 baby 16	Unknown. Main item thought to be bottom-dwelling amphipods.	Eastern No. Pacific. Summer feeding grounds off Alaska coast. Winter breeding grounds off California. Travels singly or in pairs.	13 months. One calf every two years.	At threshold of extinction.
Minke	M 27 F 27 baby 9	Krill in southern seas. Cod, herring, whiting, mackerel in northern seas.	Cosmopolitan. Highly migratory. M in deep waters. F & young in coastal waters. Travels in schools of up to 20.	12 months. One calf every two years.	At lowest ebb in history. Bordering threatened.

FIGURE 10-10 A Partial Appendix to a Formal Report

EXERCISES

1. (a) In a paragraph, explain the role and importance of format in any report. Why are headings especially important? (b) In a second paragraph, explain the role and importance of supplements.

2. *In class:* Look through your textbooks and library journals to find a chapter with an effective format. Bring the sample to class and, in small groups, compare and contrast individual choices, discussing the strong and weak points of each choice. Which of the samples discussed by the group has the most effective format? Why? Subject matter aside, what makes one sample more readable than another? Concentrate on titles, headings, and appearance.

3. The following titles are intended for investigative, research, or analytical reports. Each should be clear, accurate, comprehensive, specific, concise, and appropriately phrased. Revise all inadequate titles according to the criteria and examples discussed in this chapter.

    a. "The Effectiveness of the Prison Furlough Program in Our State"

    b. "Home Solar Heating: Boom or Bust?"

    c. "The Effects of Nuclear Power Plants"

    d. "Woodburning Stoves"

    e. "An Investigation and Comparative Analysis of the Mechanical and Other Advantages of the Rotary Combustion (Wankel) Engine over the Otherwise Popular and Conventionally Manufactured V-8 Engine"

    f. "An Analysis of Vegetables" (for a report assessing the physiological effects of a vegetarian diet)

    g. "Will Wood Save the Day as a Fuel Source?"

    h. "Oral Contraceptives"

4. Using the format principles discussed in this chapter, revise something you have written for one of your earlier assignments (a summary, classification, partition, expanded formal definition, or research report). (You might exchange rough drafts with fellow students in order to share specific suggestions for format revision.) Submit the revision to your instructor.

5. Prepare a title page, a letter of transmittal, a table of contents, and an informative abstract for a report you have written earlier (perhaps a research report). If you need to revise the heading system in your report's text, follow the instructions in this chapter.

11

Visual
Aids

DEFINITION

PURPOSE OF VISUAL AIDS

TABLES
  Levels of Complexity
  Construction

FIGURES
  Graphs
    Bar Graphs
    Line Graphs
    Construction
  Charts
    Pie Charts
    Organizational Charts
    Flowcharts
  Diagrams
    Diagrams of Mechanical Parts
    Exploded Diagrams
    Diagrams of Procedures
    Schematic and Wiring Diagrams
  Photographs
  Samples

CHAPTER SUMMARY

REVISION CHECKLIST

EXERCISES

DEFINITION

A visual aid is any pictorial representation used to clarify a discussion. The common visual aids used in report writing are (1) tables and (2) figures. Different types of figures include graphs, charts, diagrams, photographs, and samples of material.

PURPOSE OF VISUAL AIDS

Visuals attract attention and increase reader understanding by emphasizing certain information. Translate prose into visuals whenever you can, *as long as the visuals make your point more clearly than the prose does.* Use visuals to clarify your discussion, not simply to decorate it. And keep them simple.

Visual aids work in several ways to improve your report:

1. They increase reader interest by providing a view more vivid and clear than its prose equivalent. They are easier to follow and grasp. A visual satisfies the reader's demand to be shown.

2. They set off and emphasize significant data. A bar graph showing that the cost of a loaf of bread is three times as great as it was a certain number of years ago is more dramatic than a prose statement. Some readers, in fact, might only skim the prose in a report and concentrate on the visuals.

3. They condense information. A simple table, for instance, can often replace a long and difficult prose passage.

4. Certain types of visual aids (such as tables, charts, and graphs) are useful for pulling together diverse data on the basis of their similarities or contrasts. Thus they are easy to interpret.

Suppose you are researching the comparative nutritional value and cost of several kinds of sandwiches — bologna, hamburger, tuna salad, egg salad, and peanut butter. From various sources you collect the following data:

1. A bologna sandwich (3½ oz. bologna, 1 T. mustard on white bread) contains 436 calories, 17 grams of protein, 30 grams of fat, at an estimated cost of $0.51.

2. A hamburger (¼ lb. cooked beef, 1 T. catsup on bun) contains 331 calories, 25 grams of protein, 16 grams of fat, at an estimated cost of $0.47.

3. A tuna salad sandwich (3½ oz. tuna, 1 T. mayonnaise on white bread) contains 422 calories, 33 grams of protein, 21 grams of fat, at an estimated cost of $0.50.

4. An egg salad sandwich (1 large egg — cooked, 1 T. mayonnaise on white bread) contains 313 calories, 11 grams of protein, 19 grams of fat, at an estimated cost of $0.22.

5. A peanut butter sandwich (1 oz. on white bread) contains 296 calories, 12 grams of protein, 16 grams of fat, at an estimated cost of $0.19.

Clearly, this prose version is repetitious and difficult to interpret. When arranged in a table, like Table 11-1, the data become much more readable. Visuals can be great time-savers for readers.

A full-scale study of visual aids would require a course in drafting and technical illustration. Therefore, this chapter discusses only the most common and most easily composed.

TABLES

Tables are displays of data that can be numerical (as in Table 11-1) or nonnumerical (as in Table 6-4 on page 118). The data are arranged in vertical columns under category headings so they can be easily compared and contrasted.

Levels of Complexity

A table can be as simple as Table 11-2, which includes only one basis of comparison. Notice the explanatory notes that limit and qualify the meaning of the categories.

A more complex table, like Table 11-3, contains several bases of comparison. Here the bases are listed in the far left column to make the table fit the page.

Although not as visually dramatic as a graph or chart, a table is best for displaying numbers and units of measurement that must be illustrated precisely.

TABLE 11-1   Five Popular Sandwiches Classified on the Basis of Caloric, Protein, and Fat Content, and Estimated Cost

Sandwich	Calories	Protein (in grams)	Fat (in grams)	Estimated Cost
Bologna (3½ oz., 1 T. mustard)	436	17	30	.51
Hamburger (4 oz. cooked, 1 T. catsup)[a]	331	25	16	.47
Tuna salad (3½ oz., 1 T. mayonnaise)	422	33	21	.50
Egg salad (1 large egg, 1 T. mayonnaise)	313	11	19	.22
Peanut butter (1 oz.)	296	12	16	.19

Source: Figures are based on Bureau of Labor Statistics estimates found in the *Retail Food Price Index, October 1975* (Washington, D.C.: U.S. Department of Labor).

[a] The hamburger is served on a bun. All other sandwiches are on white bread.

TABLE 11-2   1977 American Subcompacts Classified in Descending Order on the Basis of Gas Mileage

Make of Car[a]	Miles per Gallon[b]
Midgo II	33.4
Vampira ST	32.9
Locomoto	32.5
Zoomer	32.3

[a] All models tested were two-door sedans with three-speed manual transmissions.
[b] Mileage figures are based on EPA averages for combined city and highway driving.

TABLE 11-3   Stand Characteristics of White Pine and Mixed Oak

	White Pine			Mixed Oak
	#1	#2	#3	
Total stems per acre[a] over 1 inch diameter at breast height[b]	260.0	390.0	390.0	325.0
Basal area per acre (sq. ft.)	104.0	224.9	183.9	85.3
Average diameter at breast height[b]	6.2	9.9	7.8	6.0
Average age, dominant and codominant trees (yrs.)	26.0	37.2	38.2	58.6
Average height, dominant and codominant trees (ft.)	52.1	63.5	65.4	59.7
Average crown space occupied (%)	96.1	94.2	95.1	90.9
Stem density[c]	1919.0	3855.0	3673.0	1940.0

Source: James H. Brown, Jr., and Thomas W. Hardy, Jr., *Summer Water Use by White Pine and Oak in Rhode Island* (Kingston, R.I.: University of Rhode Island, 1975), p. 3. Reprinted by permission.
[a] Acreage values projected from one-tenth acre samples.
[b] Diameter at 4.5 feet above ground.
[c] Stem density is the summation of tree diameters at breast height.

Construction

To make a table, follow these guidelines:

1. Number each table in order of its appearance (for easy reference), and give it a clear title that shows exactly what the table contains.

2. Begin each vertical column with a heading that identifies the types of items listed (e.g., "Make of Car") and specific units of measurement and comparison (e.g., "Miles per Gallon," "Grams per Ounce," "Percentage"). Use only the approved abbreviations and symbols listed in Appendix A. Give all items in the same column the same units of measurement (inches, sq. ft., etc.) and keep decimals vertically aligned.

3. Use footnotes to explain or clarify certain entries. Whereas footnote notation in your discussion is in arabic numerals (1, 2, 3), in your table it is in small letters (a, b, c).

4. Set your table off from the discussion by framing it with adequate white space all around. Be sure that the table does not extend into the page margins.

5. Try to keep the table on a single report page. If it does take up more than one full page, write "continued" at the bottom and begin the second page

with the full title and "continued." Also, place the same headings at the tops of each column as appear on the first page of the table. If you need to total your columns, begin second-page columns with subtotals from the first page.

6. If your table is so wide that you need to turn it to the vertical plane of your page, place the top against the inside binding.

7. Relate your table to the surrounding discussion. Refer specifically to the table by number and title in the report text. Introduce it and discuss any special features about the data. Do not leave readers with the job of interpreting raw data.

8. If the table clarifies a part of your discussion, place it in that area of your text. If, however, it simply provides supporting information of interest only to some readers, place it in an appendix so those readers can refer to it if necessary. Avoid cluttering up your discussion.

9. Identify your data sources below the table, beginning at the left margin. If the table itself is borrowed, so indicate. And list your sources even if you make your own table from borrowed data.

FIGURES

Any visual aid that is not a table is classified as a figure and should be so titled (e.g., "Figure 1: An Aerial View of the Panhandle Building Site"). The most common figures are graphs, charts, diagrams, photographs, and samples.

Graphs

A graph is made by plotting a set of points on a coordinate system. It provides a picture of the relationship between two variables and is used to show a comparison, a change over time, or a trend.

When you decide to use a graph, choose the best type for your purpose: bar graph or line graph.

Bar Graphs

A bar graph, as shown in Figure 11-1, illustrates comparisons. In this case, the visual impact of the bar graph makes it a clear choice over a prose or tabular version. Percentage figures are recorded above each bar to increase clarity. The independent variable[1] range extends only from 0 to 40 percent. This range

---

[1] In all graphs, the horizontal line (abscissa) lists those items whose value is fixed (independent variables); the vertical line (ordinate) lists those values that change (dependent variables). The dependent variable changes according to the specific activity of the independent variable (e.g., an increase in percentage over time, as shown in Figure 11-1).

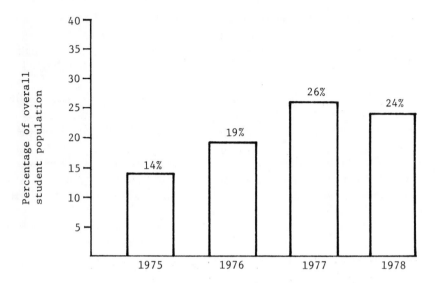

FIGURE 1. Percentage of Students Making the Dean's List at
          X College, 1975–78

FIGURE 11-1   A Bar Graph

creates enough space between vertical increments to dramatize the comparison
without taking up too much of the page. Units of measurement on the vertical
line are clearly identified.

The choice of scale in a bar graph (such as 10 percent per inch) is crucial.
Try different scales until your graph represents all quantities clearly and in
proper proportion. If, for instance, the vertical scale in Figure 11-1 were extended to 100 percent, the bars would seem dwarfed and much space would
be wasted. Other distortions would occur if the vertical increments, for example, were increased to 5 percent per inch or decreased to 30 percent per
inch.

A bar graph can also contain multiple bars (up to three) at each major point
on the horizontal line, as in Figure 11-2. In a multiple-bar graph, include a
legend to explain the meaning of the various bars.

Another common type of graph is the segmented-bar graph, which breaks
down each bar into its components. Notice that the vertical scale in Figure
11-3 is large enough to show clearly relative proportions for comparison.

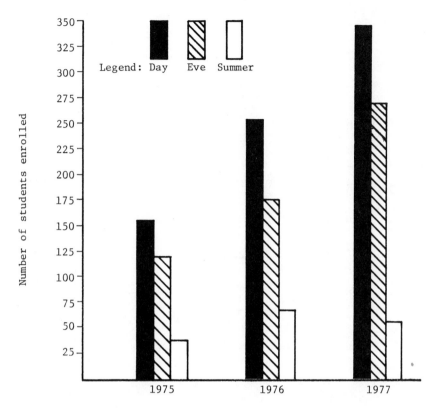

FIGURE 2. The Number of Students Enrolled in Technical Writing,
1975–77.

FIGURE 11-2   A Multiple-Bar Graph

When a horizontal quantity such as distance traveled is being compared, a horizontal bar graph can be used.

Make all graphs on graph paper so that the lines and increments will be evenly spaced. Always begin a bar graph directly on the horizontal line. To express negative values, simply extend the vertical line below the horizontal, following the same incremental division as above it, only in negative values. Make all bars the same width so readers won't be confused about the relative value of each.

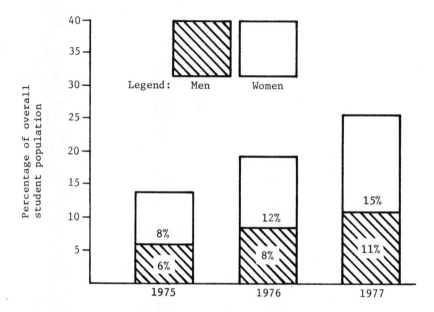

FIGURE 3. Breakdown, by Sex, of Students Making the Dean's List, 1975–77.

FIGURE 11-3   A Segmented-Bar Graph

Line Graphs

Whereas a bar graph provides units of measurement for visual comparison, a line graph, like Figure 11-4, shows a change, or a trend, over a given period. Unlike a bar graph, which must begin on the horizontal line, a line graph can begin at any intersecting point on the coordinate grid. Select a readable scale and always identify the specific units of measurement (e.g., building permits issued).

A line graph is particularly useful for illustrating a comparison of trends or changes among two or three dependent variables that are related, as in Figure 11-5. These pictorial data give an instant overview of daily shopping patterns in various locations. In this kind of multiple-line graph, your choice of scale is

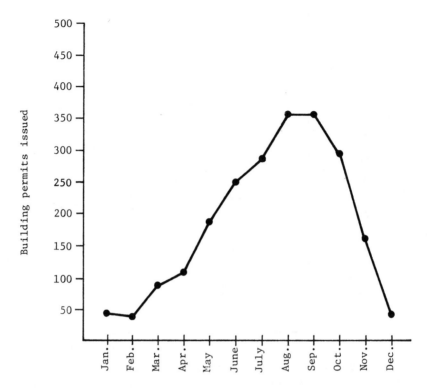

FIGURE 4. Building Permits Issued in Dade County in 1977.

FIGURE 11-4   A Line Graph

crucial. Imagine, for example, that the vertical scale in Figure 11-5 were condensed to $1500 per increment. The result is shown in Figure 11-6. With this reduced scale, the graph becomes almost impossible to interpret. Conversely, an overly expanded vertical scale would yield another kind of distortion. Figure 11-7 shows the same data on a graph whose scale had been increased to $250 per increment and whose vertical line begins at $1000 instead of 0 in order to save space. This error distorts the quantitative relationship between lines: the high for the Midwest is $2500, and for the East, $4500 (roughly 180% higher); yet the visual relationship between these lines suggests that the sales volume for the East is roughly 230% higher. Remember that the visual relationships should parallel the actual numerical relationships.

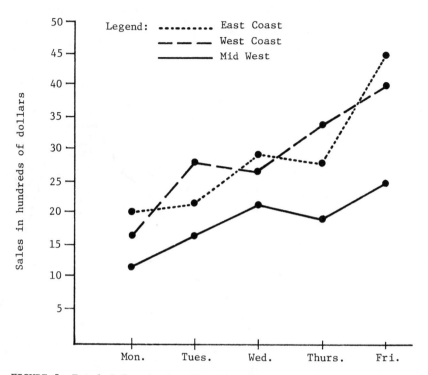

FIGURE 5. Total Sales in Our Three Major Outlets for the Week of
          June 2, 1978.

FIGURE 11-5   A Multiple-Line Graph

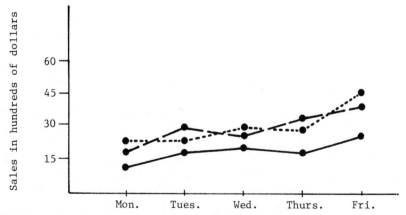

FIGURE 6. Total Sales in Our Three Major Outlets for the Week of
          June 2, 1978.

FIGURE 11-6   A Poorly Scaled Graph (Increments Condensed)

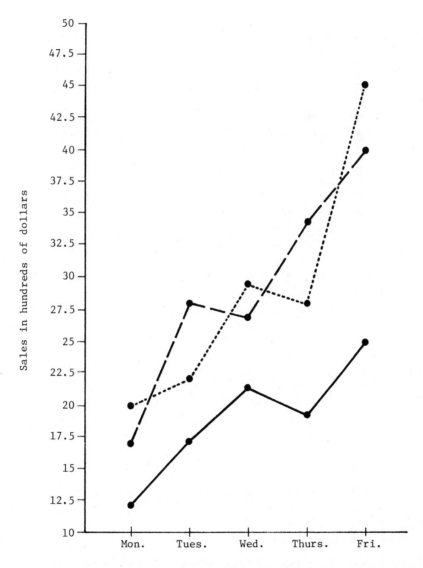

FIGURE 7. Total Sales in Our Three Major Outlets for the Week of
          June 2, 1978.

FIGURE 11-7   A Poorly Scaled Graph (Increments Expanded)

Construction

1. Number the graph in order of its appearance and give it a clear title.

2. Label the items on your horizontal and vertical lines. State units of measurement along your vertical coordinate (hundreds of dollars, pounds per square inch, etc.). In a multiple-bar or line graph include a legend identifying each bar or line.

3. Experiment with various scales until you find the one that works best.

4. Keep the graph simple and easy to read. Never plot more than three different lines or types of bars.

5. Because you are using graph paper, plan carefully for integrating a graph into your discussion. If the graph is only a few inches high, you might trim the excess graph paper and paste or glue the graph in the appropriate section of your discussion. Otherwise, place the full graph page immediately after the related discussion page or in an appendix.

6. Introduce, discuss, and interpret your graph, referring specifically to its number and title. Do not leave the reader with a page full of raw data.

7. If the graph must be presented on the vertical plane of your page, place the top against the inside binding.

8. Credit your data sources two spaces below your figure number and title.

Charts

The terms *chart* and *graph* are often used interchangeably. We define a chart as a figure that illustrates relationships (quantitative or cause-and-effect) but is not plotted on a coordinate system. Therefore, no graph paper is used. The most common types of charts are the pie chart, the organizational chart, and the flowchart.

Pie Charts

A pie chart partitions a whole into its parts and provides a pictorial image of the parts-whole relationship. The parts of a pie chart must add up to 100 percent as shown in Figure 11-8.

Follow these guidelines in making your chart:

1. Number it in order of its appearance with other figures, and give it a clear and precise title. Place figure number and title two spaces below your chart.

2. Use a compass to draw a perfect circle and to locate its center. Use a protractor for precise segmentation.

3. Begin segmenting your chart by locating your first radial line at twelve o'clock. Move clockwise, in descending order, from largest to smallest segments.

4. Use at least three, but no more than seven, segments. Combine several small segments (1 percent to 5 percent each) under the heading "Other."

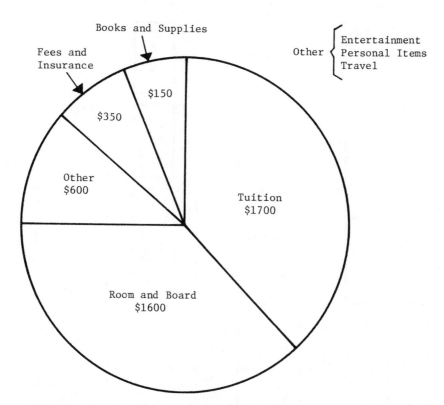

FIGURE 8. Yearly Cost Breakdown for Attending Calvin College.
(Total Cost = $4400/Year)

FIGURE 11-8   A Pie Chart

Include a parenthetical explanation of these combined items, as shown in
Figure 11-8.

5. Write all section headings, quantities, and units of measurement horizontally.

6. Place your pie chart where it belongs in your discussion. Introduce it,
explain it, and credit data sources.

Though not as precise as a tabular list, a pie chart draws your reader's attention to certain dramatic elements more effectively than a list of numbers would.

Organizational Charts

An organizational chart partitions the administrative functions of an organization. It ranks each member in order of authority and responsibility as that

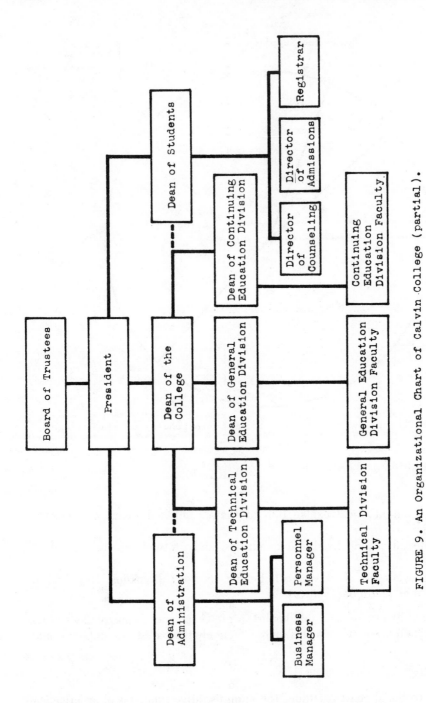

FIGURE 9. An Organizational Chart of Calvin College (partial).

FIGURE 11-9   An Organizational Chart

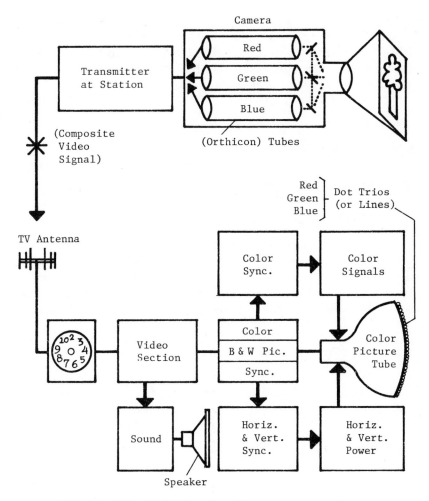

FIGURE 10. How Color Television Works.

FIGURE 11-10   A Flowchart

member relates to other members and departments. Figure 11-9 shows the partial organizational chart of a typical management structure of a college.

Flowcharts

A flowchart traces a process from beginning to end. In outlining the specific steps of a manufacturing or refining process, it moves from raw material to finished product. In illustrating how a phenomenon occurs, it moves through the specific steps that make the phenomenon possible, as shown in Figure 11-10. This chart is shown in its full textual context in the process analysis in Chapter 9. The rules discussed earlier for placement, combination, and source credit apply for organizational and flow charts.

Diagrams

Diagrams are sketches or drawings of the parts of an item or the steps in a process. Because diagrams have a broad range of types and complexity, we will discuss only some simpler types.

Diagrams of Mechanical Parts

A description of a mechanism should always be accompanied by diagrams that show its parts and illustrate its operating principle, such as Figure 11-11. Always describe the specific perspective from which you have drawn the item: frontal view, lateral view, anterior, superior, cross-sectional, and so on. Thus your reader will know what is being depicted and from what angle.

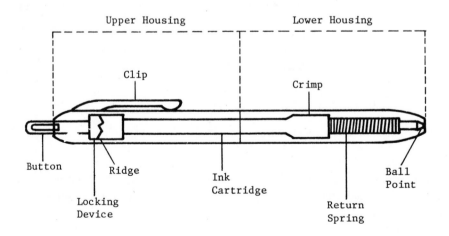

FIGURE 11.   A Retractable Ball Point Pen with Point Retracted
             (in Cutaway)

FIGURE 11-11   A Diagram of Mechanical Parts (Cutaway)

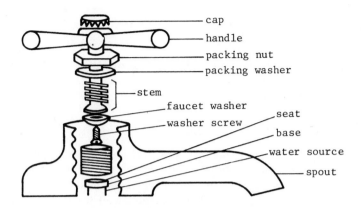

FIGURE 12.   A Typical Faucet in Exploded View

FIGURE 11-12   An Exploded Diagram

Exploded Diagrams

Exploded diagrams, like the one in Figure 11-12, show how parts of an item are assembled and often appear in repair or maintenance manuals.

Diagrams of Procedures

Diagrams are especially useful for clarifying instructions by illustrating how certain steps should be performed. Figure 11-13 is an example of this type of diagram.

FIGURE 13.   Frontal View of a Camera Held Correctly.

FIGURE 11-13   A Diagram of a Procedure

Schematic and Wiring Diagrams

Science and engineering majors will find ample illustrations of electrical diagrams in physics textbooks.

Photographs

Photographs provide an accurate overall view, but they can sometimes be too "busy." By showing all details as more or less equal, a photograph sometimes fails to emphasize the important areas within the visual field.

When you use photographs, keep them distinct, well focused, and uncluttered. For a complex mechanism, you probably should rely on diagrams instead, unless you intend simply to show an overall view. Lend a sense of scale by including a person or a familiar object (such as a hand) in your photo.

Samples

If your report discusses certain materials, such as clothing fabrics, types of paper, or paint colors, you might include actual samples. The same is true when you are discussing business forms or contracts. For things like fabrics and paints, glue or paste a small sample — titled and numbered — to your report page.

CHAPTER SUMMARY

A visual aid is any pictorial device that clarifies your discussion. In report writing, the most common visuals are tables and figures. Tables display data in vertical columns under category headings. They are best used to display precise numbers and units of measurement.

In making a table, follow these guidelines:

1. Number it chronologically and give it a clear title.
2. Begin each column with a clear heading and list all items in that column according to the same unit of measurement.
3. Use footnotes to explain complex data.
4. Leave plenty of white space between the table and your text.
5. Try to keep the table on a single page.
6. Introduce, discuss, and interpret the table.
7. Place the table appropriately in your report — in the text or in an appendix.

8. Place the top of an excessively wide table on the vertical plane against the binding.

9. Identify any data sources.

Figures include such items as graphs, charts, diagrams, photographs, and samples. Any visual aid that is not a table is a figure.

In a graph, numbers are plotted as a set of points on a coordinate system to give a picture of the relationship between data. You might choose a bar graph to show comparisons or a line graph to show change over time.

In making a graph, follow the general guidelines used for tables, with these additions:

1. In a multiple-bar or line graph, include a legend to identify various bars or lines.

2. Experiment with various scales until you find the clearest and most accurate.

3. Never plot more than three variables on one graph.

4. Use graph paper; trim away the excess; paste or glue the figure on your discussion page.

Charts also illustrate relationships, but are not plotted on a coordinate system. Pie charts, organizational charts, and flowcharts are the most common. Each partitions a whole item or process into its parts to represent the relationship of part to part and of part to whole.

Diagrams are sketches or drawings of the item or process and are usually better than photographs for emphasizing certain parts. Photographs, however, work well in presenting overall views.

In discussing certain materials, you might include actual samples.

REVISION CHECKLIST

1. Does the visual aid serve a real purpose; that is, does it clarify, not simply decorate, your report?

2. Have you chosen the best form of visual aid for your purposes?

3. Is the visual aid titled and numbered appropriately?

4. Can it stand alone in meaning, if necessary?

5. Does each vertical column in the table begin with a clear heading?

6. Are all units of measurement in the same tabular column identical (inches, grams, etc.)?

7. Are all decimal points in each tabular column vertically aligned?

8. Are explanatory notes added as needed (in lowercase letter notation)?

9. Are the margins clear?

10. Is the visual aid set off from the text by adequate white space?

11. Does the top of an excessively wide visual aid abut the inside binding?

12. Is the visual aid introduced, discussed, and interpreted as needed?

13. Is it in the best location for its purpose in your report (within the text if it clarifies the discussion; in an appendix if it merely supports it)?

14. Are all sources of data identified?

15. In a graph, are the independent variables plotted along the horizontal line and the dependent along the vertical?

16. Does a multiple-bar or line graph have a legend to identify each bar or line?

17. Does the graph have a clear and accurate scale?

18. Is the graph restricted to three or fewer lines or types of bars?

19. Does the segmentation in the pie chart begin at twelve o'clock?

20. Does the pie chart have at least three, but no more than seven, segments?

21. Do the segments add up to 100 percent?

22. Are any small segments in the pie chart (1 percent to 5 percent each) integrated under the heading "Other"?

Now list those elements of your visual aids that need improvement.

EXERCISES

1. The following statistics are based on data gathered from three competing colleges located in a large western city. They give the number of applicants to each college over the last six years.

– In 1972, X College received 2341 applications for admission, Y College received 3116, and Z College received 1807.

– In 1973, X College received 2410 applications for admission, Y College received 3224, and Z College received 1784.

– In 1974, X College received 2689 applications for admission, Y College received 2976, and Z College received 1929.

– In 1975, X College received 2714 applications for admission, Y College received 2840, and Z College received 1992.

– In 1976, X College received 2872 applications for admission, Y College received 2615, and Z College received 2112.

– In 1977, X College received 2868 applications, Y College received 2421, and Z College received 2267.

Illustrate this information in a line graph, a bar graph, and a formal table. Which form seems most effective here? Include a brief prose interpretation with the most effective illustration.

2. Devise a flowchart for a process in your career field or in an area of interest. Include a full title and a brief prose discussion of your illustrated data.

3. Devise an organizational chart showing the lines of responsibility and authority in an organization where you hold a part-time or summer job.

4. Devise a pie chart to illustrate the partition of one of your typical weekdays. Include a full title and a brief prose discussion of your data.

5. Call or visit your town or city hall and ask the town accountant for a breakdown of town income and expenditures (where each part of the revenue dollar comes from; how each part of the revenue dollar is spent). Compose the most appropriate visual aids to illustrate these partitions. Include full title, labels, and prose explanations.

6. Obtain the enrollment figures for the past five years at your college on the basis of sex, age, race, or any other pertinent category. Construct a segmented bar graph to illustrate one of these relationships over the five-year period.

7. Keep track of your pulse and respiration rates taken at thirty-minute intervals over a four-hour period of changing activities. Record your findings in a line graph, noting both times and specific activities below your horizontal coordinate. Write a brief prose interpretation of your graph and give it a full title.

8. In your textbooks, locate each of the following visual aids: a table, a multiple bar graph, a multiple line graph, a diagram, and a photograph. Bring the samples to class and discuss the effectiveness of each illustration. Is it clear, readable, meaningful? Is it introduced and discussed? Is it properly titled, numbered, and labeled? Is it necessary? Choose the most effective illustration and write a prose evaluation, discussing each of its strong points.

9. We have discussed the importance of choosing an appropriate scale for your graph and choosing the most effective form for presenting your data visually. Study the following presentation carefully:

Strong evidence now indicates that not only the nicotine and tar in cigarette smoke can be lethal. Experts have learned that a high percentage of cigarette smoke is composed largely of carbon monoxide, and the public is unaware of the danger. The bar graph in Figure 11-14 (next page) lists the ten leading U.S. cigarette brands according to the carbon monoxide given off per pack of inhaled cigarettes.

Is the scale effective? If not, why not? Can these data best be presented in a bar graph? What other form of visual aid would be more effective? Present the same data in the form that seems most effective. In a short but detailed paragraph evaluate what you have learned from this assignment.

10. Choose the most appropriate visual aid for illustrating each of the following general data areas. Justify each choice in a short paragraph.

a. a comparison of three top brands of Fiberglas ski, according to cost, weight, durability, and edge control
b. a breakdown of your monthly budget

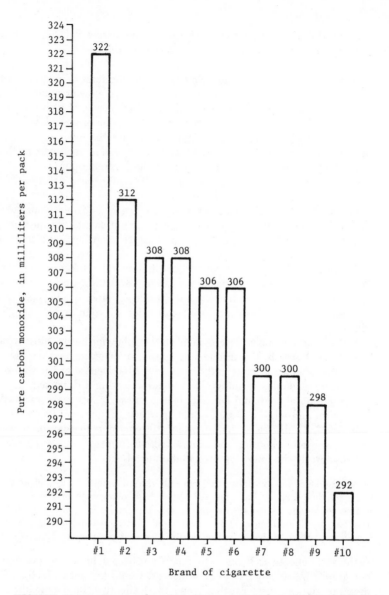

FIGURE 11-14   Ten Leading U.S. Cigarette Brands in Order of CO Content

c. an illustration of the changing cost of an average cup of coffee, as opposed to that of an average cup of tea, over the past two years

d. an illustration of the percentage of college graduates finding desirable jobs within three months after graduation, over the last ten years

e. an illustration of the percentage of college graduates finding desirable jobs within three months after graduation, over the last ten years — on the basis of sex

f. an illustration of automobile damage for an insurance claim
g. a breakdown of the process of radio-wave transmission
h. a comparison of five breakfast cereals on the basis of cost and nutritional content
i. a comparison of the average age of students enrolled at your college, in summer, day, and evening programs, over the last five years
j. a comparison of monthly sales for three models of an item produced by your company

12

Researching Information

DEFINITION

PURPOSE OF RESEARCH

IDENTIFYING INFORMATION SOURCES
  Personal Experience
  The Library
    The Card Catalog
    Reference Books
    Periodical Indexes
    The Reference Librarian
    Access Tools for U.S. Government
      Publications
  Informative Interviews
    Identification of Purpose
    Choice of Respondent
    Preparation
    Maintaining Control
    Recording Responses
  Questionnaires
  Letters of Inquiry
  Organizational Records and Publications
  Personal Observation

TAKING EFFECTIVE NOTES
  Purpose
  Procedure

CHAPTER SUMMARY

EXERCISES

DEFINITION

Research is the effort to discover any fact or set of related facts: the cost of building the first x-ray machine or the price range of half-acre building lots in Boville in January 1981. Any kind of significant research is for some purpose — to answer a question, to make an evaluation, to establish a principle. We set out to discover whether diesel engines are efficient and dependable for a reason: we're thinking about buying a car equipped with one, or we're trying to decide whether to begin producing them, or the like. Research is the way to find your own answers, to submit your opinions to the test of fact.

Depending on its information sources, research may be classified as *primary* or *secondary*. Primary research is a firsthand study of the subject; its sources are memory, observation, questionnaires, interviews, letters of inquiry, and records of business transactions or scientific and technological activities. Secondary research is based on information that other researchers — by their primary research — have compiled in books, articles, reports, brochures, and other publications. Most research combines primary and secondary approaches.

PURPOSE OF RESEARCH

The purpose of all research is to arrive at an informed opinion, to establish a conclusion that has the greatest chance of being valid.

We might have uninformed opinions about political candidates, kinds of cars, controversial subjects like abortion and capital punishment, or anything else that touches our lives. Opinions are beliefs that are not proven but seem to us to be true or valid. Without a basis in fact, opinions are uncertain, disputable, and subject to change in the light of new experience. Sometimes we forget that many of our opinions don't rest on any objective data. Instead, they are based on a chaotic collection of the beliefs that are reiterated around us, notions we've inherited from advertising, things we've read but never checked the validity of, and so on. Television commercials, for instance, are especially designed to manipulate the consumer's uninformed opinion about the quality of certain products.

Any claim is valid only insofar as it is supported by facts. A fact is a truth known by actual observation or study. Therefore, an opinion based on fact is more valuable than an uninformed opinion. In many cases we must consider a variety of facts. Consider for instance a commercial claim that Brand X toothpaste makes teeth whiter. Although this claim may be factual, a related fact may be that Brand X toothpaste contains tiny particles of ground glass, thereby harming more than helping teeth. The second fact may change your opinion about whether you want to use Brand X toothpaste. Similarly, a United States senator may claim that he is committed to reducing environmental pollution, but he may have voted against all ecology-based legislation. Therefore, if you want to know whether to believe his claim, you need to establish the facts. You want your opinion to rest on objectively verifiable information, not merely on the basis of his claim.

Clearly, we respect some opinions more than others. Assume, for example, that you are a personnel executive interviewing candidates for a position with your banking firm. Which of the following responses would you find more impressive?

1. *Q.* What is your opinion of the condition of our national economy?
   A. I think the economy is improving.
   Q. Why?
   A. Well, I just have that feeling. . . .

2. *Q.* What is your opinion of the condition of our national economy?
   A. I think the economy is improving. My opinion is based on the following indicators:
   a. Our gross national product has risen steadily over the past five months for a total 6.5 percent increase.
   b. The Dow Jones has climbed slowly and steadily for the past four months, and has now exceeded the 1,000-point mark.
   c. The wholesale price index has dropped an average of 1.5 percent monthly since last June.
   d. Seasonally adjusted unemployment figures have dropped from 9.1 percent to 7.6 percent in the past six months.
   e. New-home construction has doubled in the past year.
   f. Detroit reports a record-breaking year for automobile sales. Thus, the auto industry seems to have recovered from last year's slump.
   In short, inflation is decreasing, production and employment are increasing and the recession seems just about over.

The second candidate can support his opinions with fact because he has done his homework.

Facts also can affect the quality of your decisions. Before studying for a specific career, for example, be sure to investigate job openings, salary range, and requirements; better yet, interview someone who has such a job. Conduct the same kind of background research before buying a new car, emigrating to Australia, or making other major decisions. Although the facts may contradict your original opinion, they will enable you to make an informed decision.

IDENTIFYING INFORMATION SOURCES

Personal Experience

As a logical first step, begin with what you already know about your subject. Search your memory and use the brainstorming technique discussed in Appendix B.

The Library

A library is one of the best examples of the mind's power for classification. In a large library, you can find information on just about anything when you understand the library's classification system.

For the research work required to develop an informed opinion, the library should be your first stop. You may even learn that your problem has already been studied and solved by others, which could save you much research time.

School, public, or company libraries have one or more of these means of organizing their collections: card catalogs, reference works, indexes to periodicals, and government publications indexes.

The Card Catalog

The card catalog is the backbone of the library. In many libraries, every book, film, filmstrip, phonograph album, and tape is indexed in the card catalog under three separate designations: author, title, and subject. Thus you have at least three possible ways of locating the item.

Your library may place author, title, and subject cards in a single alphabetical file or may provide individual catalogs labeled "Author," "Title," and "Subject." In your search, first ascertain whether you are looking for a specific title, a book by a certain author, or material concerning a subject. Then check the arrangement of the card catalog to determine where you would look.

Locating the Card. Let's say that you are looking for a book on food technology written by Norman W. Desrosier. If your library has a divided catalog, locate the D cards in the author section. If not, locate the D cards in the combined catalog. Flip through the cards in the drawer until you locate those listed under "Desrosier, Norman W." Figure 12-1 shows a typical author catalog card. Each major item is labeled and explained.

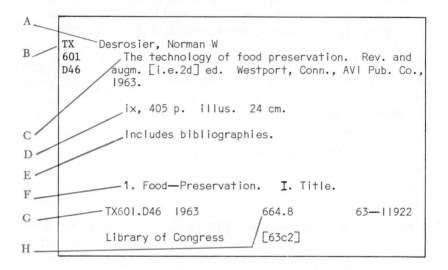

A: The author's name, listed last name first. (On some cards the author's name is followed by his date of birth — and death, if he was deceased at the time of the book's last printing.)

B: The call number: each book has a different call number under the coding system by which books are classified. Books are arranged on the shelves in the order of their respective call numbers. This number is your key for locating the book.

C: The title of one book written by this author, followed by the number of the edition, city of publication, publisher, and publication date.

D: Technical information: This book has 9 chapters, 405 pages, illustrations, and is 24 centimeters high.

E: Special information about the book: This book includes bibliographies listing related sources of information.

F: Other headings under which this book is listed in the card catalog: In the subject section, the book is listed under "Food — Preservation"; it is also listed alphabetically by title.

G: Library of Congress call number.

H: Dewey Decimal System call number.

FIGURE 12-1   A Catalog Card Classified by Author

As entry F on the author card indicates, you can find the same book listed alphabetically under its title, as shown in Figure 12-2. All other information on the title card is identical to that on the author card.

```
         The technology of food preservation.

TX     Desrosier, Norman W
601        The technology of food preservation.  Rev. and
D46        augm. [i.e.2d] ed.  Westport, Conn., AVI Pub. Co.,
1963       1963.

           ix, 405 p.  illus.  24 cm.

           Includes bibliographies.



           1. Food—Preservation.   I. Title.

         TX601.D46  1963              664.8            63—11922

         Library of Congress        [63c2]
```

FIGURE 12-2   A Catalog Card Classified by Title

If you know neither the authors nor the titles of books on a subject, turn to the subject listing. Figure 12-3 shows the card for the same book found under the subject heading "Food — Preservation." Here again, all other items on the card remain the same.

```
         Food - Preservation

TX     Desrosier, Norman W
601        The technology of food preservation.  Rev. and
D46        augm. [i.e.2d] ed.  Westport, Conn., AVI Pub. Co.,
1963       1963.

           ix, 405 p.  illus.  24 cm.

           Includes bibliographies.



           1. Food—Preservation.   I. Title.

         TX601.D46  1963              664.8            63—11922

         Library of Congress        [63c2]
```

FIGURE 12-3   A Catalog Card Classified by Subject

THE LIBRARY OF CONGRESS CLASSIFICATION SYSTEM

The books in this library are arranged on the shelves according to the Library of Congress Classification System, which separates all knowledge into 21 classes, as outlined below. Each class is identified by a letter of the alphabet, subclasses by combinations of letters, and subtopics within classes and subclasses by a numerical notation.

Classes 2nd Floor

A GENERAL WORKS (1st Floor)
(General encyclopedias, reference books, periodicals, etc.)

B PHILOSOPHY — RELIGION
B–BJ   Philosophy, including BF, Psychology
BL–BX Religion

C AUXILIARY SCIENCES OF HISTORY
CB   History of civilization (General)
CC   Archaeology
CD   Archives
CJ   Numismatics
CR   Heraldry
CS   Genealogy
CT   Biography (General)

D HISTORY: GENERAL AND OLD WORLD
(Including geography of individual countries)
D   World history, including World Wars
DA   Great Britain
DB   Austria
DC   France
DD, etc.   Other individual countries

E-F HISTORY OF AMERICA
(Including geography of individual countries)
E 1–143   America (General)
E 151–857   United States (General)
F 1–957   United States: States and local
F 1001–1140   Canada
F 1201, etc.   Other individual countries

Classes 3rd Floor

P LANGUAGE AND LITERATURE
P   Philology and linguistics
PA   Classical languages and literatures
PC   Romance languages
PD–PF Germanic languages, including PE, English
PG   Slavic languages and literatures
PJ–PL Oriental languages and literatures
PN   General and comparative literature
PQ   Romance literatures
PR   English literature
PS   American literature
PT   Germanic literatures
PZ   Fiction in English. Juvenile literature

Q SCIENCE
QA   Mathematics
QB   Astronomy
QC   Physics
QD   Chemistry
QE   Geology
QH   Natural history
QK   Botany
QL   Zoology
QM   Human anatomy
QP   Physiology
QR   Bacteriology

R MEDICINE

S AGRICULTURE

G GEOGRAPHY, ANTHROPOLOGY, FOLKLORE, ETC.

G	Geography (General)
GB	Physical geography
GC	Oceanography
GN	Anthropology
GR	Folklore
GV	Recreation

H SOCIAL SCIENCES

HA	Statistics
HB–HJ	Economics
HM–HX	Sociology

J POLITICAL SCIENCE

JA–JC	Political science
JF–JQ	Constitutional history and public administration
JS	Local government
JX	International law

K LAW

L EDUCATION

M MUSIC

M	Scores
ML	Literature of music
MT	Musical instruction

N FINE ARTS

NA	Architecture
NB	Sculpture
NC	Graphic arts
ND	Painting
NK	Decorative arts

SD	Forestry
SF	Animal culture
SH	Fish culture and fisheries
SK	Hunting sports

T TECHNOLOGY

TA	General engineering, including general civil engineering
TC	Hydraulic engineering
TD	Sanitary and municipal engineering
TE	Highway engineering
TF	Railroad engineering
TG	Bridge engineering
TH	Building construction
TJ	Mechanical engineering
TK	Electrical engineering. Nuclear engineering
TL	Motor vehicles. Aeronautics. Astronautics.
TN	Mining engineering. Mineral industries. Metallurgy
TP	Chemical technology
TR	Photography
TS	Manufactures
TT	Handicrafts. Arts and crafts
TX	Home economics

U MILITARY SCIENCE

V NAVAL SCIENCE

Z BIBLIOGRAPHY AND LIBRARY SCIENCE

BIOGRAPHY: (*1st Floor*) Lives of individuals, illustrative of any subject, are normally classified with that subject, *e.g.* Albert Einstein is classified in QC16.E5. Otherwise, they are classified with general biography in CT.

The complete Library of Congress call number for any book may be found by consulting the card catalog.

THE LIBRARIAN WILL BE HAPPY TO ASSIST YOU IF YOU ARE UNABLE TO FIND THE BOOK YOU WANT

FIGURE 12-4 A Guide to a Library's Classification System

Locating the Book on the Shelf. If your library has closed stacks, you will have to fill out a call slip with book title, author, and call number. You then wait for a staff member to find your book. If the stacks are open, you will have to find the book yourself. The call number is your key for locating the book. In the card catalog area, in stairways and elevators, and on doors to individual floors, you will see posted copies of a guide like the one in Figure 12-4.

Now simply follow the letters and numbers until you find your book. The call number for Desrosier's *The Technology of Food Preservation,* for example, is TX601. D46. 1976.[1] By locating T on the call-number map, you learn that your book is on the third floor of this library. More specifically, the card tells you that the book is in the TX section of the third floor.

Reference Books

Reference books include technical encyclopedias, almanacs, handbooks, dictionaries, histories, and biographies. These are often a good place to begin your research, because they provide background information and bibliographies that

---

[1] Instead of the Library of Congress classification system, some libraries use the Dewey Decimal System which divides all books into ten numbered categories:

000–999 General Works		500–599 Pure Science	
100–199 Philosophy		600–699 Technology	
200–299 Religion		700–799 Fine Arts	
300–399 Social Sciences		800–899 Literature	
400–499 Language		900–999 History	

Any one of these categories, in turn, forms ten smaller categories:

600–609 Technology	650–659 Business
610–619 Medical Science	660–669 Chemical Technology
620–629 Engineering	670–679 Manufactures
630–639 Agriculture	680–689 Other Manufactures
640–649 Home Economics	690–699 Building Construction

Each of these last categories can be further divided:

660– Chemical Technology	665– Oils, Fats, Waxes, Gases
661– Industrial Chemicals	666– Ceramic and Allied Industries
662– Explosives and Fuels	667– Cleaning and Dyeing
663– Beverages	668– Other Organic Production
664– Food Technology	669– Metallurgy

More specific breakdowns within each category are represented by numbers after the decimal:

664.1 Sugar Manufacturing and Refining
. . . . .
664.8 Commercial Preservation of Food

In libraries that use this system, the Dewey Decimal classification number will be found in the upper left corner of the catalog card, replacing the Library of Congress number.

can lead you to more specific information. The one drawback here is that some reference books that have not been revised recently (within five years) may be out of date. Always check the last copyright date.

Reference works will be found in a special section marked "Reference" — usually on the main floor of the building. All works will be indexed in the "Subject" card catalog, with a "Ref." designation immediately above the call number. Here is a partial list of reference works:

- *Civil Engineering Handbook*
- *Dictionary of American Biography*
- *Encyclopaedia Britannica*
- *Encyclopedia of Food Technology*
- *Encyclopedia of the Social Sciences*
- *Fire Protection Handbook*
- *Handbook of Chemistry and Physics*
- *The Harper Encyclopedia of Science*
- *Information Please Almanac*
- *McGraw-Hill Encyclopedia of Science and Technology*
- *The New Dictionary and Handbook of Aerospace*
- *The New York Times Encyclopedic Almanac*
- *Oxford English Dictionary*
- *Paramedical Dictionary*
- *The Times Atlas of the World*

There are reference works designed for every discipline. In researching the effects of food additives, for instance, you would want to start with titles such as the *McGraw-Hill Encyclopedia of Food, Agriculture, and Nutrition* or the *RC Handbook of Food Additives*. Look for your subject in the card catalogue, then check to see if there are any subheadings such as "Handbooks, Manuals, etc." or "Dictionaries." These will be books that generally get you started in your topic.

Periodical Indexes

Just as the card catalog is an index to the library's books and audiovisual materials, periodical indexes are alphabetical guides listing names, titles, and subjects of material found in journals. Through various periodical indexes, you can locate the latest information available. The most comprehensive index is the *Readers' Guide to Periodical Literature*, which indexes articles from over 150 popular magazines and journals. Because a new volume of the *Readers' Guide* is published every few weeks, you often can locate articles that are less than one month old. The *Readers' Guide* indexes items alphabetically by subject, author, and book or film title. Figure 12-5 shows a section from one of its pages.

FOOD, Organic
Energy scorecard: the nutritionist vs. the health food freak. Mademoiselle 82:163 Ap '76
Food your family eats. E. M. Whelan and F. J. Stare. il Parents Mag 51:34-5+ Jl '76
Health-food hoax. F. J. Stare. Harp Baz 109:71+ My '76
Notes from a has-been: a mother's confession about food and her family. E. Baldwin. il Org Gard & Farm 23:94+ N '76
Organic foods: today's big rip-off. il Farm J 100:66 F '76
Organic living almanac. Org Gard & Farm 23:90-1 My '76
Proxmire liberates vitamins. il Bus W p36 Mr 29 '76
Whole earth organic food fad. T. H. Jukes. Parents Mag 51:46-7+ Mr '76
    See also
Cookery—Organic food
FOOD, Raw
International chef: Hamlin, Germany steak tartare. D. Reynolds. il Travel 147:12 Ja '77
FOOD, Wild
Foraging for food provides rewards for a brave palate. L. Pringle. il Smithsonian 7:120-9 S '76
    See also
Plants, Edible
FOOD additives
Are you eating dangerously? N. Simon. Vogue 166:112-13+ Jl '76
Experts are divided about food additives. J. Mayer. il Fam Health 8:36-8 Jl '76
Food additives and federal policy: the mirage of safety, by B. T. Hunter. Review
    Consumers Res Mag 59:30 Mr '76

A —— Food additives and hyperactive kids. Sci Digest 80:13 N '76

Food additives: how safe is safe? F. Warshofsky. Read Digest 108:117-21 My '76
Testing for seeds of destruction; research on food additives as causing cancer. J. Miller. Progressive 40:37-40 D '76
    See also

B — ⌈ Coloring matter in cosmetics, food, etc.
      ⌊ Nitrosamines
FOOD and drug administration. See United States —Food and drug administration

FIGURE 12-5 A page section from *"Readers' Guide to Periodical Literature."* (Copyright © 1976, 1977 by the H.W. Wilson Company. Material reproduced by permission of the publisher.)

Locating the Entry.   Assume that you are researching the physiological effects of food additives and preservatives. As you turn to a recent issue of the *Readers' Guide,* you find many entries under the general heading "Food." Scanning the entries classified under "Food Additives," you spot an item that looks useful: "Food Additives and Hyperactive Kids" — item A. From this entry you gain the following information: because the author's name is not given, the article probably was written by a writer on the magazine staff; the name of the periodical is *Science Digest* (abbreviations are explained in the opening pages of each *Readers' Guide*); the volume number is 80; the article is found on page 13 of the November 1976 issue. Item B refers to other headings under which you might find relevant articles.

Locating the Indexed Article. Each work listed in the card catalog is actually held by that library. Therefore, unless the item has been borrowed by someone else, you should locate it easily under its call-number designation. Periodicals are more of a problem, however; not all works listed in periodical indexes are likely to be held by your library. The indexes list articles from hundreds of journals, newspapers, and magazines, but your library most likely subscribes only to a cross-section. If you find an index reference to an article not held by your library, check with other libraries or ask your librarian to see if the article is available through an interlibrary loan.

Scan the periodicals holdings list to determine which periodicals are held by your library. Copies are available in the area where indexes are shelved. Figure 12-6 shows a sample page from one library's list.

Earlier, you located an entry in the *Readers' Guide* titled "Food Additives and Hyperactive Kids," printed in *Science Digest.* Now, as you scan the periodicals holdings list you learn that your library subscribes to this journal (see item A). You also learn that all back issues to 1960 are recorded on microfilm.[2] Your article is probably not recent enough to be found in the actual journal (many libraries keep only one year of back issues as bound copies). Ask your librarian to explain the use of microfilm files and readers.

If the article is in a very recent issue, scan the periodical shelves, usually arranged alphabetically by title, to find your issue. In some libraries older issues are bound together in hard-cover bindings instead of being microfilmed. In this case, each bound volume has a call number and is listed in the card catalog under "Title."

In addition to indexes for popular, general periodicals, specialized indexes are available in nearly every discipline. A researcher in the business field, for instance, might consult the following indexes of specialized, professional periodicals:

> *Business Periodicals Index*
> *Editor and Publisher Market Guide*
> *Editorial Research Reports*
> *F and S Index*
> *Predicasts*
> *Public Affairs Index*

Here are some titles of indexes in other fields:

> *Agricultural Index*
> *Applied Science and Technology Index*

---

[2] To save space, many libraries subscribe to a microfilm service that photographs all back issues of major periodicals on microfilm. Thus they can store several periodicals on one small roll of film, and you can read the original on a microfilm reader.

-Q-

Quest   1973-

-R-

RN National Magazine for Nurses   Microfilm 1966-
RQ (ALA)   Microfilm 1970-
Ramparts (disc. micro. only)   Microfilm 1962-
                                           Sept. 1975
Readers Digest   Microfilm 1960-
Research in Education (ERIC)   1970-
Respiratory Care   1971-
Revista Rotaria (Spanish)   1970-
Revue d'Histoire Litteraire   1969-
     de la France
Rights   1971
Romanic Review   Microfilm 1960-
Rotarian   1974

-S-

Saturday Evening Post (micro. disc.)   Microfilm 1960-1969
                                           1971-
Saturday Review (Saturday Review World)   Microfilm 1924-1972,
                                           1975
Saturday Review World (formerly: 1973-1974
     Saturday Review)
Scholastic Teacher (for microfilm   1972-
     see: Senior Scholastic)
School & Society (now: Intellect)   Microfilm 1960-1972
School Libraries (now: School Media   Microfilm 1951-1972
     Quarterly)
Science   Microfilm 1959-
Science Books   1970-1975
Science Digest   Microfilm 1960-
Science News   Microfilm 1960-
Scientific American   Microfilm 1929
Sea Change (Cape Cod Community
     College Literary Magazine)
Sea Frontiers (micro.)   Microfilm 1954-1975
Secretary   1973-
Senior Scholastic   Microfilm 1960-
Sewanee Review   Microfilm 1960-

FIGURE 12-6   A Page from a Library's Periodicals Holdings List

Biological Abstracts
Chemical Abstracts
Education Index
The Energy Index
Engineering Index
Index Medicus
International Nursing Index
Psychological Abstracts

Two most valuable sources for quickly building exhaustive bibliographies (lists of publications on your subject) are the *Science Citation Index* and its counterpart, the *Social Sciences Citation Index*.

Your library may have other indexes. Ask the librarian to help you identify appropriate indexes for your subject.

Having looked in the reference books that provided a good overview of the psychological effects of food additives, you might now turn to *Psychological Abstracts* or *Index Medicus* for journal literature.

The Reference Librarian

Your best bet for help with library research is the reference librarian. This person's job — as the title implies — is to *refer* people to information sources. So don't be afraid to ask for assistance. Even professional researchers sometimes need help. The reference librarian can save hours of wasted time by showing you how to use various indexes or microfilm and microfiche readers and how to locate reference books or bibliographies. Moreover, he or she can order items from other libraries, but this may take time.

Access Tools for United States Government Publications

The federal government publishes maps, periodicals, books, pamphlets, manuals, monographs, annual reports, research reports, and a bewildering array of other information. Kinds of information available to the public include presidential proclamations, congressional bills and reports, judiciary rulings, certain reports from the Central Intelligence Agency, and publications from all other government agencies (Departments of Agriculture, Commerce, Transportation, etc.). Below is a brief sampling of the countless titles available in this gold mine of information:

Effects of New York's Fiscal Crisis on Small Business
Economic Report of the President
Major Oil and Gas Fields of the Free World
Decisions of the Federal Trade Commission
Journal of Research of the National Bureau of Standards
Siting Small Wind Turbines

Your best bet for tapping this valuable but complex resource is to ask the librarian in charge of government documents for help. If your library does not own the publication you seek, it can be obtained through an interlibrary loan.

The basic access tools for documents issued or published at government expense are discussed below.

– *The Monthly Catalog of the United States Government,* the major access to government publications, is indexed by author, subject, and title. These indexes provide you with the catalog entry number that leads you, in turn, to a complete citation for a given work (see Figure 12-7).

– *Government Reports Announcements & Index* is a listing published every two weeks by the National Technical Information Service (NTIS),[3] a federal clearinghouse for scientific and technical information — all stored in a computer base. The collection contains summaries of some 680,000 federally sponsored research reports published since 1964. About 70,000 new summaries are added annually in 22 subject categories ranging from aeronautics to medicine and biology. Copies of the full reports are available from NTIS.

Computerized information retrieval will play a major role in future research. Ask your librarian for information about NTIS and about commercial data-based information services in your field.[4]

– *The American Statistics Index,* a yearly guide to the statistical publications of the U.S. Government, is divided into two sections: *Index* and *Abstracts* (an index with summaries). The *Index* volume lists material by subject and provides geographic (U.S., state, etc.), economic (income, occupation, etc.), and demographic (sex, marital status, race, etc.) breakdowns. The *Index* volume refers you to a number and entry in the *Abstract* volume.

In addition to these three basic access tools, the government issues *Selected Government Publications,* a monthly list containing roughly 150 titles (with descriptive abstracts). These titles range from highly general (*Questions About the Oceans*) to highly technical (*An Emission-Line Survey of the Milky Way*). While this list may be of interest, it is by no means a refined research tool.

The government also publishes bibliographies on hundreds of subjects, ranging from "Accidents and Accident Prevention" to "Home Gardening of Fruits and Vegetables." Ask your librarian for information about these subject bibliographies.

[3] A branch of the U.S. Department of Commerce.

[4] For a major source of information on data bases available in various disciplines, see Martha E. Williams et al., eds., *Computer-Readable Data Bases: A Directory and Data Sourcebook* (Urbana, Illinois: American Society for Information Science, 1979).

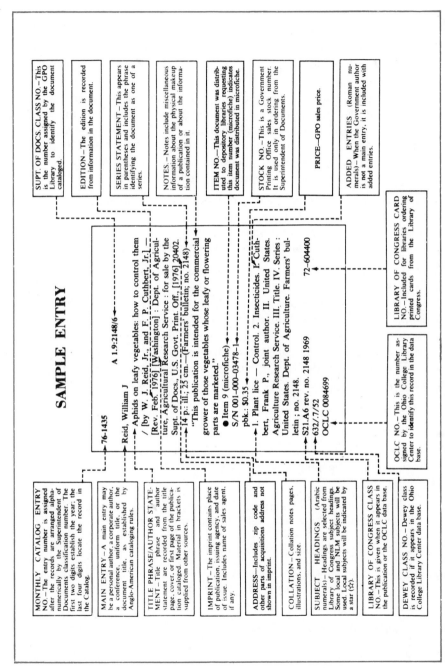

SAMPLE ENTRY

SUPT. OF DOCS. CLASS NO.—This is the number assigned by the GPO Library to identify the document cataloged.

EDITION—The edition is recorded from information in the document.

SERIES STATEMENT—This appears in parentheses and includes the phrase identifying the document as one of a series.

NOTES.—Notes include miscellaneous information about the physical makeup of a publication or about the information contained in it.

ITEM NO.—This document was distributed to depository libraries requesting this item number (microfiche) indicates document was distributed in microfiche.

STOCK NO.—This is a Government Printing Office sales stock number. It is used only in ordering from the Superintendent of Documents.

PRICE—GPO sales price.

ADDED ENTRIES (Roman numerals)—When the Government author is not a main entry, it is included with added entries.

76-1435   A 1.9:2148/6

Reid, William J
   Aphids on leafy vegetables: how to control them / [by W. J. Reid, Jr., and F. P. Cuthbert, Jr.] — [Rev. Feb. 1976] [Washington] : Dept. of Agriculture, Agricultural Research Service : for sale by the Supt. of Docs, U.S. Govt. Print. Off., [1976] 20402.
   [4 p.: ill.; 25 cm. —(Farmers' bulletin; no. 2148)
   "This publication is intended for the commercial grower of those vegetables whose leafy or flowering parts are marketed."
   ● Item 9 (microfiche)
   S/N 001-000-03478-1
   pbk.: $0.35
   1. Plant lice — Control. 2. Insecticides. I. Cuthbert, Frank P., joint author. II. United States. Agriculture Research Service. III. Title. IV. Series: United States. Dept. of Agriculture. Farmers' bulletin ; no. 2148.
   S21.A6 rev. no. 2148 1969   72—604400
   632/.7/52
   OCLC 0084699

72—604400

LIBRARY OF CONGRESS CARD NO.—Included for libraries ordering printed cards from the Library of Congress.

MONTHLY CATALOG ENTRY NO.—The entry number is assigned after the records are arranged alphanumerically by the Superintendent of Documents classification number. The first two digits establish the year; the last four digits locate the record in the Catalog.

MAIN ENTRY—A main entry may be a personal author, a corporate author, a conference, uniform title, or the document title, as established by Anglo-American cataloging rules.

TITLE PHRASE/AUTHOR STATEMENT—Title phrase and author statement are recorded from the title page, cover, or first page of the publication cataloged. Material in brackets is supplied from other sources.

IMPRINT—The imprint contains place of publication, issuing agency, and date of issue. Includes name of sales agent, if any.

ADDRESS—Includes zip code and other parts of acquisitions address not shown in imprint.

COLLATION—Collation notes pages, illustrations, and size.

SUBJECT HEADINGS (Arabic numerals)—Headings are selected from Library of Congress subject headings. Some local and NLM subjects will be used. Local subjects will be indicated by a star (☆).

LIBRARY OF CONGRESS CLASS NO.—This is given when it appears in the publication or the OCLC data base.

DEWEY CLASS NO.—Dewey class is recorded if it appears in the Ohio College Library Center data base.

OCLC NO.—This is the number assigned by the Ohio College Library Center to identify this record in the data base.

FIGURE 12-7   A Sample Entry in *The Monthly Catalog*

Informative Interviews

Although libraries are a good secondary research source, you may also have to consult firsthand, or primary, sources. One excellent primary source can be the personal interview.

Conducting a good interview is not easy. Some of my own students who are reporters for the campus newspaper have complained that their informative interviews are often disappointing and embarrassing. They tell of not knowing what questions to ask and having to suffer through long, terrible silences. Afterward, they become confused about writing up their findings. Much of this frustration results from one or more of the following errors:

1. the wrong choice of interviewee (respondent)
2. inadequate preparation for the interview
3. failure to exercise control in conducting the interview
4. failure to record vital information immediately after the interview

Each of these errors derives from the interviewer's inability to identify a clear purpose and to formulate a plan. You must know *exactly* what you are looking for! After all, an informative interview is not a casual bull session. When well planned and carefully directed it can be pleasurable, rewarding, and a solid example of your ability to engage in professional dialogue. The following suggestions should help you conduct a good interview: identify your purpose; choose the most knowledgeable interviewee; prepare well; control the interview; record your responses.

Identification of Purpose

Keep in mind the statement of purpose discussed in earlier chapters. Suppose you are thinking of opening a motorcycle shop. Determine what information you hope to obtain:

> The purpose of my interview with Ms. Jones, city clerk, is to learn how licensing requirements, fees, and zoning restrictions would affect the operation of my proposed motorcycle shop. Also, I need to know how to proceed in satisfying the legal requirements for this business.

When you are sure enough about your purpose that you can write it clearly in one or two sentences, you are ready for your next step.

Choice of Respondent

Choose your respondent carefully. Don't waste time interviewing someone who knows little about the subject. As part of your motorcycle-shop feasibility study, for instance, you might select these respondents: the head of your chamber of commerce, a representative for the brand of machine you expect to sell, a member of your local retail licensing authority board, the loan officer at your local savings bank. Any of these people should provide more vital information than, say, neighboring merchants, and they might give helpful leads for interviews with other key people.

One way to ensure that you will interview a knowledgeable person is to state the purpose of your interview when you call for an appointment:

> I am studying the possibility of opening a motorcycle shop in Winesburg, and I would appreciate the opportunity to meet with you, at your convenience, because I feel that you could answer some important questions about the future of such a business in our town.

If you communicate your purpose clearly, your intended respondent can immediately tell you whether he or she knows enough to make an interview worthwhile. If you're planning several interviews, begin with the least crucial so you can sharpen your skills for the more important ones. Your next step is to do your homework.

Preparation

Preparation is the step that requires imagination. After you have identified your purpose and chosen your respondent you have to plan questions that will get the information you need. Begin by collecting all available information. In order to ask useful questions, you have to know your subject! Complete your background investigation *before* the interview, and when your homework is done, write out your questions, *on paper*. Too many beginners set out for an interview armed only with a few blank notecards, a pencil, and a confident feeling that "the questions are all in my head." This mistake leads to embarrassing silences. Don't expect your respondent to keep the ball rolling.

Write out your questions on three-by-five notecards, one question per card, and arrange them in order. You then can jot notes about specific responses on the appropriate card.

Make your questions specific and direct. When interviewing a motorcycle wholesale representative, for example, you might phrase a question in one of two ways:

> 1. Does your company have a provision for buying back unsold stock from the retailer?
> 2. What provisions, if any, does your company have for buying back unsold stock from the retailer?

Version 1 can yield a simple yes or no answer, whereas version 2 requires a detailed explanation of the yes answer. Likewise, a question from a problem-solving interview can be worded in one of two ways:

1. Do you think this problem can be solved?
2. What steps would you recommend to solve this problem?

Notice that version 2 again requires an answer based on hard information. In any case, the answer you receive will be only as good as the question you ask.

Also, be sure that your questions cover all important parts of your subject. Brainstorming can help here (as discussed in Appendix B). Your questions for an interview about financing for your shop, for instance, might cover these subtopics: interest rate, required down payment, other required security or collateral, credit background, other provisions (such as cosigner), application procedure, time required to process the application, prepayment penalties, length of loan period, total amount to be repaid (including yearly interest), amount of monthly payments, emergency loan provisions, and so on.

Avoid questions with simple yes or no answers and questions whose answers can be found in books, articles, newspapers, and town bylaws. Finally, phrase your questions so that you do not influence the answer:

1. Don't you think this is a good plan?
2. What do you think are the principal (faults, merits) of this plan and how should they affect my success?

Version 1 suggests that you want your respondent to agree with you, whereas version 2 encourages a candid response.

Organize your questions by following the outlining procedure in Chapter 7. Move logically from one consideration to another. When your questions are specific, comprehensive, and written out in logical order, you can face your next step with confidence.

Maintaining Control

You — not your respondent — are responsible for conducting a professional and productive interview. These suggestions should help:

1. Dress neatly and arrive for your appointment on time.
2. Begin by thanking your respondent, in advance, for giving valuable time.
3. Restate the purpose of your interview.
4. Tell your respondent why you feel he or she can be helpful.
5. Discuss your plans for using the information.
6. Ask if your respondent objects to being quoted or taped. (Although

taping is the most accurate means of recording responses, it makes some people uncomfortable.)[5]

7. Avoid small talk.

8. Ask your questions clearly and directly, in the order you prepared them.

9. Be assertive but courteous. Don't be afraid to ask pointed questions, but remember that the respondent is doing you a favor by spending time with you.

10. Let your respondent do most of the talking. Remember that you are not there to instruct or to sound off, but to learn. Keep your opinions to yourself.

11. Guide the interview carefully. If your respondent begins to wander from the main points, politely bring the conversation back on track (unless, of course, the additional information is useful).

12. Be a good listener. Don't stare out the window, doodle, or ogle the secretary or office boy while your respondent is speaking.

13. Be prepared to explore new areas of questioning, if the need arises. Sometimes your respondent's answers will uncover new directions for your interview.

14. Keep your note taking to an absolute minimum. Record all numbers, statistics, dates, names, and other precise data, but don't transcribe responses word-for-word. This will be time-consuming and boring for the respondent, and may cause some tension. Simply record key words and phrases that later can refresh your memory.

15. If interviewing several people about the same issue, standardize your questions. Ask each respondent the same questions in the same order and in the same phrasing. Avoid making random comments that could affect one person's responses.

16. Don't hesitate to ask for clarification or further explanation of a response.

17. When your questions have all been answered, ask your respondent for any additional comments that might be helpful.

18. Finally, thank your respondent for his or her time and assistance and leave promptly.

Recording Responses

As soon as you leave the interview, find a quiet place to write out a summary of the responses. Don't count on your memory for preserving the main points

---

[5] In an interview about your prospective business, it would be inappropriate and unnecessary to tape the responses of the bank representative, sales representative, or licensing official. You might, however, wish to tape an interview with a member of the Small Business Advisory Services who was giving you detailed advice about how to proceed with your plan.

of the interview beyond this time. Get it all down on paper while the responses are still fresh in your mind.

If you are seeking opinions on a highly controversial subject for a newspaper or magazine article, ask your respondents to review the final draft of your text before you quote them in print.

Here is an interview text for the report titled "Survival Problems of Television Service Businesses" in Chapter 17. Notice how well-planned questions generate high-information responses that lead to more specific questions.

APPENDIX A: FIRST INTERVIEW

(This interview was with Al Jones, the owner of Al's TV Shop, which grosses under $100,000 per year.)

Q. I see that you have an excellent location. Is location of a service shop important?

A. Sure. If I ever get back to selling it will be very important. Doing just service minimizes the importance since the customer will search you out if he really wants you. I suppose the same holds true if I start selling.

Q. Well, you seem to be selling car radios and stereos. Isn't that "selling"?

A. Yes. But by selling I mean TVs. Floor plans, etc.

Q. Why don't you sell now? What's a floor plan?

A. The floor plan is why I won't sell now. I used to sell with a floor plan. As you can understand quite easily, a floor plan means a TV distributor will sell me, say, $10,000 worth of TVs for only 10 percent down payment. I pay the balance as I sell the sets. The problem is that the temptation to use the distributor's money is very great. These large amounts of money in the checking account give you a false sense of security.

Q. In other words you have had a bad experience?

A. Not only me. Most TV businesses fail because they start using someone else's money. Where can a small business come up with six, eight, and even ten thousand dollars after all the stock has been sold?

Q. So you just sell car radios and stereos and parts as needed for repairs?

A. That's right. I pay for them as I use them and stay out of trouble. I realize it's better to have thirty-day charge accounts and use the distributor's money to capitalize my business, but this can also cause trouble.

Q. OK. What about advertising?

A. I use the yellow pages. That's all I can afford. Sometimes the local radio stations offer attractive deals which cost me half the usual rate. I take advantage of these. Radio is good for the stereo radio sales.

Q. Wouldn't it be a good idea for a group of you technicians to get together and do some group advertising?

A. You already know the answer to that. You helped organize the Cape Cod Chapter of the ETG [Electronic Technicians Guild]. You know what happened to that. TV technicians don't want to get together. They just don't have the time.

Q. Tell me something about your manpower situation.

A. OK. Here's something. My TV man came up to me a few days ago and told me he was giving me thirty days' notice. He wants to go into business for himself. It's always like this. You hire and train a technician for a number of years and when he gets the itch, off he goes. Better give him a copy of your report; maybe he can avoid the mistakes I made over the years.

Q. Can you think of any mistakes you made along manpower lines?

A. I don't think I'll ever hire a beginner anymore. They just aren't productive enough. From now on they must have at least formal training and the technician's license.

Q. What do you mean by production?

A. I have to get so much production for every dollar I pay a technician. Unfortunately it has taken me many years to come to a good interpretation of what a reasonable return is. The minimum return I must have is $3.00 for every wage dollar.

Q. Your sign on the wall says "Minimum Labor Charge, $15.00." Does that mean you pay your technicians at least $5.00 per hour?

A. That's close enough. Sometimes technicians take longer than an hour to fix a unit. But it averages out to that.

Q. So your customers realize that they must be charged this amount so that you can cover your wages?

A. [Laughter] Customers don't realize anything. They complain no matter what we charge and no matter how long we take. I sincerely believe that customers start complaining even before their sets break.

Q. Have you ever tried discussing these notions with your customers?

A. Sometimes. I think they are beginning to learn that good shops and good technicians are getting scarcer. It pays for them to be cordial as it does for us.

Q. Can you think of anything else you can say about customers?

A. Yes. As you know, most homes have more than one TV set. This has created an unusual problem for us. With money being tight it seems as if the customer leaves one set at home and one in the shop. I have a time getting people to pick up their merchandise. This puts a money crunch on me since I have invested labor and money in the repair.

Q. Have you thought of any solution to this problem?

A. I have tried taking substantial deposits and threatened to sell the set for the price of the repairs (which my attorney says I can't do). As you can see the place is filled with unclaimed merchandise. This is getting a little out of hand, to say the least.

Q. How about the money situation? How much did you gross last year?

A. A little over $60,000. This wasn't too good at all. I have two technicians full time with occasional part-time help. When I'm not busy doing desk work I help out with pickups, deliveries, and antenna work. At $15.00 per hour for forty hours per man that comes out just right to cover their labor, but there are many other expenses to be covered. Also my labor wasn't covered. The figure should have been closer to $90,000.

Q. How much of this was gross margin?
A. Gross margin? Let's see. That's what's left over after the cost of the parts used in the repairs or sales. Oh, about $40,000.
Q. And how much of this turned out to be net income?
A. Very little! I only paid a couple hundred dollars in taxes this year.
Q. One last question. Is it worth it? Do you think you are getting a return on your investment?
A. It beats working for someone else. I believe I have made all the mistakes there are to be made in this business. I don't intend to make them again. Don't forget to show me the completed report.

Questionnaires

An interview has some advantages over a questionnaire: first, in a face-to-face discussion, you can tell whether or not your respondent understands your questions; moreover, your questions are more likely to be answered. Finally, you may obtain unexpected information. However, an interview sometimes seems like a formal interrogation. Respondents may feel threatened by certain questions. In this respect, a questionnaire is often a better alternative.

A questionnaire has several advantages. First, it saves time and is an inexpensive way of surveying a large cross-section. Also, respondents can answer the questions privately and anonymously — and thus more candidly. Furthermore, they have plenty of time to think about their answers.

These advantages, however, can be offset by the problem of getting people to take time to fill out and return a questionnaire. Whether you mail out or otherwise distribute a questionnaire, expect less than a 50 percent response, perhaps far less — and give plenty of time for your responses to return. You can increase your chances for a healthy response, however, by following these suggestions:

1. Introduce your questionnaire, so that your respondents will understand the importance of their answers and the purpose they will serve. Also, thank the respondent.

> Your answers to the following questions about your views on proposed state handgun legislation will be appreciated. Your state representative will tabulate all responses so that she may continue to speak accurately for *your* views in legislative session. Thank you.

2. Keep it short. Try to limit both questions and response space to two sides of a page. Also, phrase your questions concisely.

3. Place the easiest or more interesting questions first in order to attract attention.

4. Make each question specific and precise. *This is a crucial step.* If your readers can interpret a question in more than one way, their responses are

useless — or worse, misleading. Consider this question: "Do you favor foreign aid?" It is not specific or precise enough because foreign aid can mean military, economic, or humanitarian aid, or all three. Some people might support one form or another without supporting all three. Rephrase the question to allow for these differences:

> Do you favor:
> a. Our current foreign aid program of both military and humanitarian assistance to other countries?
> b. A greater emphasis on military assistance in our foreign aid program?
> c. A greater emphasis on economic and humanitarian assistance in foreign aid programs?
> d. No foreign aid program at all?

5. Avoid influencing your reader's responses with leading questions:

> Do you think foreign aid is a waste of taxpayers' money?
> Do you think foreign aid is a noble gesture by a wealthy nation toward the less fortunate?

Notice that each of these questions is phrased to elicit a definite response.

6. When possible, phrase your questions so that you can easily classify and tabulate responses: yes-or-no; multiple-choice; true-false; fill-in-the-blank.

7. Include a catch-all section ("Additional Comments") at the end of your questionnaire.

8. Enclose a stamped, self-addressed envelope along with the questionnaire.

9. If possible, do the questioning by telephone to save time and to ensure a response.

Figure 12-8 is a questionnaire designed to assess employee job satisfaction. Notice that each question is clear and easy to answer, and multiple-choice questions list *all* possible choices. In many ways, such a questionnaire is more effective than an interview.

Caution: Although questionnaires can provide valuable data, be cautious in interpreting statistical results. With a political questionnaire, for example, more people with strong views (likes, dislikes) will respond than people with moderate views. Therefore, your data may not truly reflect a cross-section. Supplement your questionnaire responses with other sources whenever possible.

Include the full text of the interview or questionnaire (with questions, answers, and tabulations) in an appendix (see Chapter 10). You can then refer to these data in your report discussion when necessary.

Letters of Inquiry

Letters of inquiry are handy for obtaining specific information. These are discussed in Chapter 14.

EMPLOYEE ATTITUDES QUESTIONNAIRE

This questionnaire is designed to record your attitudes about working conditions at Tanzor Electronics. Our firm, Obion Management Consultants, will incorporate this information into a plan for improving staff-management relations. Individual responses will be kept confidential. Please help us by taking a few minutes to answer the following questions. Choose only one answer to each question. Thank you.

1. How would you rate your own job competence?

   ____excellent   ____satisfactory   ____poor
   ____good   ____needs improvement ____don't know.

2. How would you rate the competence of other employees in your division?

   ____excellent   ____satisfactory   ____poor
   ____good   ____needs improvement ____don't know

3. How would you rate your workload?

   ____too heavy   ____too light
   ____satisfactory   ____don't know

4. How would you rate your salary for the job you do?

   ____excellent   ____satisfactory   ____poor
   ____good   ____needs improvement ____don't know

5. How would you rate Tanzor's policy for promotion and incentives?

   ____excellent   ____satisfactory   ____poor
   ____good   ____needs improvement ____don't know

6. How would you rate the communication between employees and management?

   ____excellent   ____satisfactory   ____poor
   ____good   ____needs improvement ____don't know

FIGURE 12-8 A Sample Questionnaire

7. How would you rate management's effectiveness in moti-
   vating employees?

   ____excellent   ____satisfactory   ____poor
   ____good   ____needs improvement   ____don't know

8. Are you satisfied with your job?

   _____yes   _____no   _____don't know

   Please explain your above response:

9. Are you satisfied with management's role?

   _____yes   _____no   _____don't know

   Please explain your above response:

10. Do you favor an organized union at Tanzor?

    _____yes   _____no   _____don't know

Please make any additional comments or suggestions that might
help us. Use the back of this page if necessary.

FIGURE 12-8   (*Continued*)

Organizational Records and Publications

Company records are a good primary data source. Most organizations also publish pamphlets, brochures, annual reports, or prospectuses for consumers, employees, prospective investors, voters, and so on. Your local nuclear power company, for example, may publish several pamphlets explaining the basic principles of nuclear power-generating systems and listing the advantages of this energy source. If you were writing a report analyzing the safety measures employed or omitted in a nuclear plant, you would want information representing both sides of the controversy. Along with the power company's literature you will need the publications of federal, state, and local environmental groups to discover their assessment of these safety measures. Sometimes a pamphlet can be a good beginning for your research.

Personal Observation

If possible, amplify and verify your research findings with a firsthand look at your subject. Conclude your motorcycle-shop feasibility study, for instance, by visiting possible store locations. Measure consumer traffic by observing each spot at the same hour on typical business days.

Observation should be your final step, because you now know what to look for. Observation not based on your understanding of the issues, problems, and possibilities could be a waste of time. A visit to the local nuclear power plant, for example, will be useful and informative if you have done your homework (if you understand how meltdown can occur or how safety systems work). Have a plan. Know exactly what to look for and jot down any key observations immediately. You might even take photos or make drawings, but don't rely on your memory.

Sometimes your observations can pinpoint a serious problem. Here is an excerpt from a report investigating poor management-employee relations at an electronics firm. In this case, the researcher's observations play a crucial role in defining the problem:

> The survey of employees did not reveal that they were aware of any major barriers to communication. The 75 percent of employees who have positive work attitudes said they felt free to talk to their managers, but the managers questioned said that only 50 percent of the employees felt free to talk to managers.
>
> The discrepancy appears to arise from the problems discussed in the section on morale: 50 percent of the employees don't know how management feels about them, and managers feel they don't hear everything that employees have to say. From questioning both sides, this writer concludes that both employees and management are assuming that messages have gotten

across, instead of checking to make sure they have. Managers are assuming that employees know they want to hear complaints and suggestions, so they aren't asking for them. Employees, on the other hand, mention that they would like to know how they are doing, but they never ask managers for their evaluations.

The problem involves perception and distortion from misinterpretation. Because managers don't ask for complaints, employees are afraid to make them, and because employees never ask for an evaluation, they never get one. Both sides have improper perceptions of what the other wants, and because of ineffective communications they don't know that their perceptions are wrong.

Effective research requires *educated* observation.

TAKING EFFECTIVE NOTES

Purpose

Your finished report will only be as good as the notes you take. Notecards are best because they are easy to organize. The point is to retain control at each step of the process. Begin by devising a system for recording your findings (an organized set of notes). For this purpose, a pack of notecards is a good investment. Put one idea on each card and give each card a code number in the upper right corner. By shuffling them, you can organize cards into related groups. You can pretty well figure out the organization of your report before you actually write it by arranging the cards in logical order on a large table top.

Instructions for keying your cards to your outline are discussed in the next chapter, "Writing and Documenting the Research Report."

Procedure

1. Begin by making bibliography cards, listing the books, articles, and other works you plan to consult. Using a separate notecard for each source, write the complete bibliographical information (as shown in Figure 12-9). This entry is identical to the bibliography entry you will include in the final draft of your report. (See Figure 13-1 for sample bibliography entries.) You can now begin recording information on your other notecards.

2. Skim the entire book, chapter, article, or pamphlet, noting the high points of information. (Review Chapter 4 for summarizing techniques.)

3. Go back over your reading and decide what to record. Use one card for each major item.

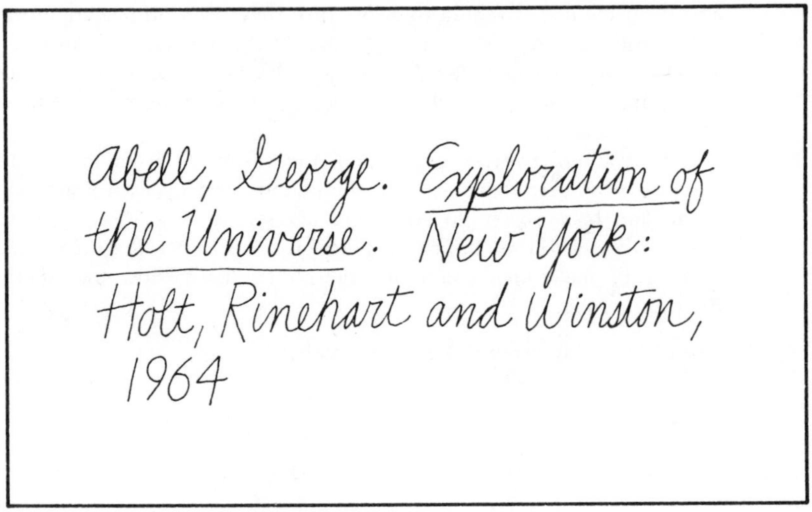
Abell, George. Exploration of the Universe. New York: Holt, Rinehart and Winston, 1964

FIGURE 12-9 A Bibliography Card

4. Decide how to record the item: as a quotation or as a paraphrase. In quoting, copy the statement word for word (as shown in Figure 12-10). Place quotation marks around all directly quoted material, even a phrase or a word

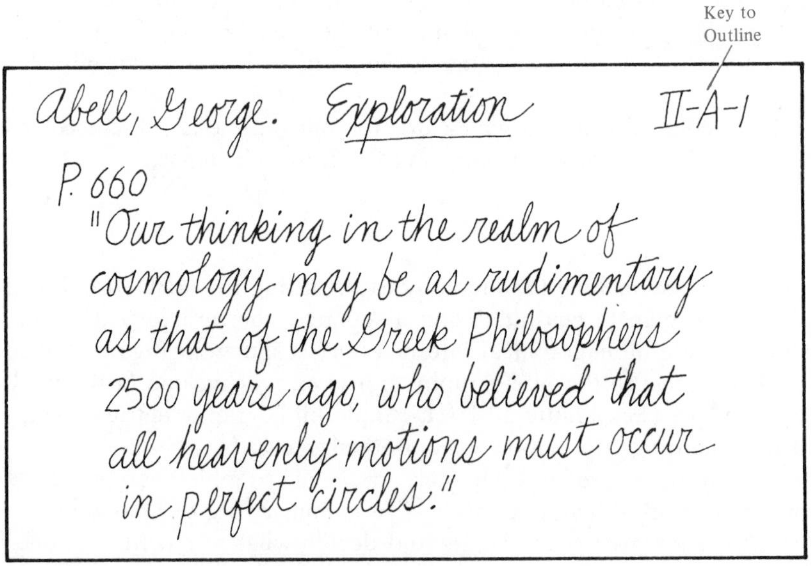
Key to Outline
Abell, George. Exploration II-A-1
P. 660
"Our thinking in the realm of cosmology may be as rudimentary as that of the Greek Philosophers 2500 years ago, who believed that all heavenly motions must occur in perfect circles."

FIGURE 12-10 Sample Notecard for a Quotation

used in a special way. Otherwise, you could forget to give proper credit to the author, and thereby face a charge of plagiarism (borrowing someone else's words without giving credit, either intentionally *or* unintentionally). Sometimes you will quote only sections of a sentence or paragraph. In this case, use an ellipsis — three dots (...) indicating that words have been left out of a sentence (as discussed in Appendix A). If you leave out the last part of the sentence, the first part of the next sentence, or one or more whole sentences or paragraphs, use four dots (....):

> The word *technical* can be misleading. ... Factual as well as abstract subjects can best be understood from a "technical" point of view.

If you insert your own comments within the quote, place brackets around them so as to distinguish your words from the author's (as discussed in Appendix A):

> "This job [aircraft ground controller] requires undivided attention for hours on end."

Avoid using direct quotes too frequently because your report will then be simply a collection of borrowed words. Instead, synthesize and condense information by paraphrasing and summarizing.

Figure 12-11 shows a paraphrased notecard entry based on the following original. The researcher condenses the message in his own words.

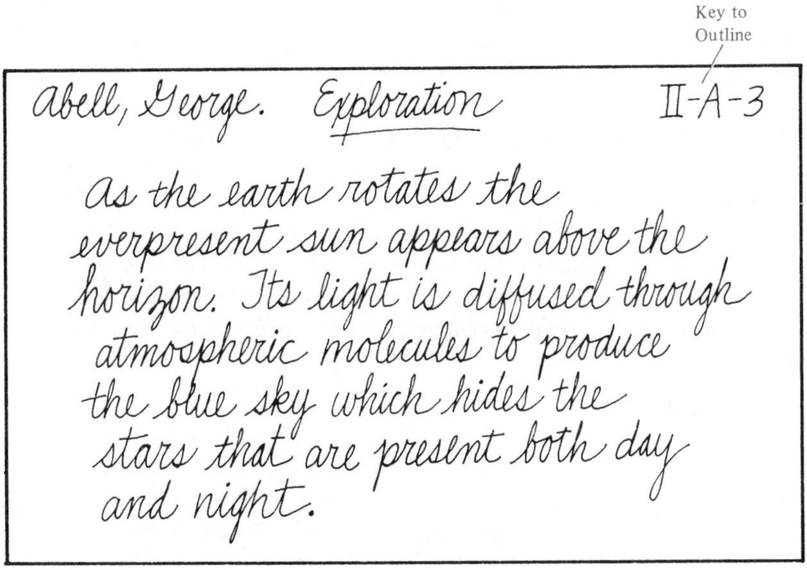

FIGURE 12-11   Sample Notecard for a Paraphrase

The sun is always present at some position on the celestial sphere. When the apparent rotation of the sphere carries the sun above the horizon, the brilliant sunlight scattered about by the molecules of the earth's atmosphere produces the blue sky that hides the starts that are also above the horizon. The early Greeks were aware that the stars were there during the day as well as at night.[6]

Most of your notes will be paraphrased, but some of your notecards may have both paraphrases and direct quotations. Paraphrased material needs no quotation marks, but it *must be footnoted* to indicate your debt to a particular source.

5. Be selective about what and how much you write in your notes. The summarizing techniques discussed in Chapter 4 should help. As you read, keep your original purpose in mind. Make notes of the main points related to your purpose, along with crucial statistics, figures, and other precise data and conclusions. Follow these guidelines: (a) Preserve the original message when quoting. Don't distort it by omitting vital information. Your aim is not to prove yourself correct but rather to uncover the facts. If the data disprove your theory, do not ignore them. (b) When you get an idea of your own, write it on a notecard immediately. Keep a few cards in your pocket or purse to record observations, questions, or ideas as they pop into your head.

CHAPTER SUMMARY

Research is designed to uncover facts. Anyone wishing to separate fact from opinion and to arrive at valid and convincing conclusions will do research.

There are many sources for research data. Primary sources include memory, personal observation, questionnaires, interviews, letters of inquiry, and business or technological records; secondary sources include books, articles, reports, brochures, films, and other publications. Overall, the library is your most valuable secondary source, so learn early how to use the library. Ask your reference librarian for help with the card catalog, reference works, indexes, government publications, computer-based information — and any other problem. To retain control of your research, be sure to take effective notes.

---

[6] This passage and those in Figures 12-10 and 12-11 (adapted) from *Exploration of the Universe,* second edition, by George Abell. Copyright © 1964, 1969 by Holt, Rinehart and Winston, Inc. Reprinted by permission of Holt, Rinehart and Winston.

EXERCISES

1. Each sentence below states a fact or an opinion. Rewrite all statements of opinion as statements of fact. (Remember that a fact can be measured, duplicated, or defined.)

   a. This set of instructions is useless.
   b. My vacation was too short.
   c. The salary for this position is $18,000 yearly.
   d. This desk calculator is priced reasonably.
   e. Our reforesting team planted 12,000 red pine seedlings last week.
   f. Our new delivery trucks get great gas mileage.
   g. This course has been very helpful.
   h. Electric heating is less efficient than oil heating.
   i. This room is much too small for our drafting and duplicating equipment.
   j. Our new company car is the most expensive of all compact sedans.

2. a. Find three articles about the same newsworthy event. Read each article carefully, separating fact from opinion. Underline all facts in blue and all opinions in red. Combine the articles and write a summary based exclusively on factual information. (Your instructor may provide the articles.)
   b. Locate a technical article from your field and underline the facts in blue and the opinions in red on your own photocopy. Submit the article to your instructor.

3. Make a list of all periodicals in your library that relate to your major field and arrange these titles in descending order of technicality or in descending order of immediate relevance to your work.

4. Assume that you belong to a statewide association of TV technicians. Summarize the interview on pages 304–306 in three to four hundred words for publication in the association's monthly newsletter. (Revised Chapter 4.)

5. Choose one situation from your life in which research could help you solve a problem or answer an important question. In a memo to your instructor, describe the problem or question and five sources you would consult to obtain suitable answers.

6. Revise the following questions so as to make them adequate for inclusion in a questionnaire:

   a. Do you think that a woman president could do the job as well as a man?
   b. Don't you think that euthanasia is a crime?
   c. Do you oppose increased government spending?
   d. Do you feel that welfare recipients are too lazy to support themselves?
   e. Do you think that teachers are responsible for the decline in literacy among American students?

f. Aren't humanities studies a waste of time?

g. Do you prefer Dipsi Cola to other leading brands?

7. *In class:* As a group, choose a familiar campus or community issue and compose a set of interview questions that could be given to people involved in the issue. Also, compose a questionnaire for sampling a broad range of attitudes about the same issue.

8. Arrange an interview with a successful person in your field. Make a list of general areas for questioning: current job opportunities, chances for promotion, salary range, training requirements, outlook for the next decade, working conditions, level of job satisfaction, and so on. Next, compose an organized set of interview questions, conduct the interview, and summarize your findings in a memo to your instructor.

9. Using the library and other sources, locate and record the required information for *ten* of the following items, and clearly document each source. (See pages 324–331 for documentation practices.)

a. the population of your home town

b. *The New York Times'* headline on the day you were born

c. the world pole-vaulting record

d. the operating principle of a diesel engine

e. the name of the first woman elected to the United States Congress

f. the exact value of today's American dollar in French francs

g. the comparative interest rates on an auto loan charged by three local savings banks

h. the major cause of small-business failures

i. the top-rated *compact* car in the *world* today

j. the origin of the word *capitalism*

k. the distance from Boston to Bombay

l. the definition of *pheochromocytoma*

m. an effective plant for keeping insects out of a home garden

n. the originator of the statement: "There's a sucker born every minute"

o. the names and addresses of five major United States paper companies

p. Australian job opportunities in your major field

q. the titles and specific locations of five recent articles on terrorism

r. the half-life of plutonium

s. a brief description of your intended occupation

13

Writing and Documenting the Research Report

DEFINITION

PLANNING THE REPORT
  Choice of Topic
  Focus of Topic
  Working Bibliography
  Statement of Purpose
  Background List
  Working Outline

GATHERING INFORMATION
  Beginning with a General View
  Skimming
  Selective Note Taking
  Designing Interviews, Questionnaires, or
        Letters of Inquiry
  Direct Observation

WRITING THE FIRST DRAFT
  Outline Revision
  Section-by-Section Development

DOCUMENTING THE REPORT
  Choosing a System
  Footnotes
    Footnoting Books
    Footnoting Articles
    Footnoting Miscellaneous Items
  Footnotes plus a Bibliography
  Lists of References
    Author-Year Designation
    Numerical Designation

WRITING THE FINAL DRAFT

CHAPTER SUMMARY

EXERCISES

DEFINITION

A research report records and discusses the findings of your search into primary and secondary sources. Its purpose is to provide the facts that will enable readers to derive an informed opinion about the subject. The information might well provide a basis for decision making, but your job in writing a research report is to *inform,* not to recommend, unless requested. Recommendations are part of the proposal or the analytical report (see Chapters 16 and 17). As we will see, research data often account for a substantial portion of a proposal or an analytical report. Our concern in this chapter, however, is only with the research report.

In Chapter 12, we discussed information sources and methods for recording data. Here we review and expand that discussion and offer a plan for completing the research report.

PLANNING THE REPORT

Choice of Topic

Your instructor will probably ask you to choose your own research topic. Instead of leaning on general topics that have been given exhaustive treatment (abortion, euthanasia, life on Mars, or the Bermuda Triangle), choose a more specific topic that concerns you or your community: job possibilities for graduates in your major, ways of improving working conditions on your part-time job, a survey of student opinions, and so on. The world is full of problems to solve and questions to answer. Make your research a worthwhile contribution.

Focus of Topic

Narrow your topic so you can discuss it effectively for your intended audience. Assume, for example, that you live in a seaside resort area whose residents are concerned about environmental damage that might be caused by proposed offshore oil-drilling rigs. Your broad topic would be "Offshore Oil Drilling." Since this topic is much too expansive to cover in a single research paper, you narrow it to "The Potential Environmental Effects of Offshore Oil Drilling at Georges Bank." But this topic also is too broad because it includes all of the following:

- effects of oil spills on the Cape Cod shoreline
- effects on area marine life
- effects of the influx of workers on Cape Cod
- effects of the building of harbor terminals and refineries on the shoreline
- effects of other changes from a vacation area to an industrial area

A more useful approach would be to discuss *one* of these subjects: "The Potential Effects of Offshore Oil Drilling on the Marine Life of Georges Bank." You now have a topic limited enough to cover adequately in a research report for a general audience.[1]

Remember that the length of your report is not important, as long as you deliver what you promise in your title.

Working Bibliography

When you have selected and narrowed your topic, be sure that you can find sufficient resources in your library and community. Make your bibliography early to avoid choosing a subject and an approach only to learn later that not enough sources are available. You might in fact choose your topic on the basis of a preliminary search for primary and secondary sources.

Conduct a quick search of the card catalogue, the periodical indexes, other reference books, and government publications. Using a separate notecard for each work, jot down the bibliographical information, as shown in Chapter 12. Many books you locate will contain bibliographies that should lead you to additional sources. With a current topic, such as offshore oil drilling, you might expect to find most of your information in recent magazine, journal, and newspaper articles. Your bibliography, of course, will grow as you read. Assess possible interview sources and include probable interviews in your working bibliography.

---

[1] For a marine biologist, however, you would need a much more specific focus such as "Potential Haddock Depletion from Oil Drilling on Georges Bank."

Statement of Purpose

Remember that you are doing more than collecting other people's views; you are screening and evaluating facts for a specific purpose. Identify your goal and your plan for achieving it. Then, your research will have a direction.

> The purpose of this report is to analyze data on the effects offshore oil drilling might have on Cape Cod's marine life. My study will compare characteristics of the Georges Bank ecosystem with those of existing offshore drilling sites.

Background List

Make a list of facts you already know about your topic. Some time spent brainstorming (see Appendix B) might produce some good ideas for investigation.

Working Outline

Now that you have identified your research goal, you need a road map. By partitioning your topic into subtopics, you emerge with a working outline:

  I. Marine Life on Georges Bank
 II. The Effects of Potential Massive Oil Spills
III. The Effects of Chronic Low-Level Oil Spills
 IV. Physical Characteristics Affecting Oil Drilling on Georges Bank

Each of these subtopics can be further divided in accordance with instructions in Chapter 7. Your working outline is only tentative; you will change it and add subtopics as your research progresses. Depending on your findings, the major topics and the organization of your paper may shift radically by the time you make your final outline. With your working outline on paper, you can begin your second step.

GATHERING INFORMATION

Beginning with a General View

Your research will be easier to control if you move from general to specific data. Encyclopedias are a good place to begin, because they contain general background information. Or, you might read a book or pamphlet that offers a comprehensive view of your subject before proceeding to specialized articles in periodicals. Technical dictionaries and newspaper or magazine articles can also provide background. Your reference librarian can help you find sources.

Skimming

Learn to skim. As you scan your bibliography cards you may find that you have a "mountain" of material to read. Don't be discouraged if you don't have time to read each source from cover to cover. With a little practice you can learn to skim pages quickly, while remaining alert for vital information. Before reading any book, look over the table of contents and scan the index. Check the introduction for an overview, a thesis, or a particular approach. In this way you can often locate information quickly, without plodding through the entire book. In reading long articles, look for major topic headings that may help you locate specific information. Short articles and pamphlets should usually be read in their entirety.

To skim effectively you have to concentrate. If you feel yourself drifting, take a break.

Selective Note Taking

Resist the temptation to copy or paraphrase almost every word so as not to miss anything. As you read, try to develop your own view of the subject instead of leaning too heavily on the ideas of others. Your finished report should be a combined product of your insights and the collected facts you have woven together.

Record the precise points of the message (names, dates, figures) along with the major ideas. Omit all supporting materials (examples, prefaces, background, etc.). Stay on track by following your outline, but remember that it can be modified as needed.

Designing Interviews, Questionnaires, or Letters of Inquiry

Your investigation may call for primary research. If interviews, questionnaires, and letters are appropriate, prepare for them well ahead of time.

Direct Observation

Now that you have read, questioned, and pondered, go and look for yourself — if direct observation is practical. If possible, take photos and draw maps, sketches, etc. Try to save this step for the end of your research so you know what to look for.

When your information is collected you should have a stack of notecards keyed to your working outline (see Chapter 7). You are now ready to write up your findings.

WRITING THE FIRST DRAFT

Outline Revision

Throughout your investigation you have relied on your working outline. By now you probably have uncovered new ideas and categories that were not part of your original outline. Before writing your first draft, examine your outline, and make any necessary changes.

Section-by-Section Development

Concentrate on only one section of the report at a time. After gathering the raw material you must create a finished product. Students often find that this is the most intimidating part of their research: pulling together a large body of information and presenting it coherently in a polished report. "What do I do now with all this stuff?" you ask. Don't despair! Those long hours of work need not lead to chaos and hysteria. Don't harm the quality of your effort as you approach the finish line by randomly throwing everything on the page, simply to be done with it. Remember that your final report will be the *only* concrete evidence of the work you have done. You can communicate a sense of the quality of your work by controlling each step.

Begin by classifying your notecards in groups according to the section of your outline to which each card is keyed. Next, find yourself a flat working surface (a large table or desk). Take the notecards for your background section and arrange them in logical order. Now, lay them out, one after another, in rows on your desk, as you would lay out playing cards. Thus armed with your outline, your statement of purpose, your own expertise, and your logically ordered notecards, you are ready to write your first section.

As you move from idea to idea, provide appropriate commentary and transitions and footnote each source of data. When you have completed your introductory section, proceed to the others, weaving ideas together by following the outline. Eventually you will have a first draft, and most of the battle will be over.

DOCUMENTING THE REPORT

In most research projects, you must draw on the information and ideas of others. It is vital that you credit each source of direct quotations, paraphrases, and visuals by using footnotes or a list of references. Proper documentation satisfies the professional requirements for ethics, efficiency, and authority.

Documentation is a matter of *ethics* in that the originator of an idea deserves to be acknowledged whenever that idea is mentioned. This is both a moral and a legal imperative. All published material is protected by copyright laws, which, like a patent for an invention, protect the originator. Therefore, failure to acknowledge your source when quoting or paraphrasing could make you liable to a charge of plagiarism, even if your omission was unintentional.

Documentation is also a matter of *efficiency*. It provides a continuous chain through which most of our world's printed knowledge can be readily located. If, for example, you make a brief reference to an article in a professional journal, your reference will enable an interested reader to locate that source easily and efficiently. Thus you add to the ongoing classification of knowledge that has made human progress possible.

Finally, documentation is a matter of *authority*. In making claims about situations or things ("A Mercedes-Benz is a better car than a Ford Granada") you are liable to be challenged with "Says who?" If you can produce data on road tests, frequency of repairs, resale value, workmanship, owner comments, and so on, you lend authority to your claim by showing that your opinion is based on *fact*. Your credibility increases in relation to the expert references you produce to support your claims.

Choosing a System

Documentation practices vary widely. Many disciplines, institutions, and industrial organizations publish their own manuals that spell out their documentation requirements. Most are listed in the card catalog under "Technical Writing." In writing your paper on food additives, for example, you might have used the American Psychological Association's *Publications Manual* or the American Medical Association's *Advice to Authors* in response to your readers' requests.

Some of the specific style manuals include the following:

> *Style Guide for Chemists*
> *Geographical Research and Writing*
> *Style Manual for Engineering Authors and Editors*
> *IBM Style Manual*
> *NASA Publications Manual*
> *Guide for Preparation of Air Force Publications*

When a specific format is not stipulated, you can consult one of three general style manuals for guidance. The Modern Language Association's *MLA Style Sheet* outlines documentation practices in the humanities. *A Manual of Style,* published by the University of Chicago Press, covers documentation in the humanities and related fields, as well as in the natural sciences. *The Style Manual,* published by the U.S. Government Printing Office, spells out documentation practices for government writing.

For a more detailed list and discussion, see John A. Walter, "Style Manuals," *Handbook of Technical Writing Practices,* 2 vols., ed. Jordan, Kleinman, and Shimberg (New York: John Wiley and Sons, Inc., 1971) 2:1267–1273.

This chapter discusses four different systems of documentation: footnotes only, footnotes plus bibliography, author-year designation, and numerical designation.

Footnotes

In most nonscholarly writing and in some scholarly writing, footnotes generally are collected in numerical order on one or more pages at the end of the report under the heading "Notes." (See Figure 13-4.)

In the text, place your footnote number immediately following and a half-space above the quoted or paraphrased material, like this: [1] . (See Figure 13-4.)

Except for a direct quotation, a note number should refer to no more than one paragraph. Sometimes you will have a paragraph in which all the data are yours except for the final two or three lines. In this case, use a hinge sentence ("Jones has shown that prolonged tetracycline treatment can lead to renal insufficiency") to separate your own from borrowed information. Otherwise, a footnote number at the end of the paragraph might lead readers to conclude that the entire paragraph is borrowed.

Unless readers request otherwise, use the following format for footnotes.

Footnoting Books

A footnote reference for a book should contain the following information (as applicable): author, title, editor or translator, edition, facts about publication (city, publisher, date), volume number, page number(s). Here are some examples:

SINGLE AUTHOR

[1]John Jones, <u>Thermodynamics: An Introduction</u> (New York: Phaeton Press, 1978), pp. 207–212.

TWO AUTHORS

[2]Harvey Keck and Roger Cayer, <u>Principles of Geology</u> (Chicago: Amherst Publishers, Inc., 1969), p. 174.

MORE THAN THREE AUTHORS

[3]Richard Fishberg et al., <u>Tumor Formation</u> (Montreal, Quebec: Temagami Publishers, Inc., 1969), pp. 108-109.

AUTHOR(S) NOT NAMED

[4]<u>The New Fisheries Directory</u>, 4th ed. rev. (Boston: Little, Brown and Co., 1976), p. 4.

AN EDITOR

[5]Martha Washburn, ed., <u>Forest Management Programs</u> (Seattle: Pinetree Press, 1974), p. 304.

A QUOTE OF A QUOTE

When your author has quoted from another work, begin with the original:

[6]Thomas Ashburn, <u>Miracle Drugs</u> (Washington, D.C.: Patriot Press, 1974), p. 84, as cited in Lester Furly, <u>The Treatment of Rheumatoid Arthritis</u> (New York: Holman Publishers, Inc., 1975), p. 116.

A SUBSEQUENT REFERENCE TO AN EARLIER NOTED WORK

[7]Jones, p. 173.

If two works by the same author have been cited earlier, include the title of this specific work:

[7]Jones, <u>Fluid Mechanics</u>, p. 173.

BACK-TO-BACK REFERENCES

[8]Ibid., p. 42.

Footnoting Articles

A footnote reference for an article should contain the following information (as applicable): author, article title, periodical title, volume and/or number, date, page number(s). Here are some examples:

A MAGAZINE ARTICLE

[9]Barbara Bonner, "Alaskan Nursing Opportunities," Woman's Life, November 1975, pp. 18–22.

If no author is given, list all other information.

A JOURNAL ARTICLE

[10]John Thackman, "Computer-Assisted Research," The American Librarian 51, no. 3 (October 1971), p. 42.

A NEWSPAPER ARTICLE

[11]Hannah Schmidt, "The Nuclear Gamble," Boston Times, 15 March 1976, p. 15.

If no author is given, list all other information.

Footnoting Miscellaneous Items

AN ENCYCLOPEDIA, DICTIONARY, OR OTHER ALPHABETIC REFERENCE WORK

[12]Technical Encyclopedia, 4th ed., s.v. "Hydraulics."

If the entry is signed, begin with the author's name.

AN INTERVIEW

[13]John Cooper, President of Datronics, in an interview regarding effects of offshore oil drilling, December 4, 1980.

A QUESTIONNAIRE

[14]Figures for the tabulation of responses to a questionnaire given to 150 freshmen at Idaho College, May 12, 1980.

A LECTURE

[15]Notes from Professor Harold Starkey's lecture on enzyme inhibitors, given at Marshall College, October 15, 1980.

Footnote other items (corporate or government pamphlets, reports, dissertations, and other unpublished work) as follows:

[16]Author (if known), Title (in quotes), Sponsoring Organization, date, page number(s).

In all cases, provide readers with enough information to locate the original. Each footnote is single spaced and separated from others by a double space.

Footnotes plus a Bibliography

Notes or footnotes list only those sources *cited* in your report. A bibliography may list *all* sources you consulted or only the works you've cited, arranged alphabetically by the author's last name (as the first item of the entry). To some extent, a bibliography repeats information contained in the footnotes; however, its purpose is to provide readers with a quick, alphabetical reference.

Figure 13-1 shows a partial bibliography for the report on pages 334–340. The entries include books, newspaper and magazine articles, research reports, and an interview. When the author's name is unknown, the title is listed alphabetically according to its first word (excluding *a, an,* and *the*). The bibliography page follows the note page in a report.

The footnotes-plus-bibliography format is used only in the humanities and a few other fields. For more information on this dual format, consult the *MLA Style Sheet.*

Lists of References

Documentation in technical writing often takes the form of an alphabetical or numerical list of references at the end of the report. These two simplified documentation formats are discussed below.

Author-Year Designation

In the author-year system, list your references cited in the report (along with references only consulted, if you wish). Arrange your items alphabetically, by authors' last names. For a book, give author, date, title (underlined), city of publication, and publisher. For an article, give author, article title (not underlined), periodical title (underlined), volume, number, and pages where the article appears. Capitalize only the first letter of work titles.

Figure 13-2 shows a sample author-year reference list that includes two works by the same author and an unpublished work.

For illustrations and other typical entries, see the University of Chicago Press *Manual of Style,* 12th ed., pages 384–388 or a style manual in your discipline.

In your report text, when you refer to the work, insert the author's name and publication date in parentheses:

BIBLIOGRAPHY

Blocke, John. "Geologist Balderdashes Hopes for Offshore
    Drilling." The Cape Codder, 1 August 1974, p. 3.

Boesch, Donald F., et al. Oil Spills and the Marine Environ-
    ment. Cambridge: Ballinger Press, 1974.

"'Clean Atlantic' Vows Protection Against East Coast Oil
    Spills." Cape Cod Times, 25 February 1976, p. 1.

Dillin, John. "Offshore Oil: America's Trillion-Dollar
    Decision." Christian Science Monitor, 28 June 1974,
    p. 2.

Flanagan, Dennis. "The Energy Resources of the Earth."
    Scientific American, 16 August 1971, pp. 19-24.

McBride, Arthur. Former Oil Geologist with Exxon. Interview
    regarding effects of offshore drilling. Harwich, Mass.,
    on May 7, 1976.

Offshore Oil Task Group. The Georges Bank Petroleum Study:
    Impact on New England Environmental Quality of Hypo-
    thetical Regional Petroleum Development. Boston: Mass.
    Institute of Technology, Vol. II, Report No. MITSG 73-5,
    1973.

Oil and The Environment: The Prospect. Houston, Shell Oil
    Co., 1972.

"Oil Drilling." The Harpers Encyclopedia of Science. New
    York: Harper and Row Publishers, 1963.

Raven, Peter H., and Helena Curtis. Biology of Plants. New
    York: Worth Publishers, 1970.

Vorse, Heaton. "Southwind." Provincetown Advocate, 16 May
    1975, p. 5.

"We Need Guarantees." Vertical File on Oil, Cape Cod Com-
    munity College Library, March 18, 1975.

FIGURE 13-1   A Sample Bibliography

REFERENCES

Albey, J. 1975. <u>Modern career choices</u>. San Francisco:
    Hamilton Publishers.

Crashaw, H. et al. 1975. <u>Technology and careers</u>. Boston:
    Little, Brown and Co.

_____. 1977. <u>Careers in the natural sciences</u>. Dallas:
    Bovary Press, Inc.

Donne, M. 1979. Job prospects for college graduates. <u>Edu-
    cation Digest</u> 28, no. 2:86–89.

Marsh, A., and Smith, C., eds. 1972. <u>Advice for the job
    seeker</u>. Boston: Arngold Press.

Peters, Claire. 1978. Wage scales for women in civil engi-
    neering. Master's thesis, Brandon University.

FIGURE 13-2   An Author-Year Reference List

> Seventy-five percent of chemistry technicians interviewed expressed a desire for further training (Albey, 1975).

With this system, you can also include page numbers in the text:

> Seventy-five percent of chemistry technicians interviewed expressed a desire for further training (Albey, 1975:114).

When you mention the writer's name in the text of your discussion, do not repeat it within the parentheses:

> Crashaw et al. (1977:93–94) claim that medical technologists "feel challenged by their work."

In cases where you are citing two works published by the same author(s) in the same year, insert an "a" and a "b" immediately after the dates, both in your reference list and in your textual citation.

> (Jones, J., 1975a)

As an illustration in actual practice, the sample report on pages 497–517 has an author/year list of references and citations.

Numerical Designation

In the numerical reference system, you number your references alphabetically or in the order first cited in the text. (Use one arrangement or the other, consistently.) Otherwise the general format is similar to the author-year system. Figure 13-3 shows some typical entries for a reference list arranged in order of first citation in the text.

With each reference numbered upon first citation, you then can use the same number in your text whenever you cite that particular work:

> Seventy-five percent of chemistry technicians interviewed expressed a desire for further professional training. (2:83)

Adding page numbers helps clarify the reference.

As an illustration in actual practice, the sample report on pages 519–538 has a numbered list of references and citations.

Whichever system of documentation you choose, be sure to use the same format consistently throughout your report.

WRITING THE FINAL DRAFT

When you have written and fully documented your first draft, you are ready to write your final draft. Polish your early draft by adding transitions, dia-

REFERENCES

1.   Donne, M. Job prospects for college graduates. Educa-
     tion Digest 28, no. 2 (1979):86–89.

2.   Albey, J. Modern career choices. San Francisco:
     Hamilton Publishers, 1975.

3.   Crashaw, H. et al. Careers in the natural sciences.
     Dallas: Bovary Press, Inc., 1977.

4.   Marsh, A. and Smith, C., eds. Advice for the job
     seeker. Boston: Arngold Press, 1972.

5.   Crashaw, H. et al. Technology and careers. Boston:
     Little, Brown and Co., 1975.

FIGURE 13-3   A Reference List Arranged in Order of First Citation

grams, examples, commentaries, and topic and area headings. Revise sentences that are awkward or unclear, or that provide little information.

Figure 13-4 shows the first section (introduction) of a student-written research report. Study it carefully to see how the writer has woven multiple sources together, using footnotes for documentation. (Normally, of course, the footnotes would be placed at the end of the report, not after the introduction.)

AUDIENCE ANALYSIS

Carol Connolly, a member of the Association for the Preservation of Cape Cod, is preparing the report in Figure 13-4 for concerned citizens of that area and all of New England as well. Because she is addressing a most diverse and general audience, she writes at the lowest level of technicality.

Her introduction is designed to provide a comprehensive background to the issue by discussing the local geography, the energy crisis, and the potential benefits and hazards of drilling on Georges Bank.

To ensure that all readers will understand the report, Carol begins by defining key terms (*oil well, continental shelf*). Then she sparks reader concern by giving sobering facts about energy consumption in the United States (and New England). By citing various authorities, Carol lends credibility to her report and thus readers are apt to take it more seriously.

Abiding by the rules of ethical research, the author is rigorously *objective* and gives a balanced assessment of "pros" and "cons." She allows no bias on her part to influence her readers' attitudes. She lets the facts speak for themselves, thus allowing her readers to make *educated* judgments. Aware of the fact that the primary purpose of a research report is to *inform*, not to *persuade*, she supplies hard information and presents it in such a way as to hold the interest of her audience. In other words, more than simply presenting bare facts, she consolidates her data to provide coherent and sensible explanations.

CHAPTER SUMMARY

Follow these steps in planning your report:

1. Choose a topic that is interesting and sufficiently narrow.
2. Make a working bibliography before you proceed.
3. Compose a clear statement of purpose and a working outline.
4. Move from general sources to particular ones, skimming your material to save time.
5. Take selected notes, keyed to your outline, with all sources clearly identified.
6. Plan early for interviews, questionnaires, and inquiry letters.
7. If possible, conclude by having a look for yourself.

THE POTENTIAL EFFECTS OF OFFSHORE OIL DRILLING

ON THE MARINE LIFE OF GEORGES BANK

INTRODUCTION

In offshore oil drilling, a well is drilled into the suspected oil-containing sediments (bottom residue) lying under the ocean's floor. An oil well is a steel-encased hole that serves as a pipeline from the underground petroleum source to the surface.[1]

The United States has potential oil reserves under its continental shelf. This shelf surrounds the entire land mass of North America and is an extension of, and connected with, the adjacent coastal plain.[2] It extends from the low-water line to the area where there is a marked increase in steepness toward the deep sea floor.[3] Off the East Coast of the United States, the shelf is approximately seventy-five miles wide.[4] Directly east of Cape Cod lies one of the potentially richest oil reserves anywhere under the continental shelf. This area, known as Georges Bank, also has some of the best fishing grounds off the United States coast.

In 1973, citing our energy squeeze, President Nixon

FIGURE 13-4   An Introduction to a Research Report

.2

called for immediate exploration of available energy sources,
including increased drilling on the shelf.[5] By 1976, the
United States was using 30 percent of the oil produced world-
wide.[6] New England alone consumed 40 percent of the nation's
oil products.[7] The oil we are consuming today is a non-
renewable source of energy.[8] Projections of the world's
available oil show that the supply will be exhausted by the
year 2075,[9] making it imperative that we develop alternative
energy sources. The oil that can be extracted from the continental shelf will help to lengthen and stabilize our long
period of transition from fossil fuels (oil and gas) to new
energy sources. Until such a transition can take place, our
dependence on foreign oil -- more than 50 percent of United
States oil is imported[10] -- makes us vulnerable. Oil
drilling on the continental shelf could help decrease our
dependence on oil imports.

There are mixed opinions about whether a marketable
quantity of oil exists under Georges Bank. John Brown, a
geologist in Dennisport, Massachusetts, feels that the
existence of extensive reserves is possible but not
probable.[11] Donald Marshall, an oil geologist in Hyannis,

FIGURE 13-4 (*Continued*)

3

Massachusetts, is uncertain about whether oil exists under Georges Bank.[12] On the other hand, the U.S. Department of the Interior has reported that the Atlantic continental shelf may contain 5.5 billion barrels of oil.[13]

If oil is indeed present, drilling on Georges Bank would exert a positive economic influence on the New England area, according to Mr. Marshall. Charter fishing should increase as a result of increased marine life, and the fishing industry, in general, would be revitalized. He cited the thriving fishing industry in the North Sea and elsewhere as evidence:

> Once the oil-drilling activity began, the fish population -- which had nearly disappeared due to overfishing -- began to return: especially cod, in the North Sea, the Java Sea, and the Persian Gulf.[14]

The legs of drilling platforms function as artificial reefs, where marine life seems to flourish.[15]

Some critics argue that the benefits derived from increased marine life would be wiped out in the event of a major drilling accident. Defenders of the project, however, argue that the waters and sands of Padre Island, the nation's largest national seashore, remain perfectly clean; this area forms the western end of the Gulf of Mexico, which contains

FIGURE 13-4 (*Continued*)

4

the world's most extensive offshore oil-drilling facilities.[16]

To date, offshore oil production has accounted for 2.1
percent of all oil pollution sources.[17] The oil companies
have a 99.95 percent safety record for all offshore drilling
activities. Out of approximately 18,200 producing oil wells
in the United States coastal waters, nine have caused "major"
spills.[18]

Against this general background, the purpose of this report is to analyze information on the possible effects of offshore oil drilling on Cape Cod's marine life. It will compare
characteristics of the Georges Bank ecosystem with those of
already existing offshore oil drilling locations. The report
is intended to inform the general reader about the most recent
data on this subject.

My information sources include research reports, newspaper and magazine articles, interviews with two local geolo-
gists, a statement by the Association for the Preservation of
Cape Cod, scientific encyclopedias, and a botany textbook. A
glossary of terms is included at the end of this report.

FIGURE 13-4   (*Continued*)

5

The major topics to be discussed include:

   I. Marine Life on Georges Bank

  II. The Effects of Potential Massive Oil Spills

 III. The Effects of Chronic Low-Level Oil Spills

 IV. Physical Characteristics Affecting Oil
      Drilling on Georges Bank

FIGURE 13-4 (*Continued*)

6

NOTES

[1] *The Why and How of Undersea Drilling* (Washington, D.C.: The American Petroleum Institute, 1970), p. 4.

[2] *McGraw-Hill Encyclopedia of Science and Technology*, 2nd ed., s.v. "Oil."

[3] *The Harpers Encyclopedia of Science*, s.v. "Oil Drilling."

[4] McGraw-Hill Encyclopedia.

[5] Andrea Stevens, "Biggest Oil Spills Caused by Tankers," *Cape Cod Standard Times*, 20 September 1973, p. 5.

[6] John Brown, geologist, in an interview on the effects of offshore drilling, May 7, 1976. See Appendix A for full text of interview.

[7] Charles Koeler, "Geologist: Offshore Oil Drilling 'Has to Happen,'" *Cape Cod Standard Times*, 28 July 1975, p. 1.

[8] Brown interview.

[9] M. King Hubbert, "The Energy Resources of the Earth," *Scientific American* 224, no. 3 (May 1969), p. 69.

[10] Donald Marshall, geologist, in an interview on the effects of offshore oil drilling, May 7, 1976. See Appendix B for full text.

[11] John Blocke, "Geologist Balderdashes Hopes for Offshore Oil," *The Cape Codder*, 1 August 1974, p. 3.

[12] Marshall interview.

[13] Paul Kemprecos, "Fishing," *The Cape Codder*, 4 April 1974, p. 8.

FIGURE 13-4   (*Continued*)

7

[14]"We Need Guarantees," The Chronicle, 5 May 1975, p. 2.

[15]Marshall interview.

[16]"Oil Rigs Off Cape Cod May Aid Fishing Yields," Cape Cod News, 1 August 1973, p. 3.

[17]John Dillin, "Pollution Challenge to World Underlined,' Christian Science Monitor, 28 June 1974, p. 2.

[18]"Favors Oil Drilling," Cape Cod Standard Times, 24 March 1975, p. 3.

FIGURE 13-4   (*Continued*)

Follow these steps in writing your report:

1. Revise your working outline.
2. Follow an introduction-body-conclusion structure.
3. Concentrate on only one section at a time.
4. Credit each data source using an acceptable documentation format.
5. Write your final draft.

EXERCISES

1. *In class:* Assume that an area within twenty miles of your college is being considered as a site for a nuclear power plant. Working in small groups, narrow the topic "Nuclear Energy" to one that will serve a real and specific information need. Next, write a collective statement of purpose, based on an intended audience that your group has identified. Finally, as a class, compare the purpose statements from individual groups for usefulness and quality of focus.

2. *Organizing the research and writing the report:* Prepare a research report by completing the following steps. (Your instructor might establish a timetable for completion of the phases outlined below.)

Phase One: Preliminary Steps

a. Choose a topic of *immediate practical importance,* something that concerns you or your community directly. (The list of research possibilities on pages 342–343 should give you some ideas.)
b. Identify a specific audience and its intended use of your information.
c. Narrow your topic, and check with your instructor for approval and advice.
d. Make a working bibliography to ensure that your library and area contain sufficient primary and secondary resources. Don't delay this step!
e. Make a list of things you already know about your topic.
f. Write a clear statement of purpose and submit it in a proposal memo (pages 436–438) to your instructor.
g. Make a working outline.

Phase Two: Collecting Data

a. In your research, move from the general to the specific; begin with broad discussions in general reference works in order to develop a good overview.
b. Skim your material, looking for high points. In this way, you will cover the most ground in the least time.

c. Take selective notes. Don't write everything down! Use notecards.

d. Plan and administer (or distribute) questionnaires, interviews, and letters of inquiry.

e. Whenever possible, conclude your research with direct observation.

Phase Three: Organizing Your Data and Writing Your Report

a. Revise and adjust your working outline, as needed.

b. Compose an audience-and-use analysis like the sample shown on page 333.

c. Follow the introduction-body-conclusion format in writing your report.

d. Concentrate on only *one* section of your report at a time.

e. Fully document all sources of information.

f. Write your final draft.

g. Proofread carefully and add all needed supplements (title page, letter of transmittal, abstract, summary, appendix, glossary).

Due Dates:

List of possible topics due:

Final topic due:

Proposal memo due:

Working bibliography and working outline due:

Notecards due:

Copies of questionnaires, interview questions, and inquiry letters due:

Revised outline due:

First draft of report due:

Final draft of report with all supplements and full documentation due:

Here are some possible research projects for Phase One:

a. Trace the history of your major department through its stages of planning, formation, staffing and growth as part of your college. Do your findings show a steady growth, periods of decline, or some other trend? What is the outlook for your department within the institution?

b. Trace the history of your major as an academic specialty. Which institutions offered the first courses or the first major? To what needs were institutions responding in offering this major? Where are the highest-ranked programs now offered? What is the outlook for further growth in this discipline?

c. Identify a modern discovery in your field that dramatically advanced the state of the art (as the discovery of antibiotics revolutionized medical treatment). Establish how the discovery was made and trace its beneficial effects along with any negative effects (antibiotics can cause severe allergic reactions and organic complications, and have caused resistant strains of organisms to develop).

d. Conduct a survey of student opinions about some proposed changes in the curriculum or another controversial campus issue. If possible, compare your findings with national statistics about student attitudes on this issue.

e. Discover how faculty hiring and tenure decisions are made at your school. What are the primary criteria, and how are they ranked (e.g., research activity, publication, quality of teaching, service to the department and college, community service)? Trace the history of tenure decisions over the last ten years. Do any trends emerge?

f. Find out how the quality of your local water supply measures up to national standards. Has the quality increased or decreased over the past ten years? What are the major factors affecting local water quality? What is the outlook for the next decade?

g. Trace the modern history of zoning ordinances in your town. How has residential zoning affected the character of the town? How has business zoning affected the character and economy of the town? What is the outlook for future zoning changes?

h. Find out which major department on your campus attracts the most students, and why. Has this department always been so popular? Will it continue to be? Trace its recent history.

i. Discover what qualities most employers look for in a job candidate. Have employers' expectations changed over the last ten years? If so, why?

j. Find out which geographic section of the United States is experiencing the greatest economic and population growth. What are the major reasons for this growth? Trace the recent history of this change.

k. Some observers claim that productivity in American industry is on the decline. Find out whether this claim is valid. If it is, identify the reasons. Include both primary and secondary research data.

l. Assume that you and a business partner have developed a desk-top duplicating machine (or some other product) that can be manufactured at low cost. What are the intricacies involved in applying for and getting a patent for your product? Include both primary and secondary research data.

m. Trace the typical promotional route for people in your specialty. What are the possibilities and essential requirements for being promoted to an executive position?

n. What is the present status of women and minorities in your specialty? Assess their hiring and promotional opportunities, relative salaries, and percentages in executive positions. Do comparative data over the past decade suggest a trend toward equal opportunity? (The article on pages 73–76 should serve as a model.)

o. What fields will offer the best job opportunities through the next decade? Where will you and your fellow majors fit into this market?

p. What are the prospects for a lifetime career in your field? What percentage of your colleagues now change careers, and why?

SPECIFIC
APPLICATIONS

14

Writing Effective Letters

DEFINITION

PURPOSE OF LETTERS

ELEMENTS OF AN EFFECTIVE LETTER
Introduction-Body-Conclusion Structure
Required Major Parts
Heading
Inside Address
Salutation
Letter Text
Complimentary Closing
Signature
Specialized Parts
Typist's Initials
Enclosure Notation
Distribution Notation
Postscript
Appropriate Format
Accepted Letter Form
Plain English
"You" Perspective
Clear Purpose

WRITING VARIOUS TYPES OF LETTERS
Letter of Inquiry
Introduction
Body
Conclusion
Revision and Final Touches

Letter of Complaint
Introduction
Body
Conclusion
Letter of Instruction
Introduction
Body
Conclusion

WRITING THE RÉSUMÉ AND JOB
APPLICATION LETTER
Job Prospecting: The Preliminary Step
Being Selective
Launching Your Search
Being Realistic
The Résumé
Name and Address
Career Objectives
Educational Background
Work Experience
Personal Interests, Activities, Awards,
and Special Skills
References
Composing the Résumé
The Job Application Letter
Your Image
Targets
The Solicited Letter
The Unsolicited Letter
The Prototype

SUPPORTING YOUR APPLICATION
  Your Dossier
  Interviews
  The Follow-up Letter
  The Letter of Acceptance
  The Letter of Refusal

CHAPTER SUMMARY

REVISION CHECKLIST

EXERCISES

DEFINITION

Whereas a report may be compiled by a team of writers and read by many readers, a letter usually is written by one writer for one or more definite readers. A letter is more personal than a report. Therefore, your attitude, as suggested by the tone of your letter, is a major ingredient of your message. Because most letters are written to elicit a definite response, you want the reader to be on your side. Also, because your signature certifies the statements in your letter (which may be used later as a legal document), the need for precision in letter writing is crucial.

PURPOSE OF LETTERS

Whether you work for yourself or for a company, in any nonmenial job you will have to write letters. Here are some common types of letters you will write on the job. The broad purpose of each type is to inform and persuade the reader.

  – sales letters designed to create interest in a product or service
  – letters of instruction outlining a procedure to be carried out by the reader
  – letters of recommendation for friends, fellow workers, or past employees
  – letters of transmittal (cover letters) to accompany reports and other documents that you mail out
  – general business letters describing progress on a project, requesting assistance, ordering parts or tools, confirming meeting times, and so on
  – letters of inquiry, asking about the cost or availability of a product,

requesting advice for solving a problem, soliciting comments about a job applicant, and so on

– claim letters written to complain about disappointing service or faulty products and to request adjustment

You may also need to write letters of response to those letters received by your company.

Most people write letters well before beginning their careers. As a student, for instance, you may write to request data for research projects or to learn about certain schools or jobs. You might also write letters to apply to colleges, to compete for scholarships or foreign study programs, or to join a campus organization. These application letters are considered important for good reasons: they provide evidence of your verbal sophistication, your talent for clear self-expression, your level of confidence, your sensitivity to your audience, your attention to detail, your level of maturity, your ability to recognize important points, your mastery of logical reasoning, and your level of personality development. From time to time, as a consumer, you might have complaints about defective items or disappointing service. Your complaint letter will be designed to express your dissatisfaction and to secure a fair adjustment. Finally, as a job applicant, you will write letters that may well be a key to your success.

A full discussion of letter writing would more than fill a textbook. In this chapter, therefore, we will discuss only four common types of letters: the letter of inquiry, the letter of complaint, the letter of instruction, and the letter of application, along with its accompanying résumé. Other useful types — the letter of transmittal and letter reports — are discussed in Chapters 10 and 15. Regardless of type, however, all letters share certain elements.

ELEMENTS OF AN EFFECTIVE LETTER

Remember this rule: *never send a letter until you genuinely feel good about signing your name to it* because your signature certifies that you approve of the contents. Use the guidelines below for judging the adequacy of your letters before you sign them.

Introduction-Body-Conclusion Structure

Structure all letters to include (1) a brief *introduction* paragraph in which you attract your reader's attention by identifying yourself and stating your purpose; (2) one or more *body* paragraphs containing the specific details of

your inquiry, application, complaint, or instructions; (3) a *conclusion* paragraph in which you tie your message together, offer to provide more details
or appear for an interview, and courteously encourage your reader to act. For
readability, keep your paragraphs short (usually fewer than eight lines). If
your body section is long, divide it into shorter paragraphs, as shown in
Figure 14-1. Notice that this body section is broken down into four specific
questions for easy answering. Figure 14-2 shows the response to the preceding
letter.

Required Major Parts

All letters have six parts, which, in order from top to bottom, appear as
follows: headings, inside address, salutation, the letter text (with introduction,
body, and conclusion), complimentary closing, and signature.

Heading

If your stationery has a company letterhead, simply include the date, two
spaces below the letterhead, as shown in Figure 14-2. On plain stationery,
include your own heading, as shown in Figure 14-1 and here:

```
Street Address           154 Seaweed Lane
City, State  Zip Code    East Harwich, Massachusetts 02163
Month Day, Year          December 24, 1981
```

Write out in full all items in the heading. Avoid abbreviations throughout
your letter. Depending on the length of the letter, place your heading eight to
ten spaces below the top of your page and far enough to the left that the
longest line abuts your right margin.

Inside Address

Four to six spaces below your heading, and 1½ inches from your left margin,
is your letter's inside address.

```
Name and Title of Reader (and position)   Dr. Marsha Mello, Dean
Company Name                              Western University
Street Address (if applicable)            Muncie, Indiana 13461
City, State  Zip Code
```

Whenever possible, address your letter to a specifically named reader, using
that reader's appropriate title (Attorney, Major, etc.), but don't be redundant
by writing "Dr. Marsha A. Mello, Ph.D."; use Dr. or Ph.D., Dr. or M.D.
Abbreviate only titles that are routinely abbreviated (Mr., Ms., Dr., etc.).
Titles such as Captain are written out in full.

154 Seaweed Lane
East Harwich, Massachusetts 02163
July 15, 1981

Land Use Manager
Eastern Paper Company
Waldoboro, Maine 04967

Dear Manager:

Mr. Melvin Blotter, your sales representative, has told me
that Eastern Paper Company makes available certain parcels
of its lakefront property in northern Maine for leasing to
the general public. Because I am interested in such a
leasing arrangement, please answer the following questions
for me.

    1. Does your company have any lakefront parcels in
        highly remote areas?

    2. What is the average size of a leased parcel?

    3. How long does a lease remain in effect?

    4. What is the yearly leasing fee?

I will greatly appreciate your answers to these questions,
along with any other details you might send along.

                      Yours truly,

                      John M. Lannon

                      John M. Lannon

P.S. I am planning a trip to Maine the week of August 5-11
      and would be happy to stop by your office any time you
      are free.

FIGURE 14-1   A Letter of Inquiry

Eastern Paper Company
Waldoboro, Maine 04697

July 25, 1981

Mr. John M. Lannon
154 Seaweed Lane
East Harwich, Massachusetts 01263

Dear Mr. Lannon:

This is in answer to your recent inquiry about leasing
lakeside lots in Maine.

We have no lands for lease in what you would classify as
"highly remote areas." The most remote area is on the
northern shore of Deerfoot Lake where presently you would
have to reach the camp lot by boat. Leases on these 30,000
square foot parcels are renewable on a yearly basis each
June. The leasing fee is approximately $250.00 per year. We
have a limited number of leases available, but you would have
to visit our office to get meaningful information about exact
locations. If you are in the area, you may drop by any weekday morning before 11 o'clock.

The Land Use Regulation Commission in Augusta, Maine,
regulates campsite leasing and you would need to apply to
this body for a building permit including soil tests, etc.,
before you would be allowed to build.

Thank you for your inquiry.

Yours truly,

EASTERN PAPER COMPANY

A. B. Coolidge
A. B. Coolidge
Townsite Manager

ABC/de

cc: Mr. Blotter

FIGURE 14-2   A Letter of Response to an Inquiry

Salutation

Your salutation, aligned two spaces below the inside address, is a greeting to your reader. Begin your salutation with "Dear" and end it with a colon. Include your addressee's full title.

```
Dear Ms. Jones:
Dear Professor Smith:
Dear Dr. Brown:
Dear Senator Smiley:
```

If you don't know your addressee's name or sex, use the person's position: "Dear Manager." When addressing several people at once, use "Gentlemen" or "Ladies" or "Ladies and Gentlemen."

Letter Text

Begin your letter text two spaces below your salutation; use single spacing within paragraphs and double spacing between them.

Complimentary Closing

Place your complimentary closing two spaces below the concluding paragraph of your letter text and aligned with your heading. Any conventional closing that is polite and not overly intimate or "gushy" will do. Select your closing on the basis of your relationship to the addressee. The following possibilities are ranked in decreasing order of formality.

```
Respectfully,
Yours truly,
Sincerely yours,
Sincerely,
Best wishes,
Warmest regards,
Cordially,
Best,
```

The complimentary closing is always followed by a comma.

Signature

Type your full name and title four spaces below and aligned on the left with your complimentary closing. Sign your name in the space between the two.

Sincerely yours,

Martha S. Jones

Martha S. Jones
Personnel Manager

Your signature is your personal stamp, indicating your approval of and re-sponsibility for the contents of the letter (even if it's typed by a secretary). If you are writing as a representative of a company or group that bears legal responsibility for your letter, type the company's name in full caps two spaces below your complimentary closing; place your typed name and title four spaces below the company name, and sign in the space between.

Yours truly,

LEVEL BROTHERS COMPANY

Leslie L. Foye

Leslie L. Foye
Collection Officer

Specialized Parts

Each letter contains the six major parts. In addition, some require one or more of the following specialized parts. Examples of these parts can be seen in the sample letters throughout the chapter.

Typist's Initials

If your letter is typed by another person, your initials (in capital letters) and your typist's (in lower-case letters) should be indicated two spaces below your typed signature and abutting the left margin.

JJ/pl

Enclosure Notation

When other documents accompany your letter, add an enclosure notation one space below the typist's initials and abutting the left margin.

Enclosure (or Enclosures 2, etc.)

Distribution Notation

If carbon copies of your letter are to be distributed to other readers, so indicate one space below the enclosure notation.

```
cc:  Office file
     Mr. Blotter
```

Postscript

A postscript is designed to draw your reader's attention to a point you wish to emphasize. Place your postcript two spaces below any other notations, and against your left margin.

```
P.S.  This product has outstanding potential for capturing the
adolescent market.
```

Use the postscript sparingly.

Appropriate Format

Follow the format instructions in Chapter 10 for each letter. Here is a brief summary of those instructions: (1) Use high quality 20-lb. bond, 8½-by-11 plain white stationery with a minimum fiber content of 25 percent. (2) Type neatly, avoid erasures, retype a smudged page, and make sure you have clear typewriter keys and a fresh ribbon. (3) Use uniform margins, spacing, and indentation: frame your letter with a 1½-inch top margin and side and bottom margins of 1 to 1¼ inches; unless your letter is very brief (three or four lines) single space within paragraphs and double space between them; avoid hyphenating at the end of a line.

If your letter extends to more than one page, begin the second page five spaces from the top, abutting your left margin:

```
P. 2, Walter James (addressee), June 25, 1981
```

Begin the text of your second page two spaces below this notation. Place at least two lines of your paragraph at the bottom of page one, and at least two lines of your final text on page two.

Your 9½-by-4⅛-inch envelope should be of the same high quality as your stationery. Center your reader's address horizontally and vertically, and single space if it occupies three lines or more. Again, use only accepted abbreviations. Place your own single-spaced address in the upper left corner, as shown in Figure 14-3.

```
Marvin Hanley
154 Seaweed Lane
East Harwich,
   Massachusetts 02163


                         Dr. Marsha Mello, Dean
                         Western University
                         Muncie, Indiana  13461
```

FIGURE 14-3   A Sample Envelope

Accepted Letter Form

Although several acceptable letter forms exist, and your own company may have its own requirements, we will discuss two most common forms: the semiblock form, with no indentations (Figure 14-4), and the modified block form, with the first sentence of each paragraph indented five spaces (Figure 14-5). The letters of inquiry and response earlier in this chapter are examples of semiblock and modified block forms, respectively.

Plain English

First, we should distinguish between the *conventions* of letter writing and *clichés* or *catchy phrases*. The conventions of letter writing concern format, form, letter parts, and general structure — the mechanical elements that govern the appearance of your letter and the parts of your message. Letter writing conventions exist because, over the years, people have agreed on a set of rules for acceptable practice in composing neat, attractive, and informative letters. One convention, for example, requires that you type your name (and position) under your signature. This procedure became part of accepted and required practice since it made for a handy way of clarifying your identity, especially if you wrote in a scrawl. Other conventions came into being because they too were useful.

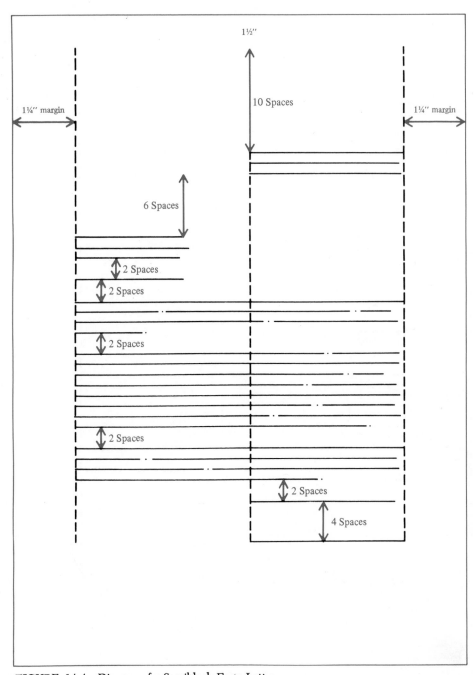

FIGURE 14-4   Diagram of a Semiblock Form Letter

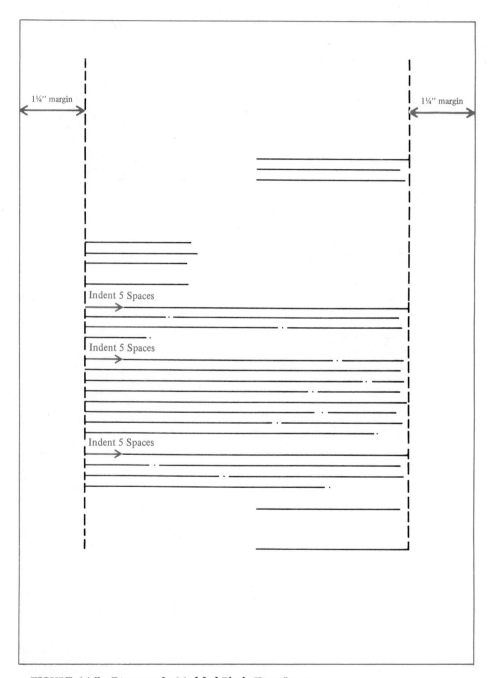

FIGURE 14-5   Diagram of a Modified Block Form Letter

Clichés and catchy phrases, on the other hand, are time-worn, pompous, and overblown phrases that some writers think they need in order to make letters seem important. Here is a typically overwritten closing sentence:

```
Humbly thanking you in anticipation of your kind cooperation, I
remain
```

Faithfully yours,

Marvin Hanley

Marvin Hanley

Although no one speaks this way, some writers lean on such heavy prose instead of simply writing, "We will appreciate your cooperation." Not only is the earlier example overworked; it is so clearly full of exaggeration that it sounds insincere. Unfortunately, statements like this and the others listed below are popular because they are easy to use. Like TV dinners they are effortless but not very impressive. Although you might save yourself some gray hair by throwing such predigested stuff into your letter, your reader might well conclude that you're predigested too. Don't borrow instant phrases.

Here are a few of the many old standards that make letters seem unimaginative and boring:

Letterese	Translation into Plain English
As per your request	As you requested
Having received your letter, we . . .	We received your letter.
Enclosed please find my résumé.	My résumé is enclosed.
I regret to advise you that I must delay payment.	I must delay payment.
It is imperative that you write at once.	Please write at once.
I am cognizant of the fact that my payment is overdue.	I know that my payment is overdue.
At the earliest possible date	As early as possible
I beg to differ with your estimate.	I disagree with your estimate.
Please be advised that my new address is . . .	My new address is . . .
This writer	I
At the present time	Now

I humbly request that you consider my application.	Please consider my application.
I beg to acknowledge receipt of your check.	I received your check.
In the immediate future	Soon
We are in hopes that you succeed.	Good luck.
In accordance with your request	As you requested
Our situation is such that we cannot immediately pay our complete bill.	We cannot immediately pay our complete bill.
Due to the fact that	Because
Herein enclosed	Enclosed
Please be kind enough to grant me an interview.	May I have an interview?
I wish to express my gratitude.	Thank you.
At this point in time	Now

Be natural: write as you would speak in a classroom discussion. Avoid the temptation to copy a textbook example word for word. Use the samples in this book as models only. Anyone who reads letters regularly can spot a borrowed letter.

Figure 14-6 shows a sample request letter that draws heavily on clichés and borrowed phrases. As you can see, the reader will find it difficult to extract the message buried under the redundancies and trite expressions.

Figure 14-7 shows the same message written in plain English. It seems more sincere. The message remains unchanged, but the writer withholds the starch. Now the letter should achieve its purpose through its clarity, conversational tone, and its "you" perspective.

"You" Perspective

In speaking face-to-face with someone, you unconsciously modify your statements and manner of expression as you read signals from the listener: a smile, a frown, a raised eyebrow, a nod, etc. Even in a telephone conversation your listener provides cues that signal approval, dismay, anger, confusion, and so on. Writing a letter, however, has one major disadvantage: because you face a blank page, it is easy to write only to please yourself, forgetting that a flesh-and-blood person will be taking the time to read your letter.

234 Idle Way
Hoboken, New Jersey 34567
September 10, 1981

Marvin Mooney
Registrar
Calvin College
Plains, Georgia 38475

Dear Mr. Mooney:

Pursuant to your notice of September 6, I regret to advise
you that my tuition payment will be delayed until January 21,
when I receive my scholarship check.

I humbly request you to be cognizant of the fact that this
writer's tuition for all five prior semesters has been paid
on time. At the present time, my first, and hopefully last,
late payment is due to the fact that a computer breakdown in
the NDEA offices has occasioned a delay in the processing of
all scholarship renewal applications for two weeks. Enclosed
please find a copy of a recent NDEA notice to this effect.

I am in hopes that you will be kind enough to grant me an
extension of my tuition-due date for this brief period of
time. Thanking you in anticipation of your cooperation, I
remain

                                    Gratefully yours,

                                    Charles Jones

                                    Charles Jones
                                    Class of '84

Enclosure

FIGURE 14-6   An Example of "Letterese"

234 Idle Way
Hoboken, New Jersey 34567
September 10, 1981

Marvin Mooney
Registrar
Calvin College
Plains, Georgia   38475

Dear Mr. Mooney:

I received your tuition-due notice of September 6 and regret
that my payment will be delayed until January 21, when I
receive my scholarship check.

Your payment records should show that my tuition bills for
all five prior semesters have been paid on time. This first,
and I hope last, late payment is the result of a computer
breakdown in the NDEA offices, which has delayed the processing of scholarship renewals for two weeks, as explained
in the enclosed copy of the NDEA notice.

May I have this brief extension of my tuition-due date, without causing you or the college great inconvenience? Your
patience in this difficult time would be a great help.

Respectfully,

Charles Jones

Charles Jones

Enclosure

FIGURE 14-7   An Effectively Phrased Letter

The "you" perspective concerns your tone; by your careful word choice you show respect for your readers' feelings and attitudes. Put yourself in their place; ask yourself how they will respond to the statements you have just written. A letter creates a relationship between you and your reader, and the words on the page are the only basis for that relationship. If you bury your readers in "letterese" and clichés, they are bound to conclude that you have a low regard for your writing task and for them — so low that you haven't taken the time to express yourself with care. Figure 14-7 is written from a "you" perspective, whereas Figure 14-6 is not.

Remember that even a single word, carelessly chosen, can offend readers. In a letter complaining about your new camera, for example, you have the choice of saying "The shutter mechanism of my new KL 50 is defective," or "The shutter mechanism . . . is lousy." Clearly, "lousy" would be a poor choice here because of its implied insult to the manufacturer. Or put yourself in the reader's place in the following example: Imagine that you manage the complaint department for a large mail order company. Which of these versions requesting repair, replacement, or refund for a faulty item would you tend to honor quickly and efficiently?

> 1. I demand that you bums immediately send me a replacement for this faulty desk calculator, and I only hope that it won't be as big a piece of junk as the first! (the belligerent attitude)
> 2. This new desk calculator does not seem to work and I wonder if you might kindly consider the possibility of sending me a replacement, if that is all right with you. Please accept my humble thanks in advance. (the pardon-me-for-living attitude)
> 3. I beg to advise you that I am appalled by the patent paucity of workmanship in this calculator and find it imperative that you refund the full purchase price at the earliest possible date. (the pompous-indignation attitude)
> 4. After I laid out all my bread on this bogus calculator it blew a fuse and blew my mind. If you are hep to my displeasure, put your money where your mouth is and send me a refund. (the pass-me-the-joint attitude)
> 5. Please send me a replacement for this defective calculator as soon as possible. (the courteous, confident, and direct attitude)

Statement 5 would most likely achieve results. This statement is neither antagonistic nor apologetic. Instead, it is courteous, suggests confidence in the reader's integrity, and makes a direct request in plain English. The earlier versions express only the writer's need to sound off (except 2 where the "you" perspective is carried to a ridiculous extreme), but version 5 creates a sense of respect, trust, and understanding. Even one or two carelessly chosen words can be offensive, so choose your words carefully.

Clear Purpose

Like any effective writing, good letters do not just "happen." Each is the product of a step-by-step process. In fact, most effective letters are *rewritten;* words rarely tumble out on your page to form a perfect message on your first try. As you plan, write, and revise your letter, answer these questions about purpose and content:

1. *What purpose do I wish to achieve?* (get a job, file a complaint, ask for advice or information, answer an inquiry, give instructions, ask a favor, share good news, share bad news, etc.)

2. *What facts does my reader need?* (measurements, dates, costs, model numbers, enclosures, other details)

3. *To whom am I writing?* (Do you know the reader's name? Whenever possible, write to a person, not a title: "Dear Ms. Robinson," not "Dear Madam.")

4. *What is my relationship to my reader?* (Is the reader a potential employer, an employee, a person doing you a favor, a person whose service or products are disappointing, an acquaintance, a business associate, a stranger?)

Answer those four questions *before* drafting the letter. Then, after you have written a draft, think about the answers to the next three questions, which pertain most directly to the *effect* of your letter on readers. Will they be encouraged to respond favorably?

1. *How will my reader react to my statements as phrased?* (with anger, hostility, pleasure, confusion, fear, guilt, warmth, satisfaction, etc.)

2. *What impression of me will my reader get from this letter?* (intelligent, courteous, friendly, articulate, obnoxious, pretentious, illiterate, confident, unctuous, servile)

3. *Am I ready to sign my letter with confidence?* (This bears some thought!)

Don't mail your letter until you have answered each question to your full satisfaction. Revise as often as needed to achieve your purpose.

WRITING VARIOUS TYPES OF LETTERS

This section covers three of the most common types of letters: the letter of inquiry, the letter of complaint, and the letter of instruction. The résumé and job application letter are discussed in a separate section.

Letter of Inquiry

Letters of inquiry may be solicited or unsolicited. You often write the first type as a consumer requesting information about an advertised product. You can expect such a letter to be welcomed by your addressee. After all, he or she stands to benefit from your interest. In this case, you can afford to be brief and to the point. Figure 14-8 shows an unsolicited letter of inquiry.

Many of your inquiries will be unsolicited, that is, not in response to an ad, but simply requesting information for a report or a class project. Here, you are asking a favor of your addressee, who must take the time to read your letter, consider your request, collect the information, and write a response. Therefore, you need to apologize for any imposition, to express your appreciation of your reader's generosity, and to state a reasonable request clearly. Begin your letter with something a bit more cordial and less abrupt than "I need some information."

Before you can ask specific questions you need to do your homework. Don't expect your respondent to read your mind. A general question ("Please send me all your data on . . .") is likely to be ignored. Only when you know your subject can you refine your questions.

Don't wait until the last minute to write your letter. Write at least three weeks before your report is due, politely indicating the due date in your letter.

Here is a typical inquiry situation: Imagine that you are preparing an analytical report on the feasibility of harnessing solar energy for home heating in northern climates. During your investigation you learn that a private, non-profit research group in your state has been experimenting in ecologically efficient energy systems. After deciding to write for details, you plan and compose your inquiry, basing it on the questions in the previous section.

Introduction

Begin by introducing yourself and stating your purpose. Your reader should know who wants the information, and why. Maintain the "you" perspective by opening with a statement that will spark your reader's interest and goodwill (as shown in Figure 14-9).

Body

In the heart of your letter compose specific and clearly worded questions that can be understood easily and answered readily. Number each question and separate it from the others, perhaps leaving space for responses right on the

180 Columbus Avenue
Cleveland, Ohio 44100
January 5, 1981

Western Cedar Log Homes
14 Valley Road
Rumford, New Hampshire 13101

Gentlemen or Ladies:

   Please send me your brochure describing your models of
log homes, as recently advertised in Country Magazine. How
far in advance must an order for a specific model be placed
to ensure a June 1 delivery date?

                              Yours truly,

                              James C. Johnson

                              James Johnson

FIGURE 14-8   A Solicited Letter of Inquiry

234 Western Road
Arlington, Vermont 05620
March 10, 1981

Director of Energy Systems
The Earth Research Institute
Petersham, Maine 04619

Dear Director:

While gathering data on home solar heating, I encountered
references (in Scientific American and elsewhere) to your
group's pioneering work. Would you please allow me to benefit from your experience?

As a technical writing student at Evergreen College, I am
preparing a report on the feasibility of harnessing solar
energy as a major source of home heating in northern climates
within the next decade. Your answers to the questions below
would help me complete my course project (April 15 deadline).

1. Have you found active or passive solar heating to
    be more practical at this stage of development?
2. Do you hope to surpass the 60 percent limit of heat-
    ing needs supplied by the active system? If so, what
    efficiency do you expect to achieve, and how soon?
3. What is the estimated cost of building materials for
    your active system, per cubic foot of living space?
4. What metal do you use in collectors so as to get the
    highest thermal conductivity at the lowest mainte-
    nance costs?

Your answers, along with any recent findings you can share,
will enrich a learning experience I will put into practice
next summer by building my own solar home. I would be glad
to send you a copy of my report, along with the house plans
I have designed. Thank you.

Sincerely yours,

Alan Greene

Alan Greene

FIGURE 14-9   An Unsolicited Letter of Inquiry

page. If you have more than five or six questions you might place them in an attached questionnaire.

Conclusion

Conclude by telling your reader how you plan to use the information and, if possible, how he or she might benefit. Offer to send a copy of your finished report. Close with a statement of appreciation; it will encourage your reader to respond.

Revision and Final Touches

Revise your letter until its tone and content measure up to the quality of a letter you would like to receive. (See the revision checklist at the end of this chapter.) When you feel good about your letter, sign it. Because you are asking a favor, include a stamped, return-addressed envelope for your reader's convenience. Remember that all letters need at least one revision, and the job-application letter, discussed later, may need several.

When completed, your letter might look like Figure 14-9. An inquiry composed with care, courtesy, and detail will yield positive results. For further study review Figure 14-1.

Letter of Complaint

Complaint letters are a challenge, not because of what you have to say, but because you have to find a reasonable way of saying it. Although a complaint is an expression of your resentment, your dissatisfaction, your frustration, it is usually a mistake to begin a complaint letter with the sole intention of "telling someone off." Everyone likes to sound off now and then, but it is less important to express your dissatisfaction than to achieve a desired result: a refund, a replacement, improved service, better business relations, or even an apology.

Imagine that you have recently bought an expensive stereo component system, with top-of-the-line speakers, from a dealer in New Jersey. Three weeks after your purchase you moved to Wisconsin and five weeks later you noticed an increasing distortion of heavy base sounds in your speakers. Your first impulse might be to write a nasty letter; however, you wisely control your temper, planning and writing your letter in line with your responses to the questions we presented earlier.

Introduction

First, identify yourself and your reason for writing. Maintain the "you" perspective by stating your claim *objectively* (this exercise in patience will mean that you have to choose your words carefully). Also remember that an apologetic and meek complaint letter is no more effective — sometimes even less — than a belligerent one.

Body

In your body section, present the details of your complaint. Identify the faulty item clearly, giving serial and model numbers. Describe the deficiency and explain how it has caused you inconvenience, expense, loss of time, and so on. Propose what you consider a fair adjustment, phrasing your statement so that your reader will feel inclined to honor your request.

Conclusion

Conclude with a courteous but firm statement indicating your goodwill and confidence in the reader's integrity.

Figure 14-10 shows how your final revision might read. You will get more positive results by addressing the reader not as a buffoon, but as a competent and responsible businessperson. If your first letter elicits no favorable response, however, phrase your follow-up letter a bit more strongly.

Letter of Instruction

A letter of instruction provides directions for a certain procedure. Instructions are best written in the imperative mood (other criteria for good instructions are discussed in Chapter 9). Be careful not to take your reader's knowledge for granted. Remember that although the procedure is familiar to you and thus easy for you to perform, it is new to the reader. Thus, you need to provide *all* necessary information.

Unclear, incomplete, or inaccurate directions can be frustrating as, for example, when people give poor directions for reaching a certain location because they assume that the addressee knows more about the area than should be expected. Generalities make instructions meaningless: "Down the road a piece" (how far, exactly?); "Turn right at the light" (stop light, blinking light?); "You will see a house on the corner" (color, size, right, left?). Deliver every detail your reader needs.

Imagine that you have obtained a lease on the wilderness property on Deerfoot Lake from the Eastern Paper Company (Figures 14-1 and 14-2). To apply for a building permit from the Maine Land-Use Regulation Commission, you must have the site evaluated (soil depth, drainage, water table, and other elements) by a state-registered engineer. After speaking with the engineer by phone you send him a letter of instruction for reaching your property, which has boat access only. You also have to pinpoint the location of your proposed log cabin. Figure 14-11 shows how you might compose your letter.

534 Hartford Way
Madison, Wisconsin 20967
March 20, 1981

Manager, Stereo Components, Inc.
143 Main Street
Newark, New Jersey 10311

Dear Sir:

On December 10, 1977, I bought a stereo component system
(sales receipt #114621) from your outlet. Three weeks later
I moved to Wisconsin, and after four weeks of stereo use I
noticed increasing distortion of heavy bass sounds in my
speakers.

As a classical music lover I bought your top-of-line
speakers (Toneway 305's, #3624 and 3625) because of their
extra-wide bass range. However, their distortion of lower
ranges of percussion and keyboard sounds is increasing to
the point of actual vibration, making my expensive system
useless.

My speaker guarantee states that items for repair or replacement must be returned to the original retailer. But because
we are now hundreds of miles apart, such an arrangement would
cost me a great deal of time and money and would further
delay the use of my equipment. Under these circumstances,
could you kindly arrange for your fellow retailer in the
Madison area to honor my guarantee directly?

Your store's reputation for good service among my friends
and relatives was a major reason for my purchase, and I am
sure that you will do everything possible to minimize my
inconvenience.

Yours truly,

Sara Fields

Sara Fields

FIGURE 14-10   A Letter of Complaint

154 Seaweed Lane
East Harwich,
  Massachusetts 01263
September 25, 1981

Mr. Lionel D. Kearns
Professional Engineer
3 Wright Lane
Otisfield, Maine 04572

Dear Mr. Kearns:

In our September 20 phone conversation about your upcoming
evaluation of my building site on the northern shore of Deerfoot Lake, you asked for directions to the property. Here
are the instructions for reaching lots #48 and 49. I am also
enclosing a topographical map of the area.

From the Seboomook dock, proceed roughly 4½ miles by water,
southeast. Immediately after passing through the channel between Seboomook Island and the mainland (see enclosed map),
look for the forest service camping area on your left (marked
by a log bench perched out on the rock point of the mainland).
My property lies 1000 feet east of the camping area. It is
marked by a granite ledge, roughly 30 feet long and 15 feet
high.

LANDING CAUTION: A rock shoal along the westerly frontage
extends about 30 feet from the shoreline. Approach the shoreline carefully, from the easterly end, and you will find a
landing area on a small gravel beach immediately to your
right of the ledge.

Lot boundaries are marked by yellow stakes a few feet from
the shoreline. Look for lot numbers carved on yellow-marked
trees adjacent to the yellow stakes.

If I can receive the results of your evaluation by early
November I will have a head start for spring building.
Please call me at 231-978-4568 (collect) for further infor-
mation. I appreciate your help.

Best wishes,

John M. Lannon

John M. Lannon

Enclosure

FIGURE 14-11 A Letter of Instruction

Introduction

In your opening paragraph, identify the procedure for which you are providing directions and explain the reasons for providing them. Supply any background necessary to the reader or to call attention to any enclosed material — lists of equipment or tools needed, charts or diagrams, maps, or the like.

Body

In your body section, describe each step of the procedure in the order of its performance. If your instructions are at all complicated, place each step in a separate paragraph and number the paragraphs sequentially. Include any cautions or warnings immediately before the steps to which they apply.

Conclusion

Conclude by mentioning the specific time or date by which the procedure should be completed (as applicable). Offer to provide further information, if needed. As a convenience, include your telephone number in case your reader has questions.

WRITING THE RÉSUMÉ AND JOB APPLICATION LETTER

Today's job market is, by-and-large, a buyer's market, with many applicants competing for few openings. Whether you are applying for your first professional job or changing careers in midlife, you have to wage an effective campaign for marketing your skills. The résumé and letter of application are your most vital tools; both must stand out among those of your competitors.

Job Prospecting: The Preliminary Step

Before writing a model letter and résumé, do some prospecting: search and study the job market to identify realistically the careers and jobs for which you qualify.

Being Selective

Many new graduates make the mistake of applying for too broad a range of jobs, including many for which they have no real qualifications. Such a shotgun approach is ambitious, but it may well decrease your chances — and it can be most discouraging. If you spend your time, energy, and optimism everywhere, you limit the time you have to concentrate on those openings for which you do qualify. Be selective in your search.

Launching Your Search

Launch your job campaign by doing some advance planning and homework. Don't wait for the job to come to you. Instead, take the initiative by observing these suggestions:

1. As early as your sophomore year, begin scanning the want ads; most big city newspapers publish an entire Help Wanted section in their Sunday editions. Here you will find descriptions, salary scales, and qualifications for countless jobs.

2. Ask your reference librarian to point out occupational handbooks, government publications, newsletters, and magazines or journals in your field.

3. Visit your college placement service; here, openings are posted, interviews are scheduled, and counselors can provide good advice about job hunting.

4. Make it a point to speak with someone working in your field so as to get an inside view and some practical advice.

5. Sign up at your placement office for interviews with company representatives who visit the campus.

6. Seek the advice of faculty in your major who do outside consulting or who have worked in business, industry, or government.

7. Look for a summer job in your field; this experience may count more than your education.

8. Establish contacts; don't be afraid to ask for advice. Make a list of names, addresses, and phone numbers of people who are willing to help.

9. Many professional organizations invite student memberships (at reduced fees). Such affiliations can generate excellent contacts, and look good on the résumé. If you do join a professional organization, try to attend meetings of the local chapter.

By taking the above steps well before your senior year, you may learn that certain courses make you more marketable. In many fields, for example, some knowledge of computers is desirable; certain nursing positions require counseling experience; and so on. Learn as much as you can in order to tailor your final semesters to these requirements (taking one or two introductory computer courses, taking counseling courses and doing volunteer work for a community service organization, or the like).

If you're changing jobs or careers, employers will be more interested in what you've accomplished *since* college. Be prepared to show how your work experience is relevant to the new job. Capitalize on the contacts you've made over the years. Collect on favors owed.

Whether you are a beginner or a veteran, you might want to register with an employment agency that handles jobs in your field. There is, of course, a fee payable after a job is landed, but in many cases the prospective employer pays this fee. Be sure to ask about fee arrangements *before* you sign up.

Being Realistic

When you do apply for jobs, be realistic. If the advertised requirements include several years of experience or a master's degree, look elsewhere. Sometimes, however, the gap between your own qualifications and the qualifications required by the employer may not be too great. In this case, an enthusiastic, well-written letter, along with alternative qualifications (related volunteer work or pertinent outside interests) might land you the job. Rely on your good judgment and the advice of placement counselors and faculty members. Don't hesitate to ask for advice! Besides applying for advertised openings you might write to some organizations that have not advertised recently. Both solicited and unsolicited letters are discussed later in this chapter.

Once you have a clear picture of where you and your qualifications fit into the job market, you will set out to answer the big question asked by all employers: "What do you have to offer?" Your answer must be a highly polished presentation of yourself, your education, work history, interests, and special skills — in short, your résumé.

The Résumé

Your résumé is a summary of your experience and qualifications — a personal inventory that accompanies your letter of application. Written before your application letter, the résumé provides the raw materials for your letter. For easy preparation and reading, try to limit your résumé to one page,[1] dividing it into the following classes of information:

- name and address
- career objectives
- educational background
- work experience
- personal activities, interests, awards, and special skills
- references

---

[1] If you are changing jobs or careers, your varied experience might call for more than one page.

This information provides an employer with a one- or two-page ready reference on you. Your application letter, in turn, will emphasize specific parts of your résumé and will discuss how your background is suited to that particular job.

Begin work on your résumé at least one month before your job search. You will need that much time to compose, revise, and polish until it represents your best effort. Your final version can then be duplicated for each of your targets.

First, list on separate sheets of paper the above classes of information. Then brainstorm each subject (see Appendix B) to identify the important items in your background.

Name and Address

Under your first heading, include your full name, street and mail address (if different), and full telephone number (many interview invitations and job offers are made by phone). Recent federal legislation protects you from job discrimination on the basis of sex, religion, race, national origin, or age. Therefore, you are not required to include a photograph (although you may) or information about these items.

Career Objectives

First, return to your collected information (from steps 1–9, listed earlier) and survey the *specific* jobs for which you *realistically* qualify. Resist the impulse to be all things to all people. The key to a successful résumé is the image of *you* that it projects — disciplined and purposeful, yet flexible. State both your immediate and long-range goals, including any plans for continuing your education: [2]

> My immediate goal is to join the intensive care nursing staff of an urban teaching hospital. Through on-the-job experience and part-time graduate study in crisis treatment and life support systems I hope eventually to supervise an intensive care unit and instruct student nurses.

Your statement of career objectives should show that you have a clear sense of purpose and have given serious thought to your future. Do not borrow a trite statement of objectives from a placement office brochure!

---

[2] To save space, you can omit your statement of career objectives from the résumé and include it in your letter instead.

Educational Background

If your education is more substantial than your work experience, place it first. Begin with your most recent school and work backward, listing your degrees and diplomas and schools attended *beyond* high school (unless you attended a high school vocational program in specific preparation for your career). List the courses that have directly prepared you for your career. If your class rank is in the upper 40 percent, mention it; otherwise, omit this information. Include any schools attended or courses completed while you were in military service. If you financed part or all of your education by working, say so, indicating the percentage of your contribution.

Work Experience

If you have solid work experience, place it before your education. Beginning with your most recent job and working backward, list and clearly identify each job, giving specific dates of employment and names of employers. Also, state whether the job was full time, part time (hours weekly), or seasonal. Briefly describe your specific duties in the major jobs and indicate any promotions or added responsibilities you received. Include any military experience you have had.

Personal Interests, Activities, Awards, and Special Skills

List hobbies, sports, and other pastimes; memberships in teams and organizations; offices held; and any recognition for outstanding performance. Include the dates and specific types of any volunteer work. These items describe you as an individual, giving employers a profile of such important traits as variety of interests, creative use of leisure time, concern for personal growth and community welfare, team spirit, ability to work within a group, leadership qualities, and performance capabilities. A history of volunteer work suggests that you give freely of your time without concern for material gain. Employers know that a person who seeks a well-rounded life is likely to take an active interest in his or her job. Who you are away from work largely defines who you will be on the job.

Obviously, you will want to be selective in this section. List only the items relevant to your field, along with those that reflect your competence and good character.

References

Your list of references names four or five people *who have agreed* to write strong, positive assessments of your qualifications and personal qualities. Often a reference letter is the major element in getting a prospective employer to want to meet you; thus, thought and care should enter into your selection of references.

Select references who can speak with authority about your ability and character. Avoid members of your family (everyone knows that grandmothers consider their grandchildren flawless), neighbors, and close friends not in your field. Choose instead among professors, previous employers, and respected community figures who know you well enough to write on your behalf. In asking for a reference keep these two points in mind:

1. *A mediocre or poor letter of reference is more damaging than no letter at all.* Therefore, don't simply ask, "Could you please act as one of my references?" A question phrased this way leaves the person little chance to say no. He or she may not know you well or may not be impressed by your work but, instead of refusing, may write a watery letter that will do more harm than good. Instead, make an explicit request: "Do you feel that you know me and my work well enough to write me a strong letter of reference? If so, would you please act as one of my references?" This second version gives your respondent the option to decline gracefully. Otherwise, it elicits a firm commitment to a positive letter.

2. *Letters are time-consuming to write.* Your references have no time to write individual letters to every prospective employer. Therefore, ask for only one letter, addressed *To whom it may concern.* Your reference keeps a copy for his or her files; you keep the original for your personal dossier (so you can reproduce it as necessary); and a copy is sent to your placement office for inclusion in your placement dossier. Because recent legislation permits you to read all material in your placement dossier, this arrangement will provide you with your own copy of your credentials.[3] (The dossier is discussed later in this chapter.)

If the people you select as references live far away, you may wish to make your request by letter. Figure 14-12 shows a sample letter of request.

Opinion generally is divided about whether the names and addresses of references should be included in a résumé. If saving space is important, simply state, "References available on request," thereby keeping your résumé only one page long, but, if your other résumé items take up more than one page, you probably should include names and addresses of references. (A prospective employer might recognize the name of a friend or colleague and

---

[3] Under some circumstances you may — and may wish to — waive the right to examine your recommendations. Some applicants, especially those applying to professional schools as in medicine and law, do in fact waive the right to see recommendations. They do so in concession, one supposes, to a general feeling that a letter writer who is assured of confidentiality is more likely to provide a balanced, objective, and reliable assessment of a candidate. In your own case, you might seek the advice of your major advisor or a career counselor.

203 Elmwood Street
San Jose, California 90462
March 12, 1981

Mr. John Knight
Manager, Teo's Restaurant
15 Loomis Street
Pensacola, Florida 31642

Dear Mr. Knight:

From September 1977 to August 1979 I worked under your supervision at Teo's Restaurant as waiter, cashier, and then assistant manager. Because I enjoyed my work I decided to study for a career in the hospitality field.

In three months I will graduate from San Jose City College with an A.A. degree in Hotel and Restaurant Management. Next month I will begin my job search. Do you feel that you know me and my work well enough to write me a strong letter of recommendation? If so, would you be one of my references?

In order to save your time, may I ask that you address your letter "to whom it may concern." If you will kindly send me the original, I will forward a copy to my college placement office.

To update my recent activities, I've enclosed my résumé. Thank you for your help.

Sincerely,

James David Purdy

James David Purdy

FIGURE 14-12 A Letter Requesting a Recommendation

thus notice *your* name among the crowd of applicants.) If you are changing careers, a full listing of references is especially important.

Composing the Résumé

With your vital data collected and your references lined up, you are ready to compose your actual résumé. Imagine yourself to be a twenty-two-year-old student about to graduate from a community college with an A.A. degree in Hotel and Restaurant Management. Before attending college, you worked at related jobs for over three years. You are now seeking a junior management position with a nationwide hospitality chain while you continue your education, part time. You have spent two weeks compiling and selecting information for your résumé and obtaining commitments from four references. Figure 14-13 shows how your finished product might look, with items listed in parallel form and uniformly indented. Notice that this résumé mentions nothing about salary. Wait until this matter comes up in your interview, or later. And because James Purdy is not entering a beauty contest, he has not attached a photograph. The information, specific but concise, describes what he has to offer, and requires a couple of minutes to skim. There are no statements of self-praise like "I did such a terrific job that I was promoted to manager." The facts speak for themselves.

As a final check of your résumé, look at it as a personnel director might, analyze your presentation, and if you find weaknesses strengthen them. As a guide, Figure 14-14 shows Personnel Director Mary Smith's assessment of James Purdy, in the form of a memo to her colleagues.

For an illustration of a résumé composed by a person with extensive career experience, see Figure 14-15. In that example, notice that work experience is placed before education and that names and addresses of references are included.

When fully satisfied with your résumé, have your model printed by a lithographer or printer. For about forty dollars you can obtain a better-looking copy than you could produce on a typewriter. This one prototype, in turn, will yield as many copies as you need. For clear, neat copies use a photocopying machine or offset printing; *never* send out carbon, thermofax, or mimeographed copies.

A final suggestion: avoid using a résumé-preparing service. Although professionals can use your raw materials to produce an impressive résumé, employers often recognize the source by its style; if they do, they may conclude that you are incapable of communicating effectively on your own.

Now, with your résumé fully prepared, you are ready to plan and compose the job application letter.

James David Purdy

203 Elmwood Avenue
San Jose, California 90462
Tel.: 214-316-2419

Professional Objective	To work in customer relations for a hospitality chain, to continue my education part time, and eventually to assume market management responsibilities.
Education 1979–1981	*San Jose City College, San Jose, California* Associate of Arts Degree in Hotel/Restaurant Management, June 1981. Grade point average: 3.25 of a possible 4.00. All college expenses financed by scholarship and part-time job (20 hours weekly).
Employment	
1979–1981	*Peek-A-Boo Lodge, San Jose, California* Began as desk clerk and am now desk manager (part time) of this 200 unit resort. Responsible for scheduling custodial and room service staff, convention planning, and customer relations.
1977–1979	*Teo's Restaurant, Pensacola, Florida* Began as waiter, advanced to cashier, and finally to assistant manager. Was responsible for weekly payroll, banquet arrangements, and supervised dining room and lounge staff.
1976–1977	*Encyclopedia Britannica, Inc., San Jose, California* Sales representative (part time).
1975–1976	*White's Family Inn, San Luis Obispo, California* Worked as bus boy, then waiter (part time).
Personal	*Awards* Elected captain of basketball team, 1976; received Lions' Club Scholarship, 1979. *Special Skills* Speak French fluently; expert skier. *Activities* High School basketball and track teams (3 years); college student senate (2 years); Innkeepers' Club — prepared and served monthly dinners at the college (2 years). *Interests* Skiing, cooking, sailing, oil painting, and backpacking.
References	Available on request from: Career Placement Office, San Jose City College, San Jose, California 90462

FIGURE 14-13   A Résumé for an Entry-Level Candidate

TO: Members of the   January 20, 1981
           Hiring Committee

FROM: Mary Smith,
           Personnel Director   *Ms*

SUBJECT: Follow-up on James Purdy's
           Application (copy enclosed)

This applicant shows a sense of purpose and responsible
planning for his future. His recent background provides
detailed and specific support for his stated plans.

The fact that he financed his own education yet achieved
a high cumulative average indicates that he is both a diligent and capable worker. His course of study, along with
practical experience in sales, food service, and hospitality
suggests that his career choice is based on sound knowledge
of the hotel/restaurant field and an obvious interest and
talent for direct customer contact.

His history of job promotions is clear indication that
the quality of his work has impressed his employers repeatedly.
His skiing ability and interest in cooking and sailing would
seem to make him a strong candidate for a position in one of
our northern resort facilities. His history of activities
and awards suggests that he works well with others (including
youngsters), is respected by his peers, has leadership qualities, and is willing to volunteer his time and talent without
compensation. Finally, his ability to speak a foreign
language could be an important asset to our customer relations
division.

Overall, James Purdy seems to be a well-rounded person
who knows what he wants and who can offer youth, enthusiasm,
and experience to our organization. He promises to be a
responsible employee who will continue to improve personally
and professionally. I recommend that we pursue his application.

FIGURE 14-14   A Personnel Director's Memo

RÉSUMÉ

Peter Arthur Profitt
14 Cherokee Road
Tucson, Arizona 85703
Telephone: Home: 602-516-1234
            Office: 602-567-5000

Qualifications and Career Objectives

Comptroller, designer of data processing systems, international sales, manager of large foreign office, manager of accounting firm, budget officer.

My immediate goal is to continue my career in fiscal/budget management, in a position with major challenges and responsibilities. Continuing my formal education part time, I plan to fit myself for top executive responsibilities.

Work Experience

1972-present   Comptroller, Datronics, Phoenix, Arizona — Oversee formation of fiscal policies of Datronics; develop appropriate operational procedures; maintain overall coordination of daily business activities, including, for example: (1) supervise development and operation of an accounting system including payrolls, operation and capital equipment budgets, and R&D funds; (2) advise president in forming company policies, plans, and procedures; (3) oversee receipt and control of operational revenues and expenditures; (4) prepare annual budgets and long-range fiscal policy for directors' approval.

1965-1972   Assistant Manager, then Manager, Financial Operations, Abernathy's, New York — (1) supervise maintenance of operations accounts; (2) supervise expenditure and receipt of funds (under vice-president for operations); (3) develop forms/procedures for accounting, purchasing, cost systems, and computerizing of entire financial operation; (4) establish and supervise inventory control system; (5) assist in preparation of budgets, financial data, and reports for immediate and long-term use.

1960-1965   Payroll Manager, Milene's Boston — (1) Sypervise payroll department, including preparation of all branch store payrolls, deductions, etc.; (2) responsible for state/federal payroll audits; (3) direct issuance of U.S. Savings Bonds; (4) assist in budget estimates of employee costs and promotions; (5) prepare periodic payroll reports.

1958-1960   Assistant in Fiscal Management, Milene's, Boston — (1) supervise three employees in preparation of departmental payrolls and maintenance of relevant personnel records and files; (2) interview and recommend applicants for clerical employment; (3) code and computer index file material.

Note: At both Abernathy's and Datronics I established the training programs for accounting and computer personnel — programs still being carried on.

FIGURE 14-15   A Résumé for a Person with Extensive Career Experience

Education Background

M.B.A. candidate, University of Tucson, 1974

Related graduate-level courses: Cases in Personnel Management, Advanced Cost Accounting, Statistical Analysis of Business Trends, Computerized Payroll Systems Development, and others

B.S., Accounting, Northeastern University, Boston, 1958 — graduated *cum laude*

A.S., Business Administration, Mass. Bay Community College, Watertown, 1956

Certificate, Proficiency in French, WSAFI, Stuttgart, Germany, 1953

Certificate, Data Processing Specialist, U.S. Air Force Base, Omaha, 1951

Personal Interests, Activities, Awards, and Special Skills

Interests: Native American archaeology, skiing, chamber music (I am first violin in an amateur group), gourmet cooking, whitewater canoeing, French and German literature.

Activities: 1968 Class Agent, Northeastern University; Rotary Club chapter president (1 year), Framingham, Mass.; United World Federalists chapter treasurer (3 years), Beverly Mass.,; Beverly Hospital Fund chairman (4 years); American Field Service chapter president (3 years), Tucson; United Fund (Commercial) chairman (2 years), Tucson; Sierra Club member (10 years).

Awards: Young Executive of the Year, Beverly Chamber of Commerce, 1964; Record Fund Raising Award, United Fund, Tucson, 1975; Alumni Fund Awards (for highest total), Northeastern University (2 years).

Special Skills: Written and oral fluency — French and German; conversational Spanish; computer operations; successful training programs.

References

Mrs. James Stirling Fell, President Datronics, 1142 Arroyo Grande, Phoenix, AZ 85903

Dr. Walter J. Enos, Vice-President (Operations), Milene's, Box 1000, Boston, MA 02114

Mr. Albert Fresco, President, Abernathy's, 500 Fifth Avenue, New York, NY 10014

Mr. Peter S. Pence, Chairman, United Fund, 5 Union Place, Tucson, AZ 02103

FIGURE 14-15   (*Continued*)

The Job Application Letter

Your Image

Your job application letter is one of the most important pieces you will ever write. Depending on its quality your letter will either open doors or be a waste of time. Although its text expands on your résumé, you must emphasize your personal qualities and qualifications in a way that is personable and convincing. Here you will project an image of your personality. In your résumé you merely present raw facts; in your application letter you discuss these facts. Furthermore, the tone and insight that you bring to your discussion suggest a good deal about the kind of person you are. The letter is your opportunity to explain how you see yourself fitting into the organization. Your purpose is to interpret the items on your résumé and show your employer how valuable you will be.

Many people are uncomfortable talking about themselves in letters to strangers. They often feel that they can say little without seeming conceited, but the most essential ingredient in your letter is self-confidence. After all, if you don't believe in yourself, who will? Your letter is a sales letter: it markets your greatest commodity — *you*. To be effective, it *must* stand out among other applications — without being kinky or cute.

Targets

Unlike the résumé, the letter should never be photocopied. Although you can base letters to different employers on the same model — with appropriate changes — type each letter freshly.

The immediate purpose of your letter is to secure an interview. Therefore, the letter itself should make the reader want to meet you. Make your statements engaging, precise, and *original*. Borrowed phrases from textbook examples and "letterese" will not do the job.

Sometimes you will apply for positions advertised in print or by word of mouth (solicited applications). At other times you will write prospecting letters to organizations that have not advertised but might need someone like you (unsolicited applications). In either case, your letter should be tailored to the situation.

The Solicited Letter

Imagine that you are James Purdy. In *Innkeeper's Monthly* you read this advertisement and decide to apply:

RESORT MANAGEMENT OPENINGS

Liberty International, Inc. is accepting applications for several junior management positions at our new Lake Geneva Resort. Applicants must have three years practical experience, along with formal training in all areas of hotel/restaurant management. Please apply by June 1, 1981 to:

> Elmer Borden
> Personnel Director
> Liberty International, Inc.
> Landsdowne, Pennsylvania 24135

Now, plan and compose your letter, using the questions on page 364 as a guide.

Introduction. Introduce yourself and create a tone of self-confidence by directly stating your reason for writing. Mention the specific title of the job you seek and remember that you are talking *to* someone; use the pronoun "you" often (this is especially important here, where you must talk about yourself without sounding conceited). If you can, establish a direct connection by mentioning the name of a mutual acquaintance; for example, you learn that your professor of nutrition, Dr. H. V. Garlid, is a former colleague of Elmer Borden; mention of his name will attract attention to your letter. Finally, after referring your reader to your enclosed résumé, you are ready to discuss the proof of your qualifications.

Body. Concentrate on the qualifications and qualities that you, specifically, can bring to the job. Don't come across as a jack-of-all-trades. Relate your experience to *this* job. Avoid empty flattery ("I am greatly impressed by your remarkable company"). Be specific. Replace "much experience," "many courses," or "increased sales" with "three years of experience," "five courses," or "a 35 percent increase in sales between June and October 1979." Always support your claims with *evidence* and show how your qualifications will benefit this employer. Create a dynamic tone by using the *active* rather than the *passive* voice:

Weak	Increased responsibilities were steadily given to me.
Stronger	I steadily assumed increasing responsibilities.

Trim the fat from your sentences:

Flabby	I have always been a person who enjoys a challenge.
Lean	I enjoy a challenge.

Project self-confidence with your language:

<blockquote>
Unsure   It is my opinion that I will be a successful manager because. . . .

Certain   I will be a successful manager because. . . .
</blockquote>

Also, avoid "letterese." Write as you speak in the classroom. Describe how you will fit in, and remember that an enthusiastic tone can go a long way. In fact, your attitude can be as important as your background. Show yourself to be a person with a clear sense of purpose.

Conclusion. Restate your interest in the job, emphasize your flexibility and willingness to retrain (if necessary), and review briefly other important personal qualities. If your reader is nearby, end with a request for an interview; otherwise, request a telephone call, stating times when you may be reached. Your strong and courteous conclusion should leave your reader with the impression that you are more than a name on a page; you are worth knowing.

Revision. *Never* settle for a first draft — or even a second or third! Perhaps by your fourth you will be approaching the best possible answers to our guide questions. And because, in any case, your letter is your model for letters serving a wide variety of circumstances, it must be your best effort. When you are sure that your letter has high-quality content, an appropriate tone, and an impeccable format, sign it. If you still feel unsure, revise it again.

James Purdy's letter to Elmer Borden, replying to the *Innkeeper's Monthly* advertisement, was the product of his careful attention to these principles. After making a number of revisions, he finally signed the letter in Figure 14-16. He wisely chose to emphasize his practical experience because his background is varied and impressive. An applicant with less practical experience would emphasize education instead, discussing related courses and extracurricular activities.

The Unsolicited Letter

Ambitious job seekers will not limit their search to advertised openings. The unsolicited, or "prospecting," letter is a good way of uncovering other possibilities. Such letters have advantages and disadvantages.

Disadvantages. The unsolicited approach does have two drawbacks: (1) You may waste time and energy writing letters to organizations that simply have no openings. (2) Because you don't know what the opening is — if there is one — you cannot tailor your letter to the specific requirements.

203 Elmwood Avenue
San Jose, California 10462
April 22, 1981

Mr. Elmer Borden
Personnel Director
Liberty International, Inc.
Lansdowne, Pennsylvania 24135

Dear Mr. Borden:

Please consider my application for a junior management position at your Lake Geneva resort. I will graduate from San Jose City College on May 30 with an Associate of Arts degree in Hotel/Restaurant Management. Dr. H. V. Garlid, my nutrition professor, has told me of his experience as a consultant for Liberty International, which further encouraged me to apply.

For two years I worked as a part-time desk clerk and now am the desk manager at a 200-unit resort. This experience, combined with earlier customer contact work in a variety of situations (see enclosed résumé) has given me a clear and practical understanding of customers' needs and expectations. As an amateur chef, I know of the effort, attention, and patience required to prepare fine food. Moreover, my skiing and sailing background might well be assets to your resort's recreation program.

I have confidence in my managerial ability and am determined to succeed in the hospitality field. My experience and education have strengthened the most vital skills any management professional can have: to work well with others and to respond creatively to changes, crises, and added responsibilities.

I would very much like to discuss how my placement with Liberty International would benefit us both. Would you please phone me any weekday after 4 p.m. at 214-316-2419? I hope to hear from you soon.

Yours truly,

James David Purdy

James David Purdy

Enclosure

FIGURE 14-16   A Solicited Job Application Letter

Advantages. This cold-canvassing approach does have one important advantage: for an advertised opening you will compete with legions of qualified applicants, whereas your unsolicited letter might arrive just when an opening has materialized. If it does, your application will receive immediate attention and you just might get the job! Even when there is no immediate opening, the company may file an impressive application until an opening does occur. Or the application may be passed along to a company that has an opening.

There are often good reasons for going further than the Help Wanted columns. Unsolicited letters generally are a sound investment if your targets are well chosen and your expectations are realistic.

Reader Interest. Because your unsolicited letter is unexpected, attract your reader's attention early and make him or her want to read further. Don't begin: "I am writing to inquire about the possibility of obtaining a position with your company." By now, your reader is asleep. If you can't establish a direct connection through a mutual acquaintance, use a forceful opening like the one in Figure 14-17.

```
Does your hotel chain have a place for a junior manager with
a college degree in hospitality management, a proven commit-
ment to quality service, and customer relations experience
that extends far beyond mere textbook learning?  If so,
please consider my application for a position.
```

FIGURE 14-17   An Effective Opening for a Letter of Inquiry

Unlike the usual, time-worn, and plastic openings, this approach gets through to your reader.

The Prototype

Most of your letters, whether solicited or unsolicited, can be versions of your one model, or prototype. Thus your prototype must represent you and your goals in the best possible light. As you approach your job search give yourself plenty of time to compose the model letter and résumé. Employers will regard the quality of your application as an indication of the quality of work you will do.

SUPPORTING YOUR APPLICATION

Your Dossier

Your dossier is a folder containing your credentials: college transcript, letters of recommendation, and any other items (such as a notice of a scholarship award or letter of commendation) that testify to your accomplishments. In your letter and résumé you talk about yourself; in your dossier others talk about you. An employer impressed by what you have said about yourself will want to read what others think about you and will request a copy of your dossier. By collecting your letters of recommendation in one folder you spare your references from writing the same letter over and over.

If your college has a placement office, it will keep your dossier on file and send copies to employers who request them. In any case, keep your own copy in a manila folder. Then, if an employer writes to request your dossier, you can make a photocopy and mail it out, advising your reader that the placement office copy is on the way. This is not needless repetition! Most employers establish a specific timetable for (1) advertising an opening, (2) reading letters and résumés, (3) requesting and reviewing dossiers, (4) holding interviews, and (5) making job offers. Obviously, if your letter and résumé do not arrive until the screening process is at step 3 you are out of luck. The same is true if your dossier arrives when the screening process is at the end of step 4. Timing, then, is crucial. Too often, dossier requests from employers sit and gather dust in some "incoming" box on a desk in the placement office. Sometimes one or two weeks will pass before your dossier is mailed out. And, of course, the only loser in this case is you.

Interviews

An employer impressed by your credentials will phone or write to invite you for an interview. Now you can be sure that you are one of the finalists. You may meet with one interviewer, a group, or several groups in succession. You may be interviewed alone or with several candidates at once. Interviews can last one hour or less, a full day, or even several consecutive days. The character of the interview can range from a pleasant, informal chat to grueling quiz sessions. Some interviewers may antagonize you deliberately to observe your reaction ("Whatever made you imagine that you could fill this position?"). If that happens, suppress your annoyance, look your interviewer straight in the eye, and answer calmly and confidently.

Your library has books with far more detailed interview advice than there is space for here; because self-confidence is of first importance, prepare yourself by studying the techniques of being interviewed. Better yet, get some practice. Take as many interviews as you can. You will find your confidence increasing with each one.

Prepare for your interview by learning whatever you can about the company (its products or services, history, recent growth, prospects, branch locations) in trade journals and industrial registries or indexes. If time permits, request company brochures and annual reports. Prepare specific answers to the obvious questions:

> Why do you wish to work here?
> What do you consider to be your strongest quality?
> What do you see as your biggest weakness?
> Where would you like to be in ten years?

Plan informative and direct answers to questions about your background, training, experience, and salary requirements. Project a strong sense of purpose. Prepare your own list of questions about the job and the organization; you will be invited to ask questions, and the questions you ask can say as much about you as the answers you give.

If you have done your homework, your interview should be a stimulating experience in which you and your prospective employer can learn much about each other. Remember that the purpose of the interview is to confirm the impressions about your qualifications and personality that an employer has derived from your application. Your interviewer wants to know if you are as impressive in person as you seem on paper. Consequently, your best strategy is to be yourself. You have specific skills and a unique personality to offer. Busy people are taking the time and expense to speak with you because they recognize your worth. Knowing this, you can enter your interview with confidence.

Know the exact time and location of the interview. Come well dressed and groomed. Maintain direct eye contact most of the time; if you stare at your shoes your interviewer will not be impressed. Relax in your chair but don't lounge. Don't smoke, even if invited. Don't pretend to know more than you do; if you can't answer a question, say so. Avoid abrupt yes or no answers as well as complicated life stories. Make your answers detailed but to the point.

When your interviewer hints that the meeting is ending (perhaps by checking his or her watch), don't overstay your welcome. Restate your interest in the job; ask when you might expect further word; thank your interviewer; and leave promptly.

The Follow-up Letter

Within a few days of your interview, jog the employer's memory with a letter of thanks that also restates your interest. Figure 14-18 shows the text of James Purdy's follow-up to his interview with Elmer Borden.

The Letter of Acceptance

If all goes well you will receive a job offer by phone or letter. If it is by phone, request a written offer and respond with a formal letter of acceptance. This letter may serve as part of your contract; be sure to spell out the terms of the offer you are accepting! Figure 14-19 shows the text of James Purdy's letter of acceptance.

Thank you again for your hospitality during my visit to your
Lansdowne offices. The trip and scenery were delightful.

After meeting you and your colleagues, and touring the
resort, I remain convinced that I could be a productive
member of your staff.

FIGURE 14-18   A Follow-up Letter

I am happy to accept your offer of a position as assistant
recreation supervisor at Liberty International's Lake Geneva
Resort with a starting yearly salary of $10,500.

As you requested, I will phone Ms. Druid in your personnel
office for final instructions on reporting date, physical
examination, and employee orientation.

I look forward to a long and satisfying career with Liberty
International.

FIGURE 14-19   A Letter of Acceptance

The Letter of Refusal

You may find yourself in the enviable position of having to refuse some offers.
Even if you refuse by phone, write a prompt and cordial letter of refusal, explaining your reasons, and, especially, leaving the door open for future possibilities. You may find later that you are disillusioned with the job you accepted and wish to explore old contacts. Figure 14-20 shows how James Purdy
handled a refusal.

Although I was impressed by the challenge and efficiency of
your company's operation, I am unable to accept your offer of
a position as assistant desk manager of your Beirut hotel. I
have taken a position with Liberty International because it
will allow me to complete the requirements for my B.S. degree
in hospitality management on a full-time basis.

If any later openings should materialize, however, I would
again appreciate your considering me as a candidate.

Thank you for your time and courtesy, and best wishes for
continued success.

FIGURE 14-20   A Letter of Refusal

CHAPTER SUMMARY

The letter is a more personal form of communication than the report; its tone must connect with the reader. As a rule of thumb, don't ever send off a letter until you feel good about signing your name to it.

Any effective letter has an appropriate format and form, is written in conversational language with a "you" perspective, and expresses a clear purpose. The types of letters you will write most often are:

– the letter of inquiry, requesting detailed information from your reader
– the letter of complaint, written to register your dissatisfaction and to request adjustment
– the letter of instruction, explaining how to carry out a procedure

Among your most important correspondence are your résumé and job application letter. Follow these steps in writing your application:

1. Spend some time job prospecting.
2. After brainstorming on your background, begin work on your résumé and letter early, and plan on many revisions.
3. Compose your résumé as an inventory of your qualifications. Make a perfect model from which you can run off photocopies.
4. Compose your letter, emphasizing the major qualifications from your résumé, along with your personal qualities. Project a sense of self-confidence without being pompous. Tailor your letter to fit the job and the application situation (solicited or unsolicited). Write a fresh letter (based on your prototype) for each application.

Follow these steps in supporting your application:

1. Compile your dossier and retain control of its distribution.
2. Prepare well for interviews and take as many as you can to sharpen your skills.
3. Write follow-up letters after completing interviews.
4. Request job offers in writing and respond to each with a letter of acceptance.
5. For offers that you refuse, send refusal letters that leave doors open for future contacts.

REVISION CHECKLIST

Use this checklist to refine the content, arrangement, and style of your letters.

Content

1. Does the letter contain all major parts (heading, inside address, salutation, letter text, complimentary closing, signature)?

2. Does it contain all needed specialized parts (typist's initials, enclosure notation, distribution notation, postscript)?

3. Have you given the reader all necessary information?

4. Have you identified the name and position of your reader?

Arrangement

1. Does the letter have an introduction, body, and conclusion?

2. Is the format appropriate (good paper, neat typing, uniform margins and spacing)?

3. Does the letter follow an accepted form (semiblock or modified block)?

Style

1. Is the letter phrased in conversational language (free from triteness and "letterese")?

2. Does the letter reflect a "you" perspective (words chosen to establish trust, respect, and mutual understanding with the reader)?

3. Is the opening paragraph designed to interest the reader?

4. Does the tone reflect your relationship to your reader?

5. Is the reader likely to derive a favorable impression from this letter?

6. Is the closing paragraph designed to encourage the reader to act in your interest?

7. Are all sentences clear, concise, and fluent?

8. Is the letter written in correct English (see Appendix A)?

Now list those elements of the letter that need improvement.

EXERCISES

1. Bring to class a copy of a business letter addressed to you or a friend. Working in small groups, compare letters, using the principles of effective letters discussed in this chapter. Choose the most effective and the least effective letter and, as a group, compose two separate paragraphs explaining the group's choices.

2. Write and mail an unsolicited letter of inquiry about the topic that you have investigated, or will investigate, in an analytical report or research assignment. Your letter may request brochures, pamphlets, or other informative literature, or it may ask specific questions about your subject ("What chemicals are used to clean algae, barnacles, and other marine vegetation from the cooling system's filters?" "Are these chemicals then discharged into the sea?"). Submit a copy of your letter and your addressee's response to your instructor. Plan and write this and all other letters in these exercises according to the revision checklist.

3. As a student in a state college you learn that your governor and legislature have cut next year's operating budget for all state colleges by 20 percent. This cut will cause the firing of many young and popular faculty members, a drastic reduction in student admissions, reduction in student financial aid programs, cancellation of several new college programs, and erosion of the morale of the college community and the quality of instruction. Write a complaint letter to your governor or your legislative representative, expressing your strong disapproval of this cut and justifying a major adjustment in the proposed budget.

4. Hide a slip of paper with your name on it in some remote corner of your college campus, outside of your classroom building. Write a formal letter of instruction to a classmate, explaining how to locate the hidden item within fifteen minutes. Exchange letters and launch your fifteen-minute search. When all items have been found, critique individual instructions in class.

5. Besides providing factual data about one's background, what does a letter of application say about an applicant? Explain in a short essay titled "Reading Between the Lines."

6. Write a 500-word essay explaining your reasons for applying to a specific college for transfer or for graduate or professional school admission. Be sure your essay covers two general areas: (1) what you can bring to this school by way of attitude, background, and talent and (2) what you expect to gain from this school in personal and professional growth.

7. Write a letter applying for a part-time or summer job. Choose an organization that can offer you experience that is directly related to your career goal. Be sure to identify the exact hours and calendar period during which you are free to work.

8. a. Identify the job that you would like to have in two to five years. Using newspaper, library (see your reference librarian for assistance), placement office, and personal sources, write your own full description of the job: duties, responsibilities, work hours, salary range, requirements for promotion, highest promotion possible, unemployment rate in the field, employment outlook for the next decade, need for further education (advanced degrees, special training, etc.), employment rate in terms of geography, optimum age bracket (as in football, does one fade around thirty-five?), and any other items you can think of.

   b. Using the same sources and your own good judgment, construct a profile of the ideal employee for this job. If you were the personnel director screening applicants, what specific qualifications would you require in an employee (education, experience, age, physical ability, appearance, special skills, personality traits, attitude, outside interests, and so on)? If you can, locate an actual newspaper advertisement describing job responsibilities and qualifications in detail. Better yet, assume that you're a personnel officer, and compose an ad for the job.

   c. Assess your own credentials against each item in the ideal-employee profile. Review the plans that you have made to prepare yourself for this job: specific courses, special training, work experience, etc. Assume that you have completed your preparation. How do you measure up to the requirements in part b? Are your goals realistic? If not, why not? What alternative plan should you formulate?

   d. Using your list from part c as raw material, construct your personal résumé. Revise your résumé until it is perfect.

   e. Write a letter of application for the job described in part a. Revise your letter until you feel good about signing it.

   f. Write a follow-up letter to your fictional employer, thanking him or her for your recent interview and again expressing your interest in working for the organization.

   g. Write a letter accepting the job offer you received from this same employer.

   h. Write a letter graciously declining this job offer.

   i. Submit each of these items (job description, employee profile, newspaper ad, résumé, application letter, follow-up letter, letter of acceptance, letter of refusal), in order, to your instructor.

   Note: Use the sample letters in this chapter for guidance but don't borrow specific expressions.

   9. Assume that the following advertisement has appeared in your school newspaper:

STUDENT CONSULTANT WANTED

The office of the Dean of Students invites applications for the position of student consultant to the Dean for the upcoming academic year. Duties will include (1) meeting with fellow students as individuals and groups to discuss issues, opinions, questions, complaints, and recommendations regarding all areas of college policy, (2) presenting oral and written reports of findings to the Dean of Students on a regular basis, and (3) attending various college planning sessions in the role of student spokesperson. Time commitment: 15 hours weekly during both semesters. Salary: $2000.

Candidates for this position should be full-time students with at least one year of student experience at this college. The ideal applicant will be skilled in report writing and oral communication, will have the ability to work well in groups, and will demonstrate a firm commitment to the welfare of our college community. Application deadline: May 15.

    a. Compose a résumé and a letter of application for this position.

    b. As a class, split into groups of six to form screening, interview, and hiring committees.

    c. Exchange your group's letters and résumés with another group.

    d. As an individual committee member, read and evaluate each of the six applications you have received. Rank each application, privately, on paper, according to the criteria discussed in this chapter before discussing them with the colleagues in your group. *Note:* While screening applicants you will be competing for selection by another committee who is reviewing your own application.

    e. As a committee, select the three strongest applications and invite the applicants for interviews. Interview each selected applicant for ten minutes, after you have prepared a list of standardized questions.

    f. On the basis of these interviews, rank your preferences privately, on paper, giving specific reasons for your final choice.

    g. Compare your conclusion with those of your colleagues and choose the winning candidate.

    h. As a committee, compose a memo to your instructor, giving specific reasons for your final recommendations.

    i. Write your own evaluation of this exercise. In two or three paragraphs discuss what you have learned here.

10. Most of the following sentences need to be overhauled before being included in a letter. Identify the weakness in each statement and revise as needed.

    a. Pursuant to your ad, I am writing to apply for the position of junior accountant.

    b. I need all the information you have about methane-powered engines.

    c. You idiots have sent me a faulty carburetor!

    d. It is imperative that you let me know your decision by January 15.

    e. You are bound to be impressed by my credentials.

    f. I could do wonders for your company.

    g. I humbly request your kind consideration of my application for the position of junior engineer.

    h. If you are looking for a winner, your search is over!

    i. I have become cognizant of your experiments and wish to ask your advice about the following procedure.

    j. You will find the following instructions easy enough for an idiot to follow.

    k. I would love to work for your wonderful company.

    l. As per your request I am sending the county map.

    m. I am in hopes that you will call soon.

    n. We beg to differ with your interpretation of this leasing clause.

    o. I am impressed by the high salaries paid for this kind of work.

11. Evaluate the letters in Figure 14-21 and Figure 14-22 according to the revision checklist at the end of this chapter and make any needed changes.

                                          1289 Fourth Street
                                          Madison, Wisconsin 86743
                                          October 5, 1981

Mr. James Trask
Trask and Forbes, Attorneys at Law
17 Lord Street
Bartly, Michigan 47659

Dear Mr. Trask:

    Having just graduated from law school, I am looking for
an established law firm with which to join and learn. Your
firm seems to meet my requirements and I hope I meet yours.

I had thought of going into legal services, but then decided
to go immediately into a private practice. I will be able
to perform innumerable tasks while gaining invaluable
knowledge.

Your firm is considered to be one of the finest in the region
and that is another of the aspects that attracted me. Your
firm is without a junior partner or assistant at this point
in time and I feel very qualified for the position.

Enclosed please find my educational qualifications included
in my résumé. I have just passed the bar on my first attempt
and received very high grades in law school.

Should you have any questions or comments, we could discuss
them at an interview. I am available any time during the
business week from 9:00 to 5:00. Feel free to phone me at
304-756-9759 or write, as I would like to hear from you in
the immediate future.

                                   Humbly yours,

                                   Brendan Gaines

FIGURE 14-21   A Letter for Evaluation

Mr. Arthur Marsh
Durango Chemical Corporation
Box 278
Lakeland, Wisconsin 39765

Dear Mr. Marsh:

I was reading the local paper and came across your advertisement in regards to an opening for a crushing and grinding operator's position at your plant. At the present time I am in college, but would like to fill that opening when this semester is over. I am highly qualified for this job as I have already had two years experience in this area. I have operated both crushing and grinding circuits that provide the raw ore used in the processing of phosphate products. I also have experience in operating front-end loaders, forklifts, cats, and 30, 50, and 120-ton haul units. I have held the different positions of laborer, operator, and foreman, so I have a full understanding of this type of operation. I am a very organized and safety-minded worker that can handle myself well in emergency situations. I am a responsible and punctual employee. If you need any further information concerning my work or personal background, please contact me at the address on the envelope. I thank you for considering my application and hope to hear from you soon.

Cordially,

Raymond Manning

Raymond Manning

FIGURE 14-22   A Letter for Evaluation

15

Writing Informal Reports

DEFINITION

PURPOSE OF INFORMAL REPORTS

CHOOSING THE BEST REPORT FORM
    FOR YOUR PURPOSE

COMPOSING VARIOUS INFORMAL
    REPORTS
  Typical Reports in Memorandum Form
    Survey Results
    Progress Reports
  Typical Reports in Letter Form
    Site Inspection Reports
    Work Estimates
  Typical Reports on Prepared Forms
    Periodic Activity Reports
    Consumer Complaints
  Typical Reports in Miscellaneous Forms
    Minutes
    Progress Reports

CHAPTER SUMMARY

REVISION CHECKLIST

EXERCISES

DEFINITION

Informal reports are written daily to communicate messages within and among organizations. Most of them require no extended planning, are quickly prepared, contain little background information, have no supplements (title page, abstract, etc.), and can have a variety of formats. They range in length from one sentence to several pages.

PURPOSE OF INFORMAL REPORTS

In the working world, the informal report is the one item most often written and read. Its purpose is to communicate rapidly and precisely in one of these formats: the memorandum, the letter report, the prepared-form report, or a variety of other forms that fit into none of the first three categories and which we will call "miscellaneous."

On the job, personnel must communicate with speed and precision. Your success may depend on your skill in sharing useful information with colleagues. Here are some of the kinds of informal reports you might write on any workday:

– a request for assistance on a work project
– a requisition for parts and equipment
– a proposal outlining the reasons and suggesting a plan for a new project
– a brief set of instructions for one or more colleagues
– a cost estimate for planning, materials, and labor on a new project
– a report of your progress on a specific assignment
– an hourly or daily account of your work activities

– a voucher detailing your business travel expenses

– a report of your inspection of a site, item, or process

– a statement of reasons for equipment malfunction or failure to meet a deadline

– a record of the minutes of a meeting

– a memo describing a change in company personnel policy (vacation time, promotions, etc.)

– a report of your survey to select the best prices, materials, equipment, or service among those offered by several competing firms

These are just a few samples of the wide variety of daily reports that keep companies moving. Most of these reports may be cast in a number of different forms.

CHOOSING THE BEST REPORT FORM FOR YOUR PURPOSE

Whether you decide to report your data in a memo, a letter, on a prepared form, or in any of the miscellaneous forms will depend on your purpose, reader, and situation. It is possible that the same information you cast in a memorandum to your superior will be incorporated in a letter to another company. Or, in some cases, your company might provide prepared forms for certain reports.

The memos that follow were written for readers within the writer's organization. The same information might have been written in letter form if the readers had been outside the company. Conversely, the letter examples in this chapter might have been written as memos.

The point is that in any organization a great deal of working information has to be exchanged as easily as possible. This flow of information is maintained most conveniently by a variety of report forms.

COMPOSING VARIOUS INFORMAL REPORTS

Typical Reports in Memorandum Form

The memorandum is the most common form of in-house communication. A hasty note to a colleague to arrange a luncheon meeting can be classified as a memo. So can a telephone message as shown in Figure 15-1. This kind of brief note usually is discarded immediately after it has served its purpose.

Aside from such brief pieces, most memos follow a fairly well-specified format. Unless your organization provides its own prepared form for memos,

To _Harvey Smith_

Date _Feb. 15, 1978_   Time _11:45 AM_

WHILE YOU WERE OUT

M_s_ _Alma Kelley_

of _Ludlow Electronics_

Phone _631   421-7432   341_

   Area Code   Number   Extension

TELEPHONED	✓	PLEASE CALL	✓
CALLED TO SEE YOU		WILL CALL AGAIN	
WANTS TO SEE YOU		URGENT	
RETURNED YOUR CALL			

Message _Our shipment of Beta-Transistors to Ludlow Electronics is 3 days overdue, causing a serious slowdown in the production schedule of their new mini-computer. Ms. Kelley has asked for a special delivery shipment to be sent out today. B. Daigle_

Operator

EFFICIENCY® LINE NO. 2725 AN **AMPAD** PRODUCT   **60 SHEETS**

FIGURE 15-1   A Telephone Message

follow the standard format shown in Figure 15-2. The standard memo has a heading that names the organization and identifies the sender, recipient, subject, and date. Its text follows an introduction-body-conclusion structure: first, you identify your purpose for writing the memo; next you give the specific information related to your purpose; finally, you conclude cordially with a request, a recommendation, or an offer of further assistance.

If you need a second page, list the recipient's name, the date, and the page number (Ms. Jones, 9/4/78, page 2) three spaces from the top of the page. Begin your text three spaces below. You may also need to photocopy or mimeograph your memo for distribution throughout your organization or for perma-

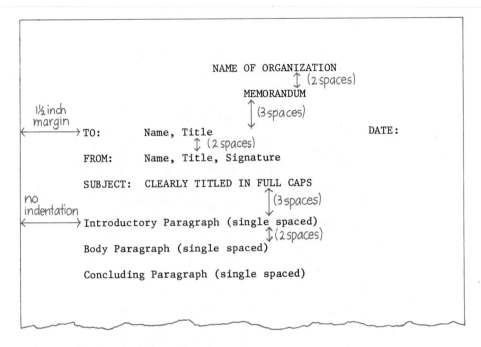

FIGURE 15-2   Standard Memo Format

nent filing. Memos can cover just about any topic and purpose. This chapter illustrates two typical categories that can be reported in memo form: survey results and progress reports. (Proposal memos are illustrated in Chapter 16, pages 431–441.)

Survey Results

Brief surveys are often done to examine conditions that affect an organization (consumer preferences, available markets, etc.). The memo in Figure 15-3, from the research director for a Midwest grain distributor, gives clear and specific information concisely. Notice the absence of background information. (An explanation of how and where these data were obtained would not be significant for this writer's purpose.) The titles of sender and recipient are stated clearly, and the subject heading, typed in full caps for emphasis, promises exactly what the memo will deliver. The writer signed the memo immediately after her typed name and title. She wisely chose to arrange her body section in a classification table to further simplify the reader's job of interpretation. The distribution enclosure (cc.) at the bottom of the page identifies all other readers receiving copies.

ACME GRAIN WHOLESALERS INC.

MEMORANDUM

TO: Charles Jones, Manager,   April 15, 1981
          Marketing Division

FROM: Margaret Spaulding,   MS
          Research Director

SUBJECT: FOOD-GRAIN CONSUMPTION IN U.S., 1969-72

Here are the data you requested on April 7 as part of your
division's annual marketing survey.

U.S. PER CAPITA CONSUMPTION OF FOOD GRAINS IN POUNDS, 1969-72

Corn Products:	1969	1970	1971	1972
Cornmeal and other	15.8	15.8	15.8	15.8
Corn syrup and sugar	20.3	20.8	21.4	21.7
Oat Food Products	3.2	3.2	3.2	3.2
Barley Food Products	1.2	1.2	1.2	1.2
Wheat:				
Flour	112.0	110.0	110.0	111.0
Breakfast cereals	2.9	2.9	2.9	2.9
Rye, Flour	1.2	1.2	1.2	1.2
Rice, Milled	8.3	6.7	7.7	7.0

If you have any questions or require additional information,
please call Ms. Smith at 316.

cc: Mr. C. B. Schultz, Vice-President in Charge of Marketing

FIGURE 15-3   A Memo Reporting Survey Results

In your own field, you may be asked to report research findings comparing the cost or quality of items that are similar. These findings may lead to contracts with certain suppliers. The student-written memo in Figure 15-4 is a good example of how data in memo form can serve as a practical basis for decision making.

Progress Reports

Large organizations depend on progress (or status) reports as a way of keeping track of activities, problems, and progress in various jobs. Daily progress reports are vital in a business that assigns several work crews to a variety of projects.

Sometimes, you will be asked to report on your progress in compiling a longer, formal report. Figure 15-5, for instance, shows a memo describing a student's progress on her term project. She has partitioned her memo into four sections: work completed, work in progress, work to be completed, and date for completion. Although this writer chose a memo format, her report — like most in this chapter — might be recast in an alternate format (letter, prepared form, or miscellaneous form) as dictated by the situation. Many companies in fact provide prepared forms for progress reports.

Typical Reports in Letter Form

Memos usually remain in-house, whereas letter reports generally go to outside readers: potential clients, colleagues, policy-making authorities, and other readers whom you may not have met. Your reader probably will file your letter as a permanent record. Therefore, be sure that your data are accurate. Also, because a letter is a more personal form of correspondence than a memo, give it a "you" perspective (see Chapter 14).

For the introductory and closing elements of your letter, follow the standard letter format discussed in Chapter 14. The only two format additions in a letter report are (1) a "subject" heading, placed two spaces below the inside address and two spaces above the salutation and (2) other headings, as needed to segment your letter into specific areas. The following sections include examples of two typical letter reports: a report of a site inspection and a work estimate. (Proposal letters are discussed in Chapter 16.)

Site Inspection Reports

Figure 15-6 shows a soil engineer's evaluation of a proposed cottage site in a wilderness area. This sample contains three kinds of data: (1) a description

CALVIN COLLEGE

MEMORANDUM

TO: Professor Smith   DATE: February 15, 1981
           Technical Writing
               Instructor

FROM: Susan Grimes, Student

SUBJECT: CONSUMER SURVEY OF COMPARATIVE RETAIL PRICES FOR
           DILANTIN TABLETS

I surveyed comparative prices for a Dilantin prescription by
calling six local pharmacies.

SIX LOCAL PHARMACIES CLASSIFIED IN DESCENDING ORDER
OF THEIR RETAIL PRICE FOR DILANTIN

	Price/ 100 tablets
Pharmacy	
Hargrove Pharmacy, Harwich	$4.14
Cascade Village Pharmacy, Hyannis	4.14
Murphy's Rexall, Sandwich	4.10
Apothecary, Dennis Village	3.89
Consumer's Pharmacy, Harwich	2.79
Dunn's Pharmacy, Hyannis	2.19

These data are important to me because I must take Dilantin
every day. The 100 tablets last only about one month and the
expense of this medicine quickly adds up. From my data I
conclude that my best choice for future Dilantin purchases
is Dunn's Pharmacy in Hyannis.

FIGURE 15-4   A Memo Reporting Survey Results

PROGRESS REPORT

TO: Dr. J. Lannon, English Professor

FROM: T. Fitzgerald, student

DATE: April 27, 1981

SUBJECT: Analytical Report
A STUDY OF THE IMPACT OF THE SANDWICH HISTORIC
DISTRICT COMMITTEE ON THE ZONING AND ARCHITECTURAL
CHARACTER OF SANDWICH

WORK COMPLETED TO DATE

March 22: Completed report on Sandwich Historic
District Committee.

April 10: Obtained maps of Sandwich, legislation,
certificates of appropriateness, exemption, and appeals
from Town Hall.

April 12, 14, 15: Investigated district area by car,
noted differences in historic and nonhistoric areas.

April 18: Photographed sampling of areas in district.
Twenty photos taken, fifteen returned by developer.

April 19: Divided master map into sections, rede-
fining district.

April 20: Completed tentative outline.

April 22: Interviewed chairman of Committee, Donald
Bourne. Received new set of rules and regulations for
district.

WORK IN PROGRESS: Drawing maps of present district and
redefined area.

WORK TO BE COMPLETED

April 28: Interview Edward Carey, committee member.

May 2: Interview George Smith, committee member.

DATE FOR COMPLETION: May 11, 1981

FIGURE 15-5   A Progress Report in Memo Form

Telephone
312-547-9758

Lionel D. Kelley, P.E.
3 Wright Lane
Otisfield, Maine 04572

December 21, 1981

Mr. John M. Lannon
154 Seaweed Lane
East Harwich, Massachusetts 01263

SUBJECT: SITE EVALUATION--COTTAGE LOT, LITTLE "W" TOWNSHIP

Dear Mr. Lannon:

Forms HHE 200 are enclosed detailing a site evaluation conducted on your leased 30,000-square-foot lot. Below is a
summary of my evaluation, along with instructions for proceeding with your plan.

Site Description and Evaluation

The site is made up of Lots 48 & 49 and a 50- by 200-foot
extension to the rear of these lots, which are owned by the
Eastern Paper Company and shown on their maps. The report
and the design are meant to comply with the intent of the
Maine State Plumbing Code and are contingent on your
obtaining the additional land at the rear of the lots.

The site has a high ledge outcrop on the lake side and drops
off at the rear. A ridge runs perpendicular to the lake and
the waste disposal systems must be constructed on the west
side of this ridge. This area is about 5 or 6 feet above
high water of the lake.

The water supply will be hand-carried from the lake to serve
your 16- by 20-foot seasonal cottage.

FIGURE 15-6   A Site Inspection Report in Letter Form

page 2, J. M. Lannon, 12/21/81

Instructions for Building a Waste Disposal System

To construct the gray water system, select an area across
the slope and running level between the two blue flags of
the test borings. Drive a steel rod along this 18-foot run
to determine that you have 36 inches from the surface to
the bedrock. Once this line has been established, dig a
trench 24 inches wide by 12 inches deep, grading level.
Backfill the trench with 3/4 to 3 inches of stone,
embedding a 4-inch diameter PVC perforated pipe, 18 feet
long, in the top 4 inches. Connect the pipe to a standard
16- by 36-inch distribution box, and connect the cottage
sink drain with a 1½-inch diameter polyethylene pipe. Cover
the 4-inch diameter pipe with orange building paper and
cover with 4 inches of topsoil.

To construct the privy pit, probe for bedrock, obtaining a
depth of 48 inches if possible. The ledge appears to go
downward from the ridge and we can anticipate that this
depth can be reached downslope from the first blue flag
toward Lot 50. Excavate a pit to a depth leaving 2 feet of
soil to the bedrock level. Then construct a mound to have a
pit of 36 inches deep and locate the privy as directed by
code section 9.13.

I have enclosed a waiver form to allow you the use of coarse
gravel in lieu of the graded stone. The lot location, being
in a heavily wooded, remote area, does not warrant the use
of stone and I recommend its substitution. Hand-dug test
holes at the boring sites revealed a well-draining high void
granular material which I consider adequate for your requirements. The code permits your applying for this waiver.

Instructions for Completing Building Permit Application

I have enclosed a separate set of HHE 200 forms for you to
include as exhibit 3 of your LURC building permit application. The check marks indicate where your signature is
required. Complete and sign all copies before submitting
them to the respective agencies.

FIGURE 15-6 (*Continued*)

page 3, J. M. Lannon, 12/21/81

A plumbing inspector has not yet been assigned to this area.
LURC will advise you about proper procedure for inspection
on issuance of your building permit.

If you have any questions, please contact me at any time.
Best of luck on your project.

Very truly yours,

Lionel D Kelley

Lionel D. Kelley, P.E.

LDK/jh
Enclosures 1

FIGURE 15-6   (*Continued*)

and evaluation of the building site, (2) instructions for building a suitable waste disposal system, and (3) instructions for completing the building permit application forms. Each section is subsumed under its own heading to increase readability. Notice the absence of extraneous information. The letter simply delivers what its subject heading promises. Both descriptions and instructions are easy to follow. The writer closes with a personal expression of good wishes.

Work Estimates

A work estimate is a proposal to provide a service or product at a certain cost. This type of report is designed to tell customers exactly what they will and will not get for their money. Because a signed estimate can be legally binding, specifications and figures must be precise. To illustrate, the second-to-last paragraph in Figure 15-7 spells out exceptions to the repairs outlined in the estimate.

Typical Reports on Prepared Forms

To streamline communications and to keep track of data, more and more companies are using prepared forms for short reports. Such forms are useful in two ways: (1) A prepared form simplifies your task by providing guidelines for recording data. If you complete the form correctly you are sure to satisfy your reader's needs. (2) A prepared form standardizes the data reported from various sources. Each reporter provides the same classes of data recorded in the same order. This fixed format allows for rapid processing and tabulation of data. The sample questionnaire in Figure 8-1 is a good example of the effectiveness of a prepared-form report; its data can be easily reviewed and tabulated. Another good example is the accident report in Figure 8-3.

The one drawback of a prepared form is its limited space for recording data. From time to time you will need to attach your own statement explaining certain items on the form. Two of the many kinds of information routinely recorded on prepared forms are illustrated and discussed below.

Periodic Activity Reports

The periodic activity report is similar to the progress report in summarizing work activities over a specified period. Your company most likely will provide a prepared form for this reporting task. The sample report in Figure 15-8 is a log of hours worked, type of work accomplished, and mileage traveled by one construction company employee during one week. Most jobs where employees carry out their responsibilities without direct supervision will require

LEVERETT LAND & TIMBER COMPANY, INC.   creative land use
                                        quality building materials
                                        architectural construction

                                                January 17, 1981

Mr. Thomas E. Muffin
Clearwater Drive
Amherst, Massachusetts 01002

Dear Mr. Muffin:

I have examined the damage to your home caused by the
ruptured water pipe and consider the following repairs to be
necessary and of immediate concern:

Exterior:
    Remove plywood soffit panels beneath overhangs
    Replace damaged insulation and plumbing
    Remove all built-up ice within floor framing
    Replace plywood panels and finish as required

Northeast Bedroom--Lower Level:
    Remove and replace all sheetrock, including closet
    Remove and replace all door casing and baseboards
    Remove and repair window sill extensions and moldings
    Remove and re-install electric heaters
    Re-spray ceilings and repaint all surfaces

Northwest Bedroom--Lower Level:
    Remove and replace all sheetrock, including closet
    Remove and replace all door casings and baseboards
    Remove and re-install electric heaters
    Remove and repair window sill extensions and moldings
    Re-spray ceilings and repaint all surfaces

    Post Office Box 185, Amherst, Mass. 01002   413-549-6239

FIGURE 15-7   A Work Estimate in Letter Form

page 2, T. E. Muffin, 1/17/81

<u>Family Room:</u>
Remove and replace sheetrock on north and west walls
Remove and replace door casings and baseboards where required
Remove and re-install electric heaters
Remove and repair window sill extensions and moldings
Re-spray ceiling and paint all new work
Repair closet door under stairs

<u>Entry and Stairwell:</u>
Repair entry sill and refit doors
Remove and replace sheetrock on north wall and partition
Remove and re-install stair rails
Repair and replace stair treads
Replace skirt boards and trim
Replace entry landing and baseboards

<u>Main Level:</u>
Remove, replace, and finish all oak floors
Remove and re-install door casings and electric heaters
Replace baseboards as required
Repair sheetrock corners as required
Paint all new work

This appraisal of damage repair does not include repairs and/or replacements of carpets, tilework, or vinyl flooring. Also, this appraisal assumes that the plywood subflooring on the main level has not been severely damaged.

Leverett Land and Timber Company, Inc. proposes to furnish the necessary materials and labor to perform the described damage repairs for the amount of nine-thousand-one-hundred-and-eighty dollars ($9,180).

Sincerely,

Gerald A. Jackson
President

GAJ/cb

FIGURE 15-7 (*Continued*)

SUMMARY OF WEEKLY ACTIVITIES

NAME   P. Daily

WEEK BEGINNING MONDAY,   April 4, 1981

DAY	JOB	TIME IN	TIME OUT	TOTAL TIME	DESCRIPTION OF WORK	TRAVEL DEST.	TRAVEL MILES
4/4/76 MONDAY	Hadley	8:00	4:00	8 hrs	frame upper level walls, block for shelving, apply plywood to east side		
4/5 TUESDAY	Hadley	8:00	4:15	8 1/4	frame upper level walls, straighten and brace walls, finish plywood sheathing, layout roof framing	Jones Lumber	10
4/6 WED.	Hadley	8:00	4:00	8	cut rafters and set north wing trusses, brace and plumb trusses for sheathing		

FIGURE 15-8   A Periodic Activity Report

DAY	JOB	TIME IN	TIME OUT	TOTAL TIME	DESCRIPTION OF WORK	TRAVEL DEST.	TRAVEL MILES
THURS.	Hadley	8:00	4:00	8	finish roof frame and setup for plywood, sheath roof		
FRIDAY	Hadley	8:00	4:00	8	apply roof trim and layout for shingles, set up staging and stock for shingling next week, block partitions for shelving and cabinets		
SAT.							
SUNDAY							
TOTAL TIME THIS WEEK				40¼		TOTAL MILEAGE	10

FIGURE 15-8 (*Continued*)

periodic activity reports from all personnel. If your company does not supply prepared forms, you can easily devise one for your own use.

Consumer Complaints

Consumers who have purchased a faulty product or received poor service should first complain directly to the company responsible. The complaint letter discussed in Chapter 14 (pages 368–369) is often the best vehicle for recording your dissatisfaction and for requesting compensation. If, however, the company fails to honor your request after one or two letters and a phone call, a formal complaint filed with your state's consumer protection agency may be necessary. Figure 15-9 is typical of the prepared forms used by state agencies for obtaining the details of a complaint.

Typical Reports in Miscellaneous Forms

Reports that do not follow the format of the memo, the letter report, or the prepared-form report we will call "miscellaneous." There are probably as many ways of setting up informal reports as there are subjects to report about. Some reports follow conventions; others do not because their data categories can be so variable that no conventions would serve as adequate guidelines. The reporting of minutes of a meeting, for instance, follows fairly standard conventions. The reporting of survey results for a broad and largely nontechnical audience, however, may require the writer to invent a suitable format. The following sections provide two common examples of reports cast in miscellaneous forms.

Minutes

Minutes are the official records of organizational and committee meetings. Copies of minutes are distributed to all members and concerned superiors as a way of keeping track of proceedings in a large organization. The person appointed secretary records the minutes.

Minutes are filed as part of an official record; therefore they must be precise, clear, highly informative, and free of the writer's personal commentary ("As usual, Ms. Jones disagreed with the committee") or judgmental words ("good," "poor," "irrelevant," etc.). When you record minutes, identify the group, date, place, and purpose of the meeting. Next, give the names of the convener (person calling the meeting) and all members present (unless there are so many that your list would be unwieldy). Finally, record each item on the agenda in chronological order, beginning with a statement that the min-

DEPARTMENT OF THE ATTORNEY GENERAL   FOR OFFICE USE ONLY
Consumer Protection Division
One Ashburton Place, 19th Floor
Boston, MA 02108

YOUR NAME _____

ADDRESS _____ CITY/TOWN _____

ZIP _____ HOME PHONE _____ BUSINESS PHONE _____

NAME OF STORE OR COMPANY _____

ADDRESS _____ CITY/TOWN _____

ZIP _____ PHONE NUMBER _____ PERSON YOU DEALT WITH _____

INFORMATION

WAS A CONTRACT SIGNED? YES _____ NO _____ DATE _____

PRODUCT OR SERVICE INVOLVED _____

DATE PURCHASED? _____ WAS DEPOSIT PAID? _____ COST _____

DID YOU PAY CASH? _____ LOAN _____ ON TIME _____ OTHER _____ AMOUNT _____

WAS THE PRODUCT OR SERVICE ADVERTISED? YES _____ NO _____

WHERE AND WHEN WAS IT ADVERTISED? _____

DID YOU COMPLAIN TO THE COMPANY? _____ TO WHOM _____

HOW DID YOU COMPLAIN? BY PHONE _____ LETTER _____ IN PERSON _____

MAY WE SEND A COPY OF THIS COMPLAINT TO THE COMPANY? YES _____ NO _____

STATE *BRIEFLY* THE *FACTS* OF YOUR COMPLAINT

FIGURE 15-9   A Prepared Form for Consumer Complaints

utes of the previous meeting were approved (or disapproved) as written or as amended. Summarize the points made during the group's discussion of each agenda item. Name the person who makes a motion, and the person who seconds it. Record the results of votes on each motion offered, along with a full description of the motion itself. If the group votes to support a specific proposal, include a description of that proposal. Figure 15-10 shows how effective minutes can provide a valuable record for future reference.

Progress Reports

As we said earlier, progress reports are often written daily as a way of keeping track of a project's status. For such regular daily reporting, many companies provide prepared forms. When they do not, the writer is expected to invent a form, as the writer in Figure 15-11 has done, to accommodate the data categories as they might change from day to day.

CHAPTER SUMMARY

Although informal reports are only a few lines to a few pages long, and usually are prepared quickly, they are the backbone of day-to-day written communication on the job. Depending on your subject, your reader's needs, and your company's policy, you might record your data in memo form, letter form, on a prepared form, or in a variety of other forms that we have called miscellaneous. Each format lends itself to certain reporting assignments.

1. The report in *memo form* follows a fixed format and is best used for in-house communication. It is easily prepared and often photocopied or mimeographed for distribution.

2. The report in *letter form* follows a standard letter format, with the addition of a subject line and headings as needed. It is designed for the special requirements of communicating outside your organization (most often with clients or prospective customers). Like all letters, it has a more personal tone than other types of informal reports.

3. The report on a *prepared form* follows specific guidelines for listing data so that they can be easily located, processed, and tabulated. Prepared-form reports are the easiest to write, but they sometimes require an additional statement to explain certain entries.

4. The report in miscellaneous form does not follow the format of a memo, a letter, or a prepared form. These reports are often in-house communication such as minutes, brief survey reports, or progress reports.

PARKS COLLEGE
Kadoka, Maine 04843

MINUTES OF THE STUDENT SENATE MEETING OF OCTOBER 7, 1981

Members Present

Kevin Ames, Charles Mott (treasurer), Janet Leroux, Donna Campbell, Steve Parks, Laurie Davis, Kate Taylor, Joe Griggs, Leslie Robett, Bill Faber, Laine Logan, Gerry Stark, Peter Jones (secretary), Kevin Oates, Tom Reid, Tricia Kelly, Ann Kearns, Mike Wills, Bob Moor, Tom Mackie, John Verdellini, Ann Reagan, Gerry Grimes (president).

Agenda

1. The meeting was called to order at 3:30 p.m. by President Gerry Grimes.

2. The minutes of the September meeting were approved as printed.

3. In his treasurer's report, Charlie Mott summarized his meeting with Dean Bailey, in which they discussed the dean's suggestion that the Student Senate use approximately $3800 from the President's discretionary fund for intramurals ($3000 for an equipment man and $800 for equipment). Charlie asked the Senate to consider the idea of using the money in this way.

4. The Black Student Union is sponsoring a play called Black Nativity by the Alma Lewis School of Fine Arts on December 10, at an admission fee of $1.00 per person. They requested $500 from the Senate to help defray costs. A motion was made by Donna Campbell and seconded by Tom Reid that the Senate allocate $500 from its reserve fund to the BSU for that purpose. (passed 15-3)

5. Arthur Burnham, bookstore manager, came before the Senate to answer questions about the bookstore's operation.

FIGURE 15-10   A Report of the Minutes of a Meeting

2

Topics discussed were: the bookstore's rate of profit, the
policy for charging books, the policy on book buy-backs, and
the ordering of individual books.

6. Members appointed to the commencement committee are
Laine Logan, Tricia Kelly, Tom Mackie, and Joe Griggs.

7. The Nursing Club requested $160 to cover bus rental
for a field trip to a cancer research center in Boston. A
motion was made by Mike Wills and seconded by Ann Reagan that
the Student Senate allocate $160 to the Nursing Club for its
trip. (passed unanimously)

8. Core requirements were discussed at the meeting.
The decision about what courses fit core requirements is
left to individual department heads. Kevin Oates suggested
that this decision be made by the Academic Curriculum Committee. Volunteers were requested in order to form a sub-
committee to further study this issue.

9. In the interest of encouraging active student parti-
cipation in upcoming local elections, a motion was made by
Kevin Ames and seconded by Laine Logan that the Senate form
a committee to establish a voter-registration drive on campus.
(passed unanimously) Committee members: John Verdellini,
Tom Reid, Bill Faber.

10. Mike Wills read his letter of resignation. A motion
was made by Kate Taylor and seconded by Steve Parks that the
Senate accept Mike's resignation with deep regret. (passed
unanimously) The meeting adjourned at 5:25 p.m.

Respectfully submitted

Peter Jones

Peter Jones, Secretary

cc: President Dithers
     Dean Bumstead
     All senators

FIGURE 15-10   (*Continued*)

DAILY CONSTRUCTION REPORT

LEVERETT LAND &   DATE: April 5, 1981
     TIMBER CO., INC.   PROJECT: Hadley Residence
BOX 185   JOB NO: 76-0141
AMHERST, MASS. 01002   SUBMITTED BY: D. Jenks, Sup't.

WEATHER: Cloudy, temp. in 50s

EMPLOYEES PRESENT: Jenks, Taylor, Barry, Smith, Reed

DESCRIPTION OF WORK: Frame upper level walls, sheath
                              walls, cut pilot rafters, cut and
                              block for mechanical equipment
                              installation. Supervise rough
                              grading.

SUBCONTRACTORS PRESENT: Masonry - W. Markowski, 4 men;
                              Excavator - W. Clark, 1 1/2 yd.
                              front-end loader with operator and
                              2 laborers

DELIVERIES & PURCHASES: Jones Lumber - roof framing materials,
                              see invoice #E-3341, LLT Purchase
                              Order #76-567. Valley Roofing
                              Supply - shingles, see invoice
                              #52-564, LLT Purchase Order #76-573
                              W. Clark - 48 yards bank run gravel,
                              9 yards 3/4" stone for drains

VISITORS TO SITE: W. Jones Lumber salesman,
                              T. Lombardi from Ashton Real Estate,
                              2 casual onlookers, owner

CHANGES OR EXTRA WORK: Enlarge master bath linen closet by
                              9" as requested by owner; send
                              memo! Realign foundation drain
                              pipes to avoid future garden area -
                              increase of 54' of 4" perforated
                              pipe, requested by owner; send
                              memo!

FIGURE 15-11   A Daily Progress Report in Miscellaneous Form

EQUIPMENT FAILURES: Change air hose fittings on small compressor (downtime 25 minutes); check spare parts inventory. Piston jammed twice on Bostitch nailgun; send in for rebuild ASAP.

COMMENTS: Rain last week causing difficulty in drainage work and rough grading; if things don't dry out by tomorrow we should delay driveway work until next week. Check schedule for co-ordinating roofing work with masons' chimney progress.

FIGURE 15-11   (*Continued*)

REVISION CHECKLIST

Use this checklist as a guide to refining and revising your informal reports.

1. Have you chosen the best form of short report for your specific purpose?
   a. a memorandum form for in-house readers
   b. a letter form for outside readers
   c. a prepared form if an appropriate one is available
   d. a miscellaneous form if your data must be supported by background
      information and definitions for readers who are not technically informed
2. Does the memo have a complete heading (name of organization, name
   and title of sender and recipient, identification of subject, date)?
3. Does the memo text follow an introduction-body-conclusion structure?
4. Have you single spaced within paragraphs and double spaced between?
5. Have you used headings, charts, or tables wherever they are needed?
6. If more than one reader is receiving copies, does the memo include a
   distribution notation (cc.) to identify all other readers?
7. Does the memo give your readers what they need to know — no more
   and no less (including all major points and omitting minor ones)?
8. Does the letter report follow a standard letter format (as in Chapter 14),
   and does it include a subject heading and any needed internal headings?
9. Does the letter embody a "you" perspective?
10. In a prepared-form report, are all required data recorded accurately,
    neatly, and clearly?
11. For a report in miscellaneous form, is the chosen format neat, attractive,
    and readable?
12. Does the arrangement of data represent the best possible choice (e.g.,
    charts, lists, etc.)?
13. Is the report written in correct standard English (grammar, mechanics,
    and usage, as discussed in Appendix A)?

Now list those elements of your report that need further attention.

EXERCISES

1. Write a memo to members of your college newspaper staff or to fellow
members of some other campus organization, announcing the time, place,
purpose, and brief agenda for an upcoming organizational meeting.

2. Ask a person working in your field to describe the kinds of short reports he or she most often writes for various purposes and audiences. Report your findings in a memo to your instructor.

3. In a brief essay, explain the difference between formal and informal reports. What elements do formal and informal reports have in common?

4. In a memo to your instructor, outline your progress to date on your term project. Describe your accomplishments, your plans for further work, and any problems or setbacks encountered. Also, request assistance if you need it. Conclude your memo with a specific completion date.

5. In a memo to your instructor, outline your progress on your career plans. Describe your achievements to date and any problems or setbacks, and request any needed advice.

6. Conduct a brief survey (e.g., of comparative interest rates from various banks on an auto loan, comparative tax and property evaluation rates in three local towns, or comparative prices among local retailers for a certain item). Arrange your data in a classification table (see Chapter 6) and report your findings to your instructor in a memo that closes with specific conclusions and recommendations for making the most economical choice.

7. Make up a periodic form that will allow you to keep a one-week record of your major daily activities. During that week, record the time spent in each activity. When your prepared-form report is completed, review the data and write an interpretation with conclusions and recommendations for better budgeting of your time.

8. Identify an area or situation on campus or in your community that is inconvenient or that poses a threat to public safety (such as endless cafeteria lines, a dangerous intersection, slippery stairs, a busy street filled with potholes, a poorly marked railway crossing, a bus stop without a weather shelter, a poorly adjusted traffic light). Observe the problem area or situation for several hours during a peak usage period. Write a site inspection report to appropriate authorities describing the deficiencies, noting your firsthand observations, making recommendations for positive change, and enlisting your reader's support through positive action. Submit your letter report to your instructor.

9. *In class:* Divide into groups of five or six students. Choose a subject composed of several related areas (for example, *Academic Policy:* final examination policy, grading policy, core requirements, the foreign language requirement, the minimum cumulative average required for graduation, the need for certain new courses, the minimum cumulative average required for first-semester freshmen, etc.). In one class period discuss each of the items related to your subject, making proposals for positive change. *Each* member of your group should act as secretary, recording appropriate minutes of the meeting.

At home: compose your minutes (see Figure 15-10) and bring them to class for comparison with the minutes written by the other members of your discussion group. Exchange copies for proofreading and revision. Are your minutes accurate, concise, and clear? Revise your report according to your proofreader's suggestions, and submit your early and final drafts to your instructor.

10. In a memorandum, compose a list of specifications and a full cost estimate of the items required for your college attendance during one full semester. Specifications should include a complete, itemized list of school fees, supplies, clothing, entertainment, travel, living arrangements (apartment, room, house, roommates), etc. Your final cost estimate will be a total of the costs of each of these items.

11. Use the prepared form for consumer complaints (Figure 15-9) as a model for your own report to the attorney general's office describing a legitimate complaint.

16

Writing
a Proposal

DEFINITION AND PURPOSE

THE PROPOSAL PROCESS

TYPES OF PROPOSALS
The Planning Proposal
The Research Proposal
The Sales Proposal

ELEMENTS OF AN EFFECTIVE
PROPOSAL
Appropriate Format and Supplements
A Focused Subject and a Worthwhile
Purpose
Identification of Related Problems
Realistic Methods
Concrete and Specific Information
Visuals
Appropriate Level of Technicality
A Tone that Connects with Readers

PLANNING AND WRITING THE
PROPOSAL
Introduction
Body
Conclusion

APPLYING THE STEPS

CHAPTER SUMMARY

REVISION CHECKLIST

EXERCISES

DEFINITION AND PURPOSE

A proposal is an offer to do something or a suggestion for action. The general purpose of a proposal is to *persuade* readers to improve conditions, authorize work on a project, accept the offer of a service or product (for payment), or otherwise support a plan for solving a problem or doing a job.

Your own proposal may be a letter to your school board to suggest changes in the English curriculum; it may be a memo to your firm's vice-president to request funding for a training program for new sales representatives; or it may be a 1,000-page document to the Defense Department to bid for a missile contract (competing with proposals from other firms). You might write the proposal by yourself or as part of a team. It might take hours or years.

Whether in science, business, industry, government, or education, proposals are written for decision makers: managers, executives, directors, clients, trustees, board members, community leaders, and the like. Inside or outside your organization, these are the people who decide if your suggestions are worthwhile, if your project will ever get off the ground, if your service or product is useful. In fact, if your job depends on funding from outside sources, proposals will be your most important writing activity. To achieve its purpose, a proposal must be convincing.

THE PROPOSAL PROCESS

The basic proposal process can be summarized simply: someone offers a plan for something that needs to be done. In business and government, this process has three phases:

1. Client *X* needs a service or product.
2. Firms *A*, *B*, *C*, etc., propose ways to satisfy the need.
3. Client *X* awards the job to the firm offering the best proposal.

The complexity of events within each of these phases will of course depend on the specific situation. Here is a typical situation:

Assume that you manage a mining engineering firm in Tulsa, Oklahoma. On Wednesday, February 19, you spot this announcement in the *Commerce Business Daily*:[1]

> **R — Development of Alternative Solutions to Acid Mine Water Contamination from Abandoned Lead and Zinc Mines** to Tar Creek, Neosho River, Grand Lake, and the Boone and Roubidoux aquifers in northeastern Oklahoma. This will include assessment of environmental impacts of mine drainage followed by development and evaluation of alternate solutions to alleviate acid mine drainage in receiving streams. An optional portion of the contract, to be bid upon as an add-on and awarded at the discretion of the OWRB, will be to prepare an Environmental Impact Assessment for each of three alternative solutions as selected by the OWRB. The project is expected to take 6 months to accomplish with anticipated completion date of September 30, 1981. The projected effort for the required task is 30 person-months. Request for proposals will be issued and copies provided to interested sources upon receipt of a written request. Proposals are due March 1, 1981. (044)
>
> Oklahoma Water Resources Board
> P.O. Box 53585
> 1000 Northeast 10th Street
> Oklahoma City, OK 73152
> (405) 271-2541

Your firm has the personnel, experience, and time to do the job, so you decide to compete for the contract. Because the March 1 deadline is fast approaching, you write immediately for a copy of the request for proposal (RFP). The RFP will contain the guidelines for developing the proposal — guidelines for spelling out your plan to solve the problem (methods, timetables, costs, etc.).

---

[1] This daily publication lists the government's latest needs for *services* (salvage, engineering, maintenance, etc.) and for *supplies, equipment, and materials* (guided missiles, engine parts, building materials, etc.). The *Commerce Business Daily* is an essential reference tool for anyone whose firm seeks government contracts.

The Grantsmanship Center News, published six times a year, contains a wealth of information about federal grants for nonprofit organizations in areas such as rural development, media, science, energy and environment, and education. This publication also lists notices of training programs in proposal writing, and offers books on proposal writing and evaluation, along with bibliographies of essential publications about the proposal process.

Two days later, you receive a copy of the RFP. It is now February 21 — only one week before the deadline. You get right to work, with the assistance of the two staff engineers you have appointed to your proposal team. Since the credentials of your staff will be a factor in gaining client acceptance of the proposal, you ask all members to update their résumés (for inclusion in an appendix to the proposal).

Several other firms will be competing for this one job. The Oklahoma Water Resources Board will award the contract to the firm submitting the best proposal, and board members will use the following criteria (and perhaps others) in evaluating individual proposals:

- understanding of the client's needs described in the RFP
- soundness of the firm's technical approach
- quality of project organization and management
- ability to complete the job by the deadline
- ability to control costs
- specialized experience of the firm in this type of work
- qualifications of the staff to be assigned to the project
- the firm's track record for similar projects

A client's specific evaluation criteria often are listed (in order of importance or according to a point scale) in the RFP. Although these criteria may vary, every client expects a proposal that is *clear, informative,* and *realistic.*

Some clients will hold a preproposal conference for the competing firms. During this briefing, the various firms are told of the client's needs, expectations, specific start-up and completion dates, criteria for evaluation, and any other matters that will lead to high-quality proposals.

The sample situation described above is only one among countless possibilities. You might encounter proposal situations that are quite different, but in every case the process will be similar: you will propose a plan to fill a need.

TYPES OF PROPOSALS

Despite their wide variety, proposals can be classified in three ways: according to *origin, audience,* or *intention.* Based on its origin, a proposal may be classified as *solicited* or *unsolicited* — that is, requested by someone or initiated on your own because you have recognized a need. Business and government proposals are most often solicited and originate from a customer's request (as shown in the sample situation on page 430).

Based on its audience, a proposal may be classified as *internal* or *external* — that is, written for members of your organization or written for clients and funding agencies. (The sample situation on page 430 would call for an external proposal.)

Based on its intention, a proposal may be classified as a *planning, research,* or *sales* proposal. These last categories by no means account for all variations among proposals. In fact, certain proposals may fall under all three categories, but these are the types you will most likely have to write. Each type is discussed and illustrated below.

The Planning Proposal

A planning proposal suggests ways to solve a problem or to bring about improvement. It might be a request for funding to expand the campus newspaper, an architectural plan for new facilities at a ski area, or a plan to develop energy alternatives to fossil fuels. In every case, the successful planning proposal answers this central question for readers:

– What are the benefits of following your suggestions?

The planning proposal in Figure 16-1 is internal and unsolicited. The writer, an ecology-minded member of the Coastal Zone Management Commission, is proposing certain changes in CZM's plan for developing the coastline. The proposed changes affect two sections of the plan, so the headings, "Philosophy" and "Costs," give readers a clear signal about the contents. The issue is controversial: Should we locate refineries and nuclear plants on the Massachusetts coastline? Several commission members feel that the answer is yes. Therefore, the writer selects a tone that is firm and confident but not antagonistic or superior. Her tone shows that she believes in her recommendations. A less confident writer might have replaced "should" with "might possibly." A more militant writer might have replaced "should" with "must." [2]

Because this proposal is internal, it is cast informally as a memo. A formal planning proposal is illustrated on pages 454–469.

The Research Proposal

Research (or grant) proposals request approval (and often funding) for a research project. A professor of microbiology, for instance, might address a research proposal to the Environmental Protection Agency to request funds for a study of toxic contaminants in local groundwater. There are numerous fed-

---

[2] "Must" is appropriate in the second paragraph under "Costs" because this item deals with the most immediate and crucial issue (siting decisions about power plants, etc.). Therefore, the wording must be as strong as the situation will bear in order to influence decisions crucial to preserving the environment. (See page 446 for a discussion of tone.)

MASSACHUSETTS COASTAL ZONE
MANAGEMENT COMMISSION

TO: Members of the Commission   February 13, 1981

FROM: Judith M. Barnet
           APCC Representative

SUBJECT: PROPOSAL FOR AMENDMENTS TO THE ENERGY CHAPTER OF
           THE MASSACHUSETTS COASTAL ZONE MANAGEMENT PREVIEW

A realistic energy policy should reflect the broadest philosophy and a detailed assessment of costs.

Philosophy

Any public energy policy written in the 1980s should begin by
encouraging conservation and development of alternate energy
sources. This commitment should be reflected in every statement made by a governmental body. To implement this attitude,
the Association for the Preservation of Cape Cod proposes that
Massachusetts CZM amend its policy as follows:

   1. Insert as one of the objectives in the energy
       chapter of the CZM plan (page 2-G/22): "to
       promote the conservation of energy by every
       consumer -- residential, commercial, industrial,
       or municipal."

   2. Insert as one of the earliest policies and recom-
       mendations (before present policy #29) a state-
       ment of support and encouragement for development
       and use of alternate, nontraditional energy
       sources, especially by small communities.

Costs

Any public energy policy written in the 1980s should include
a proper cost assessment, which in the CZM plan should be

FIGURE 16-1   A Planning Proposal

2

implemented by appropriate discussion and tables reflecting
the following:

1. Environmental costs of siting oil and nuclear
    facilities on the coast. The Program Preview
    does not address these costs.

2. Economic costs (best estimate) of alternate
    sources. These should be supplied in order to
    permit comparison with economic costs of conven-
    tional sources.

The Energy Facilities Siting Council must agree to review all
possible technologies when making siting decisions, recognizing the inherent characteristics of each. In other words,
the council should not measure alternate sources by the criteria appropriate to centralized sources (e.g., that they be
able to supply vast amounts of energy over vast networks).

Implementing these proposed amendments will help ensure a
balanced program of energy development in coastal areas.

FIGURE 16-1 (*Continued*)

eral and private agencies that solicit proposals such as the Department of Health and Human Services, the National Science Foundation, the National Institutes of Health, the Department of Agriculture, the Carnegie Foundation, the Fulbright-Hays Foundation, and others. Each granting agency has its own requirements for the format and content of proposals, but any successful research proposal answers these central questions for readers:

- Why is this project worthwhile?
- What qualifies you to undertake the project?
- What are its chances of success?

In college, you might submit proposals for independent study, field study, or a thesis project. Here is the title of a research proposal submitted to the thesis committee in a geology department:

A PROPOSAL FOR A MASTER'S THESIS PROJECT
TO INVESTIGATE THE TERTIARY GEOLOGY
OF THE ST. MARIES RIVER DRAINAGE
FROM ST. MARIES TO CLARKIA, IDAHO

A technical writing student might submit an informal proposal requesting the instructor's approval of a term project (which, in turn, may be a formal proposal). The introduction of the proposal in Figure 16-2 describes the background of the problem and justifies the need for the study. The body outlines the scope, method, and sources for the proposed investigation. The conclusion describes the goal of the investigation and encourages reader support. Note that the proposal is convincing because it answers questions about *what, why, how, when,* and *where.*

The Sales Proposal

A sales proposal is addressed to clients and offers a service or product (for payment). Sales proposals may be solicited or unsolicited. If they are solicited, several firms may be competing with proposals of their own. Because sales proposals are addressed to readers outside your organization, they are cast as letters (if they are brief), but long sales proposals, like long reports, are presented as formal documents complete with supplements (cover letter, title page, table of contents, etc.).

The sales proposal, a major marketing tool in business and industry, will be successful if it answers this question:

Why will your service or product fill our needs better than that of your competitors?

or

Why should we hire you over someone else?

TO: Dr. John Lannon   March 16, 1981

FROM: T. Sorrells Dewoody /SD/

SUBJECT: A PROPOSAL FOR DETERMINING THE FEASIBILITY OF
           MARKETING DEAD WESTERN WHITE PINE

Introduction

Over the past four decades huge losses of western white pine
have occurred in the Northern Rockies, primarily attributable
to white pine blister rust and the attack of the mountain
pine beetle. Estimated annual mortality is 318 million board
feet. Because of the low natural resistance of white pine
to blister rust, this high mortality rate is expected to continue indefinitely.

If white pine is not harvested while the tree is dying or
soon after death, the wood begins to dry and check. The sap-
wood is discolored by blue stain, a fungus carried by the
mountain pine beetle. If the white pine continues to stand
after death, heart cracks develop. These factors work together to cause degradation of the lumber (as graded by the
Western Wood Product Inspectors) and a consequent loss in
value.

Statement of Problem

White pine mortality results in a reduction in value of white
pine stumpage, since the commercial lumber market will not
accept it. The major implications of this problem are two:
first, in the face of rising demand for wood, vast amounts of
timber are not being used; second, dead trees are left to
accumulate in the woods, where they are rapidly becoming a
major fire hazard.

Proposed Solution

One possible solution to the problem of white pine mortality
and waste is to search for markets other than the conventional
lumber market. The last few years have seen a burst of

FIGURE 16-2   A Research Proposal

2

popularity and a growing demand for weathered barn boards and wormy pine for interior paneling. Some firms around the country are marketing defective wood as specialty products. (Note: These firms call the wood from which their products come "distressed." This term will hereafter be used to refer to dead and defective white pine.) There is a good possibility that distressed white pine might find a place in such a market.

Scope

To determine the feasibility of developing a market for distressed white pine, I plan to pursue six areas of inquiry:

1. What products are presently being produced from dead wood, and what are the approximate costs of production?

2. How much demand is there for distressed wood products?

3. Can distressed white pine meet this demand as well as other species meet it?

4. Is there room in the market for distressed white pine?

5. What are the projected costs of retrieving and milling distressed white pine?

6. What prices for the products can the market bear?

Methods

My primary data sources will include consultations with Dr. James Hill, Professor of Wood Utilization, and Dr. Sven Bergman, Forest Economist -- both members of the College of Forestry, Wildlife, and Range. I will also inspect decks of dead white pine at several locations, and visit a processing mill to evaluate it as a possible base of operations. I will round out my primary research with a letter and

FIGURE 16-2 (*Continued*)

telephone survey of processors and wholesalers of distressed
material.

Secondary sources will include selected publications on the
uses of dead timber, and a review of an ongoing study by Dr.
Hill concerning the uses of dead white pine.

<u>My Qualifications</u>

I have been following Dr. Hill's study on dead white pine for
two years. In June of this year I will receive my B.S. in
forest management. I am familiar with wood milling processes
and have had firsthand experience at logging. My association
with Dr. Hill and Dr. Bergman creates the opportunity for an
in-depth feasibility study.

<u>Conclusion</u>

Clearly, something should be done to reduce the vast accumula-
tions of dead white pine in our forests. The land on which
they stand is among the most productive forest land in northern Idaho. By addressing the six areas of inquiry mentioned
earlier, I can determine the feasibility of directing capital
and labor to the production of distressed white pine products.
With your approval I will begin my research at once.

FIGURE 16-2 (*Continued*)

The solicited proposal in Figure 16-3 offers a service. Because the writer is competing against other firms, the body of his proposal explains specifically *why* his machinery is best for the job, *how* the job can best be completed, *what* his qualifications are for getting the job done, and *how much* the job will cost. He will be legally bound by his estimate. Therefore, he points out possible causes of increased costs to protect himself. In your own sales proposals, *don't make the mistake of underestimating costs by failing to account for all variables* — a sure way to lose money!

The writer's introduction describes the subject and purpose of the proposal. The conclusion reinforces the confident tone throughout and encourages reader acceptance by ending with — and thereby emphasizing — two key words: "economically" and "efficiently."

The above three categories for proposals (planning, research, and sales) are neither exhaustive nor mutually exclusive. A research proposal, for example, may request funds for a study that will eventually lead to a planning proposal. The architectural proposal shown partially on pages 442–445 is a combined planning and sales proposal: if the clients accept the writer's preliminary plan, they will then hire the firm to design the new ski lodge facilities.

ELEMENTS OF AN EFFECTIVE PROPOSAL

Readers will evaluate your proposal according to how clearly you answer their questions about *what, why, how, when,* and *how much.* An effective proposal is clear, informative, and realistic and conforms to the following guidelines.

Appropriate Format and Supplements

We have seen that short proposals can be cast as memos or letters, depending on whether they are internal or external. Longer memos or letters, however, may be more easily read if headings are used. Either format may include appendixes (see pages 239–256) for support material (maps, blueprints, specifications, calculations, etc.) that would interrupt the flow of the text.

Although a short proposal should serve many of your purposes, some projects are too complex for a short proposal. Moreover, perhaps different readers inside and outside your organization will be interested in different parts of your proposal: some may need only a summary; others may already know about the problem and will want to read only your plan; still others may want all the details. Or perhaps a soliciting agency will specify a certain format and supplements. Such cases may call for a long proposal that, like the formal report, is presented with all needed supplements (cover letter, title page, summary or informative abstract, etc.). See the sample proposal on pages 454–469.

James A. Landmover
Route #1
Buhl, Idaho 83331
Tel. 208-634-5267

March 28, 1981

Greg Haver
Star Route
Bliss, Idaho 83314

SUBJECT: PROPOSAL TO DIG A TRENCH AND MOVE BOULDERS AT SITE
           TEN MILES WEST OF BLISS

Dear Mr. Haver:

I've inspected your property and would be happy to undertake
the landscaping project necessary for the development of your
farm.

The backhoe that I use cuts a span three feet wide and can
dig as deep as eighteen feet -- more than an adequate depth
for the mainline pipe you wish to lay. Because this backhoe
is on tracks rather than tires, and is hydraulically operated,
it is particularly efficient in moving rocks. I have more
than twelve years of experience with backhoe work and have
completed many jobs similar to this one.

After examining the huge boulders that block access to your
property, I am convinced they can be moved only if I dig out
underneath and exert upward pressure with the hydraulic ram
while you push forward on the boulders with your D-9 Cater-
pillar. With this method we can move enough rock to enable
you to farm that now inaccessible tract. Because of its
power, my larger backhoe will save you both time and money
in the long run.

FIGURE 16-3   A Sales Proposal

2

The job should take twelve to fifteen hours, unless we encounter subsurface ledge formations. My fee is $60 an hour.
The fact that I provide my own dynamite crew at no extra
charge should be an advantage to you since you have so much
rock to be moved.

Please phone me at any time for any additional information.
I'm sure we can do the job economically and efficiently.

                                         Yours truly,

                                         James A. Landmover

                                         James A. Landmover

FIGURE 16-3   (*Continued*)

Remember that decision makers are busy people who dislike fog. Begin with a title that is absolutely clear about the intent of your proposal:

Foggy   A PROPOSAL FOR FACILITIES
                               AT SUGARPLUM SKI AREA

What kinds of facilities are being planned — lifts, lodge parking? What is being proposed — new construction, remodeling, design?

Revised   A PRELIMINARY DESIGN PROPOSAL
                               FOR THE NEW LODGE FACILITIES
                               AT SUGARPLUM SKI AREA

Foggy   RECOMMENDED IMPROVEMENTS
                               FOR THE RAILWAY LOCOMOTIVE
                               FACILITY IN KANSAS CITY

What kinds of improvements are being recommended?

Revised   RECOMMENDED WASTEWATER TREATMENT
            FOR THE RAILWAY LOCOMOTIVE
            FACILITY IN KANSAS CITY

A concrete and specific title signals a concrete and specific proposal.

A Focused Subject and a Worthwhile Purpose

Do not try to solve all the world's problems in a single proposal. A fatal mistake in a research proposal, for instance, is to begin writing too soon — before you have zeroed in on your subject and purpose. Instead, focus on *one* specific research question that you can answer exhaustively and make your approach original enough to get the reader's attention and support.

The same need for a tight focus applies to planning or sales proposals. Readers want specific suggestions for filling specific needs. Show them immediately that you understand their problem by spelling out your subject and purpose.

Subject

The Sugarplum Corporation wishes to expand its ski area facilities so as to develop a year-round destination resort. As part of its expansion program, Sugarplum needs hotel rooms, a day care center, and larger space for its cafeteria, bar, ski-and-rental shop, ski school, ski patrol room, and parking.

Purpose

The architectural firm of Brinckerhoff and Brinckerhoff offers the following preliminary design proposal for the lodge facilities. The proposed design is based on our evaluation of site information, the master plan, and a list of required facilities and their relationships, as well as the projected circulation around and through the required spaces. Our building philosophy and design priorities also take into account local building restrictions and possibilities of future expansion.

Notice that the focus here is limited to the *preliminary design phase*. The actual plans for construction, decor, landscaping, etc., will be the subject of later proposals. This, then, can be called a *preproposal*. If accepted, it will lead to more specific proposals.

Identification of Related Problems

Do not underestimate the complexity of the project. Identify *all* related problems that readers themselves might not recognize. Here is how the architectural proposal treats one such problem:

> The Sugarplum Corporation has expressed a desire for glass walls on the north, east, and south exposures of the main lodge. Although improving the view, glass walls would increase heat loss through thermal conductivity. Concentrated areas of glass (triple-glazed) should be limited to southerly exposures for maximum use of solar energy. For further energy efficiency, exterior walls should be insulated to a value of R-19 or better, and the doors built in a double-door configuration to minimize infiltration.

Realistic Methods

Resist the temptation to propose easy answers to hard questions. Be conservative. Propose only those methods that have a good chance of success. If a certain solution is the best available — but still leaves doubt as to its effectiveness — let the readers know; otherwise, you or your firm could later be held liable for the project's failure. Avoid overstatement. Here is how the architectural proposal treats a questionable solution:

> The subsoil near the lodge is sand, with a two-foot cover of humus and no rock outcrops. Some places on the higher mountain have depths of over 10 feet. Where the bedrock is close to the surface, however, considerable runoff occurs, causing erosion around the drainage ditches and the parking lot (see the drainage map in Appendix B). A system of French drains, tile beds, and drainage pipes coupled with a landscaping program *can eliminate most, if not all, of the drainage and erosion.*

Notice how the writer qualifies his suggested solution (in the phrase italicized here) to avoid promising more than he can deliver.

Concrete and Specific Information

Vagueness is a fatal flaw in a proposal. Before you can *persuade* readers, you must *inform* them; therefore, you need to *show* as well as *tell*. (Review pages 56–57.) Instead of writing "The mountain is covered with mixed vegetation," write "The mountain's vegetation consists primarily of tamarack, cedar, and sub-alpine fir, with a ground cover of bear grass and alder bush."

Here are the specifications that will serve as design guidelines for the ski patrol suite:

> *Ski Patrol Suite*
>
> *Square Footage Requirements:* 900 sq. ft. total; first aid room = 400 sq. ft. (to include a 30-sq.-ft. restroom); dressing room/lockers = 500 sq. ft.
>
> *Occupants:* ski patrol persons; injured skiers and their friends.
>
> *Special Equipment:* ski, boot, and pole storage, lockers, boot-drying racks, good heater, storage for toboggans and first aid equipment, two cots, coat racks, blackboard, desk, exterior ski racks.
>
> *Special Considerations:* first aid room should be pleasant and comfortable; ski patrol room should be functional; cots should be well sheltered from drafts and doors; floor surface in the patrol room should be durable; first aid room should have easy slope, snow-cat, and ambulance access.
>
> *Location:* close to the parking lot.

A less competent writer might have said:

> The ski patrol suite will have to be large enough to accommodate the various personnel, patrons, and equipment, and will have to be located in an easily accessible area.

Notice how abstract words ("large," "accessible") and general words ("personnel," "patrons," "equipment") provide no useful design guidelines until they are supported with concrete and specific details.

Visuals

If they will enhance your proposal, use visuals, but be sure they are properly introduced and discussed (review Chapter 11).

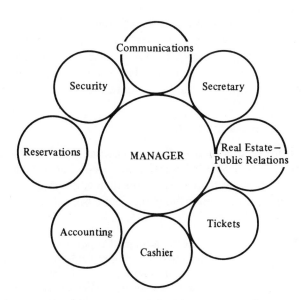

FIGURE 1   Bubble Diagram of the Administrative Complex

Spatial Relationships
The design of the administrative complex should facilitate circulation and communication among various departments, as well as allowing for centralized management. See Figure 1.

Appropriate Level of Technicality

A single proposal might address a diverse audience. A research proposal might be read by experts in the field, who then advise the granting agency whether to accept or reject it. Planning and sales proposals might be read by colleagues, superiors, and clients (who are often laypersons). Informed and expert readers will be most interested in the technical details of the project. Lay readers will be more interested in the projected results, but they will need an explanation of technical details as well.

Unless your proposal gives all readers what they need, it is not likely to move anyone to action. This is where supplements are useful (especially abstracts, glossaries, and appendixes). Let your knowledge of the audience guide your decisions about supplements. Who is your secondary audience (pages 17–18)? Who else will be evaluating your proposal?

If the primary audience is expert or informed, keep the proposal itself technical. For uninformed secondary readers (if any), provide an informative abstract, a glossary, and appendixes that explain specialized information. If the primary audience has no expertise and the secondary audience does, follow this pattern: write the proposal itself for laypersons and provide appendixes containing the technical details (formulas, specifications, calculations, etc.) that the informed readers will use to evaluate the soundness of your plan.

A Tone that Connects with Readers

The successful proposal is the one that is *accepted* by its audience, so make it reader-oriented. Express your information in a tone that is confident, encouraging, and diplomatic. Show readers that you believe in your plan; urge them to act on your proposal; anticipate how they will respond to your suggestions. Do not come across with a superior attitude.

> Superior The Sugarplum Corporation *has remained ignorant* of the *obvious* problems of drainage and erosion, snow accumulation, and hazardous circulation patterns. Moreover, the corporation *had better pay close attention* to the following design restrictions: building codes, energy conservation, the limited use of stairs, and anticipated snow loads on the structures.

Although *what* is said above is worthwhile, *how* it is said alienates readers. The italicized words create a superior and insulting tone.

On the other hand, don't come across as a milquetoast:

> Weak *It would seem that there is a possibility of* problems with drainage and erosion, snow accumulation, and hazardous circulation patterns. Also, *it might not be a bad idea to give some consideration to* the following design restrictions. . . .

The words italicized here make the writer seem unsure, and the tone will not inspire the reader's confidence.

The following version has a tone that is confident, encouraging, and diplomatic:

> Revised Design considerations must address the problems of drainage and erosion as well as snow accumulation and hazardous circulation patterns. Moreover, a successful design must abide by the following restrictions: building codes, energy conservation, the limited use of stairs, and anticipated snow loads on the structures.

PLANNING AND WRITING THE PROPOSAL

Like all good factual writing, proposals have an introduction-body-conclusion arrangement. Depending on project complexity each section contains some or all of the subsections listed in the following general outline:

I. INTRODUCTION
    A. Subject and Purpose (or Objective)
    B. Statement of Problem
    C. Background
    D. Need
    E. Qualifications of Personnel
    F. Data Sources
    G. Limitations
    H. Scope

II. BODY
    A. Methods
    B. Timetable
    C. Materials and Equipment
    D. Personnel
    E. Available Facilities
    F. Needed Facilities
    G. Cost
    H. Expected Results
    I. Feasibility

III. CONCLUSION
    A. Summary of Key Points
    B. Request for Action

These subsection headings can be rearranged, combined, divided, or deleted as needed. Although not every proposal contains all subsections, each major section must answer certain reader questions, as illustrated below.

Introduction

The introduction answers all the following questions (or all those that apply to the situation):

    – What problem are you proposing to solve?
    – In general, what solution are you proposing?
    – Why are you proposing it?
    – What are the benefits?
    – What are your qualifications for this project?

From the outset, your goal is to sell your idea, to convince readers that the job needs doing and that you (or someone else) are the one to do it. If your introduction is long-winded, evasive, or vague, readers probably will not read on. Make it concise, specific, and clear.

State the problem and give a brief background. Explain why the problem should be solved or the project undertaken. Identify any sources of data. In a research or sales proposal, state your qualifications for doing the job. If your plan has limitations, explain them. Finally, define the scope of your plan by enumerating the specific subsections to be discussed in the body section.

Here is the introduction for a planning proposal titled "A Proposal for Solving the Noise Problem in the University Library." Jill Sanders, a library work-study student, addresses her unsolicited proposal to the chief librarian and the administrative staff. Because this proposal is unsolicited, it must first make the problem vivid through concrete details that will arouse reader concern and interest. Therefore, this introduction is longer than it would be in a solicited proposal, where readers would already agree about the severity of the problem.

INTRODUCTION

Subject

During the October 1979 Convocation at Margate University, many students and faculty members complained about noise in the library. Shortly afterward, certain areas were designated for "quiet study," but the library continues to receive complaints about noise. In order to create an atmosphere for scholarly work, the library staff should take immediate action to decrease noise.

Purpose

This proposal examines the noise problem from the viewpoint of students, faculty, and library staff. It then offers a plan for ensuring that certain areas of the library will be sufficiently quiet to promote serious study and research.

Sources

The data for this proposal come from a university-wide questionnaire, interviews with students, faculty, and library staff, inquiry letters to other college libraries, and my own observations for three years as a library staff member.

Statement of Problem

This subsection examines the noise problem in terms of its severity and its causes.

Severity

Since the 1979 Convocation, the library's fourth and fifth floors have been reserved for quiet study, but students hold group-study sessions at the large tables and disturb others working alone. The constant use of computer terminals on both floors adds to the noise, especially when students working in pairs discuss the program. Moreover, people often chat as they enter or leave study areas.

On the second and third floors, designed for reference, staff members help patrons locate materials. Thus there is a constant shuffling of people and books, as well as loud conversation. At the computer service desk on the third floor, conferences between students and instructors cause additional noise.

The most frequently voiced complaint from the faculty members interviewed concerned the second floor where people using the Reference and Government Documents services converse loudly. Students complain about the lack of a quiet spot to study, especially in the evening, when even the "quiet" floors are as noisy as the dorms.

Over 80 percent of the respondents (530 undergraduates, 30 faculty, 22 graduate students) to a university-wide questionnaire (Appendix A) insisted that excessive noise discourages them from using the library as often as they would otherwise. Of the student respondents, 430 cited quiet study as their primary reason for wishing to use the library.

The library staff recognizes the problem but has insufficient personnel to cope with it. Because all staff members have assigned tasks, they have no time to monitor noise in their sections.

Causes

Respondents complained specifically about the following causes of noise (in descending order of frequency):

1. Loud study groups that often lapse into social discussions.
2. A general disrespect for the library, with the attitude of some students characterized as "rude," "inconsiderate," or "immature."
3. The constant clicking of computer terminals on all five floors, and of typewriters on the first three.
4. Vacuuming by the evening custodians.

In addition, virtually all complaints centered on the lack of enforcement by the library staff.

Because the day staff works on the first three floors quiet-study rules are not enforced on the fourth and fifth floors. Work-study students on these floors have no authority to enforce rules not enforced by the regular staff. Small, black-and-white "Quiet Please" signs posted on all floors apparently go unnoticed, and the evening security guard provides no added deterrent.

Needs

Excessive noise in the library is keeping patrons away. By addressing this problem immediately, we can help restore the library's credibility and utility as a campus resource. We must take steps to decrease noise on the lower floors and to eliminate it from the quiet-study floors.

Scope

The proposed plan includes a detailed assessment of methods, costs and materials, personnel requirements, feasibility, and expected results.

Body

The body section will receive the most attention from readers. It answers all the following questions that are applicable:

- How will it be done?
- When will it be done?
- What materials, methods, and personnel will it take?
- What facilities are available?
- How long will it take?
- How much will it cost, and why?
- What results can we expect?
- How do we know it will work?

Here you spell out your plan in enough detail for readers to evaluate its soundness. If this section is vague, your proposal stands no chance of being accepted. Besides being clear, be sure that your plan is realistic, that it promises no more than you can deliver. The main goal of this section is to prove that your plan will work.

PROPOSED PLAN

The following plan takes into account the needs and wishes of our campus community as well as the available facilities in our library.

Methods

Noise in the library can be reduced through three complementary steps: (1) improving publicity, (2) shutting down and modifying our facilities, and (3) enforcing the quiet rules.

Improving Publicity

First, the library must publicize the noise problem to the university. This assertive move will demonstrate the staff's concern. Publicity could include articles by staff members in the campus newspaper, leaflets distributed on campus, and a freshman library orientation that acknowledges the noise problem and asks the cooperation of new students. All forms of publicity should include a detailed account of the steps the library is taking to solve the problem.

Shutting Down and Modifying Facilities

After the campus and local newspapers have been notified, the library should be shut down for one week. To minimize disruption, the shutdown should occur during the week between the end of summer school and the beginning of the fall term.

During this period, the fixed tables on the fourth and fifth floors can be converted to cubicles by the use of temporary partitions (6 cubicles per table). The cubicles could later be converted to shelves as the need for book space increases.

All unfixed tables from the third, fourth, and fifth floors can be taken to the first floor where a space for group study will be set up. Plans already are underway for removing the computer terminals from the fourth and fifth floors.

Enforcing the Quiet Rules

Enforcement is the essential, long-term element of this plan. No one of any age is likely to follow all the rules all the time — unless they are enforced.

First, new "Quiet" posters can be made to replace the present, innocuous notices. A visual design student can be hired to draw up large, colorful posters that attract attention. Either the design student or the university print shop can take charge of poster production.

Next, through publicity, library patrons can be encouraged to demand quiet from noisy people. To support such patron demands, the library staff can begin periodic monitoring of the fourth and fifth floors, asking study groups to move to the first floor, and revoking library privileges of those who refuse. Patrons on the second and third floors can be asked to speak in whispers. Staff members should set an example by regulating their own voices.

Costs and Materials

The major cost would be for salaries of new staff members who would help monitor. Next year's library budget, however, would include an allocation for four new staff members.

A design student has offered to make up four different posters for $200. The university printing office can reproduce as many posters as needed at no extra cost.

Prefabricated cubicles for 26 tables sell for $150 apiece, for a total cost of $3,900.

Rearrangement of various floors can be handled by the library's custodians.

The Student Fee Allocations Committee and the Student Senate routinely reserve funds for improving student facilities. A request to these organizations would yield at least partial funding for the plan.

Personnel

The success of this plan ultimately depends on the willingness of the library administration to implement it. The program itself can be run by committees made up of students, staff, and faculty. This is yet another area where publicity is essential to persuade people that the problem is severe and that their help is needed. In order to recruit committee members an added incentive can be offered to students through the untapped resource of Contract Learning credits.

The proposed committees include an Antinoise Committee overseeing the program, a Public Relations Committee, a Poster Committee, and an Enforcement Committee.

Feasibility

On March 15, 1979, I mailed survey letters to 25 New England colleges, inquiring about their methods for coping with noise. Among the respondents, 16 stated that publicity and the administration's attitude toward enforcement were key elements in their success.

Improved publicity and enforcement could certainly work for us as well. Moreover, slight modifications in our facilities to concentrate group study on the busiest floors would automatically lighten the burden of enforcement.

Expected Results

Increased publicity will improve communications between the library and the campus community. An assertive stance will show that the library is aware of the needs of its patrons and willing to meet those needs. Offering the program for public inspection will draw the entire community into improvement efforts. Publicity, begun now, will pave the way for the formation of committees.

The library shutdown will have a dual effect: it will dramatize the problem to the community, and also provide time for the physical changes. (An antinoise program begun with carpentry noise in the quiet areas would hardly be effective.) The shutdown will be both a symbolic and a concrete measure and perhaps lead to the reopening of the library with a new philosophy and a new face.

The library's seemingly passive stance has been a major cause of the problem. Continued, strict enforcement will be the backbone of the program. It will prove that staff members are concerned enough about the atmosphere to jeopardize their own friendly image in the eyes of certain users and that the library is not afraid to enforce its rules.

Conclusion

The conclusion restates the need for the project and persuades readers to act. It answers the questions that readers will ask:

- How badly do we need this change?
- Why should we accept your proposal?
- How do we know this is the best plan?

End on a strong note, with a conclusion that is assertive, confident, and encouraging — and keep it short.

CONCLUSION AND RECOMMENDATION

The noise in Margate University library has become embarrassing and annoying to the whole campus. Positive forceful steps must now be taken to improve the academic atmosphere.

Aside from the intangible question of image, close inspection of the proposed plan will show that it will work if all recommended steps are taken and — most important — if the enforcement of quiet becomes a part of the library's services.

When a few lines or short paragraphs can answer the reader questions in each section, a short proposal will suffice, but a complex plan calls for a long, formal proposal. The following section illustrates a formal proposal accompanied by all necessary supplements.

APPLYING THE STEPS

The formal planning proposal in Figure 16-4 typifies the kind of specialized proposal that justifies a request for funding a project.

AUDIENCE ANALYSIS

A university newspaper is struggling to meet rising costs. The paper's yearly budget is funded by a college allocations committee that disburses money to all student organizations. Because of tight money, the newspaper has received no funding increase in three years.

The author of the proposal, the paper's business manager, must justify his request for a 17 percent budget increase for the coming year. This proposal, solicited by the allocations committee, is both external and internal. The primary audience will be the committee members, who will evaluate the soundness of the plan and decide whether to grant the additional funds. The secondary audience will be the newspaper staff, who will implement the plan — if it is approved by the allocations committee.

To encourage reader acceptance, the writer must propose a realistic plan showing that the newspaper staff is sincere in its intention to cut operating costs. At a time when everyone is expected to get by with less, the writer has to make an especially strong case for salary increases because a section of the plan calls for a slight increase in staff salaries to attract talented personnel.

When considering budget requests, readers invariably want to know exactly *how* the money will be spent, so the writer has to spell it out by itemizing every expense in the proposed budget. Thus the "Costs" section is the longest of the proposal.

To further justify the requested budget, the writer decides to show just how well the newspaper staff manages its funds. Under "Feasibility" he gives a detailed comparison of funding, expenditures, and size of newspaper

A BUDGET PROPOSAL
FOR
THE SMU TORCH
(1980-81)

Prepared for
The Student Fee Allocations Committee
Southeastern Massachusetts University
North Dartmouth, Massachusetts

by

William Trippe
Torch Business Manager

May 1, 1980

FIGURE 16-4   A Formal Proposal

ii

The SMU <u>Torch</u>
Old Westport Road
North Dartmouth, Massachusetts

May 1, 1980

Charles Marcus, Chairperson
Student Fee Allocations Committee
Southeastern Massachusetts University
North Dartmouth, Massachusetts 02747

Dear Dean Marcus:

No one needs to be reminded about the staggering effects of
inflation on our campus community. We are all forced with
having to make do with less.

Accordingly, we at the <u>Torch</u> have spent long hours devising
a plan to cope with increased production costs -- without
compromising the newspaper's tradition of quality service.
I think you and your colleagues will agree that our plan is
realistic and feasible. Even the "bare-bones" operation that
will result from our proposed spending cuts, however, will
call for a $4,432.14 increase in our 1980-81 budget.

We have received no funding increases in three years. Our
present need is absolute. Without additional funds, the <u>Torch</u>
simply cannot continue to function as a professional newspaper.
Therefore, I submit the following budget proposal for your
consideration.

Respectfully,

William Trippe

William Trippe
Business Manager, SMU <u>Torch</u>

FIGURE 16-4 (*Continued*)

iii

TABLE OF CONTENTS

                                                                    Page

LETTER OF TRANSMITTAL   ii

INFORMATIVE ABSTRACT   iv

INTRODUCTION   1

    Subject and Purpose   1

    Background   1

    Statement of Problem   2

    Need   3

    Scope   3

PROPOSED PLAN   4

    Methods   4

    Budget Request for 1980–81   6

    Feasibility   9

    Personnel   11

CONCLUSION   11

APPENDIX (Comparative Performances)   12

FIGURE 16-4   (*Continued*)

INFORMATIVE ABSTRACT

The student newspaper at Southeastern Massachusetts University is crippled by inadequate funding, having received no budget increases in three years. Inflation is the major problem facing the Torch. Increases in costs for printing, layout, and photographic supplies have called for a decrease in production. Moreover, our low salaries are inadequate to attract and retain qualified personnel. A nominal increase would make salaries more competitive.

This proposal outlines a plan for cutting costs by reducing page count, saving on photographic paper, reducing circulation, and hiring a new printer. The only proposed cost increase (for staff salaries) is essential.

A detailed breakdown of projected costs establishes the need for a $4,432.14 budget increase to keep the paper running weekly and with an adequate page count and distribution to serve our campus.

Compared with other college newspapers, the Torch makes much better use of its money. This comparison certifies the cost effectiveness of our proposal.

FIGURE 16-4 (*Continued*)

INTRODUCTION

Subject and Purpose

Our campus newspaper faces the contradictory challenge
of surviving ever-increasing production costs while maintaining its standard of quality. The following proposal
offers a realistic plan for responding to this crisis. The
plan's ultimate success, however, depends on the Student Fee
Allocations Committee's willingness to approve a long-overdue
increase in our 1980-81 budget.

Background

In its ten years, the Torch has grown in size, scope, and
quality. It is now the central means of communication on
campus. Roughly 6,000 copies (24 pages/issue) are printed
weekly during each 14-week semester.

Each week, the Torch prints news, features, editorials,
sports articles, announcements, notices, classified ads, a
calendar column, and letters to the editor. A vital part of
university life, the paper disseminates information, ideas,
and opinions -- all with the highest professionalism.

FIGURE 16-4   (Continued)

2

Statement of Problem

Fewer businesses are harder hit by inflation than the
newspaper business. The Torch, with much of its staff about
to graduate, faces next year with little money, rising costs
in every phase of production, and the need to replace expensive equipment that is outdated and in constant need of repair.

Our newspaper also suffers from a serious lack of student
involvement. Few students can be expected to devote full
time and energy to the paper without some kind of salary.
Most staff members do receive minimal weekly salaries: from
$10.00 for the distributor to $45.00 for the Editor in Chief,
but salaries averaging less than $1.00 per hour cannot
possibly compete with a minimum wage of $3.10 per hour. As
inflation forces more students to work, the Torch will have
to do what it can to make its salaries more competitive.

The newspaper's operating expenses can be divided into
four categories: composing costs, salaries, printing costs,
and miscellaneous (office supplies, mail, etc.). The first
three categories account for nearly 90 percent of the budget.

FIGURE 16-4 (*Continued*)

3

In the past year, costs in all categories have increased from as little as 3 percent for darkroom chemicals to as much as 60 percent for photographic paper and film. Printing costs (roughly one-third of our total budget) rose by 9 percent in the past year, and another price hike of 10 percent has just become effective.

Need

Despite the steady increase in production costs, the Torch has received no increase in its yearly allocation ($21,500.00) in three years. In these times of double-digit inflation, inadequate funding is virtually crippling our newspaper.

Scope

The proposed plan offered below includes:

1. methods for reducing production costs while maintaining the quality of our staff
2. an itemized list of projected costs for equipment, material, salaries, and services in 1980–81
3. a demonstration of feasibility through an assessment of our cost effectiveness

FIGURE 16-4 (*Continued*)

4

4. a summary of the attitudes shared by our staff

PROPOSED PLAN

The following plan is designed to trim operating costs --
without compromising the quality of our newspaper.

Methods

We can respond successfully to our budget and staffing
crisis by taking the following steps:

Reducing Page Count

By condensing free notices for campus organizations from
two pages to one page, abolishing "personals," and limiting
press releases to one page, we can reduce our average page
count per issue from 24 to 20. This reduction will mean
nearly a 17 percent saving in production costs.

Saving on Photographic Paper

A newspaper with fewer pages will call for fewer photographs. Also, we can further save money by purchasing a large
supply of photographic paper within the next month, before the
upcoming 15 percent increase becomes effective.

Reducing Circulation

By reducing circulation from 6,000 to 5,000 copies

FIGURE 16-4 (*Continued*)

5

weekly, we barely will cover the number of full-time day students, but we will save nearly 17 percent in printing costs.

Hiring a New Press

We can save money by hiring Sadus Press of Worcester to
do our printing. Bids from several other presses (including
our present printer) were at least 25 percent higher than
Sadus's price. Moreover, no other company offers the
delivery service that we will get from Sadus.

Increasing Staff Salaries

Although our staff seeks talented students who expect
little money and much good experience, our salaries for all
positions must increase by an average of $5.00 weekly. With
the minimum wage at $3.10 per hour, any of our staff could
make as much money elsewhere by working only one-third of
the time. In fact, many students could make more than the
minimum wage by working for local newspapers. To illustrate:
The Standard Times pays $20.00 to $30.00 for a news article
and $10.00 for a photo, while the Torch pays nothing for
articles and $2.00 for a photo.

A most striking example of our low salaries is the

FIGURE 16-4   (*Continued*)

6

$3.10 hourly we pay our typesetters. On the outside, trained
typesetters like ours make from $8.00 to $12.00 per hour.
Thus, our present typesetting cost of $3,038 easily could be
as much as $7,000 -- or even higher if we had the typesetting
done by an outside firm, as many colleges do.

In short, without this nominal salary increase, we cannot possibly hope to attract and retain qualified personnel.

Budget Request for 1980-81

Our proposed budget is itemized below, but the main
point is clear: the Torch must receive increased funding if
it is to remain a viable newspaper.

Projected Costs   Yearly Total

Equipment Leasing (We must continue
to meet the obligations of a leasing
arrangement with Chase Manhattan
Leasing Corporation for Compugraphic
equipment.)

    1. Execuwriter $106.25/month
    2. 7200 L and hardware $127.38/month
    3. Execuwriter II $138.63/month
                                 $372.26/month = $3,350.34/year

FIGURE 16-4 (*Continued*)

7

Composing Chemicals

   1.   Activator $105.00
   2.   Stabilizer   180.00
   3.   Processor Cleaner   55.00

                           $340.00   = $340.00/year

Photo Paper for Execuwriters

   1.   3" x 150' $711.00
   2.   7" x 150'   556.37
   3.   Headline paper   732.63
   4.   8" x 150'   227.00

                       $2,227.00   = $2,227.00/year

Layout Supplies

   1.   Font strips $160.00
   2.   Layout boards   160.00
   3.   Exacto knives and blades   80.00
   4.   No-repro pens   30.00
   5.   Rulers   12.00
   6.   Scotch tape   50.00
   7.   Folders   30.00
   8.   Reduction wheels   20.00
   9.   Wax   45.00
 10.   Construction paper   10.00
 11.   Border tape   50.00

                       $647.00   = $647.00/year

Photography

   1.   10 $1.00 photos/issue $280.00
   2.   10 $2.00 photos/issue   560.00

                       $840.00   = $840.00/year

FIGURE 16-4 (*Continued*)

8

Photography Supplies

1.	Paper	$730.00	
2.	Film	250.00	
3.	Darkroom Supplies	100.00	
		$1,080.00	= $1,080.00/year

Typing Staff

1.	35 hrs. at $3.10/hr.	$108.50/week	= $3,038.00/year

Salaries

1.	Editor in Chief	$1,400.00/year	
2.	News Editor	840.00	
3.	Asst. News Editor	420.00	
4.	Features Editor	840.00	
5.	Head Writer (features)	420.00	
6.	Sports Editor	840.00	
7.	Head Writer (sports)	420.00	
8.	Advertising Manager	1,050.00	
9.	Advertising Designer	700.00	
10.	Free Ad Designer	280.00	
11.	Layout Editor	840.00	
12.	Art Director	560.00	
13.	Photo Editor	840.00	
14.	Business Manager	840.00	
		$10,290.00	= 10,290.00/year

Fixed Printing Costs

1.	5,000 copies/week x 28	$12,399.80	= $12,399.80/year

FIGURE 16-4   (*Continued*)

9

Miscellaneous Costs

1. Graphics by SMU art
   students: 3/wk.
   at $5.00 each $420.00
2. Mail costs   550.00
3. Telephone costs   500.00
4. Print shop costs   200.00
5. Copier fees   50.00
                          $1,720.00   = $1,720.00/year

TOTAL YEARLY COSTS $35,932.14

Expected Ad Revenue
   ($1,000.00/mo. x 10) $10,000.00

Total Cost minus Ad Revenue: $25,932.14

TOTAL BUDGET REQUEST $25,932.14

Feasibility

    Beyond exhibiting our need, we feel that the feasibility

of this proposal can be measured through an objective assess-

ment of our cost effectiveness: In comparison to other school

newspapers, how well does the Torch use its funds?

    In a survey of the other four colleges in our area, we

found that the Torch -- by a sometimes huge margin -- makes

the best use of its money. Table 1 in Appendix A shows that,

of the five newspapers, the Torch costs the students least,

runs the most pages per week, and spends the least money per

FIGURE 16-4 (*Continued*)

10

page, <u>despite having a circulation two to three times the</u>

<u>size of the other papers.</u>

   The most striking comparison is between the <u>Torch</u> and

the newspaper at Fallow State College (Appendix A). Each

student at FSC pays $12.33 yearly for a paper averaging 12

pages per issue. Here at SMU, each student pays $4.06 yearly

for a paper averaging 20 pages per issue. Thus, for 33 per-

cent of FSC's cost, SMU students are getting 66 percent more

newspaper.

   The <u>Torch</u> has the lowest yearly cost of all five papers,

despite the largest circulation. With the budget increase

requested above, the cost would rise only by $0.44, for a

yearly cost of $4.48 to each student. Although Alden Col-

lege's paper costs each student $4.29, it is published only

every third week, averages 12 pages per issue, and costs

nearly $50.00 yearly per page to print -- as opposed to our

yearly printing cost of $38.25 per page.

   As the figures in Appendix A demonstrate, our management

of funds is responsible and effective.

FIGURE 16-4   (*Continued*)

11

Personnel

    Students on the Torch staff are unanimous in their de-
termination to maintain the highest professionalism. Many are
planning careers in journalism, writing, editing, advertising,
photography, or public relations. In any issue, the balanced,
enlightened coverage is evidence of our judicious selection
and treatment of articles and our shared concern for quality.

CONCLUSION

    As a forum for ideas and opinions, the Torch continues to
reflect a seriousness of purpose and a commitment to free expression. Its place in the campus community is more vital than
ever in these troubled times.

    There are increases and decreases in student fee alloca-
tions every year. Last year, for example, eight allocations
were increased by an average of $2,166. The Torch has re-
ceived no increase since 1976-77. Presumably, various increases materialize as priorities change and as special circumstances arise. The Torch staff urges the Allocation
Committee to respond to the paper's legitimate and proven needs
by increasing our 1980-81 allocation to $25,932.14.

FIGURE 16-4   (*Continued*)

12

APPENDIX A

Table 1. A Comparison of Allocations and Performance of Five Massachusetts College Newspapers

	Stonehorse College	Alden College	Simms University	Fallow State	SMU
Enrollment	1,600	1,400	3,000	3,000	5,000
Fee paid (per year)	$65.00	$85.00	$35.00	$50.00	$65.00
Total fee budget	$88,000	$119,000	$105,000	$150,000	$334,429.28
Newspaper budget	$10,000	$6,000	$25,300	$37,000	$21,500 $25,932.14[a]
Yearly cost per student	$6.25	$4.29	$8.43	$12.33	$4.06 $5.20[a]
Format of paper	Weekly	Every third week	Weekly	Weekly	Weekly
Average no. of pages	8	12	18	12	20
Average total pages	224	120	504	336	560 672[a]
Yearly cost per page	$44.50	$50.00	$50.50	$110.11	$38.25 $38.69[a]

[a]These figures are next year's costs for the SMU Torch.

Source: Figures were quoted by newspaper business managers in April 1980.

FIGURE 16-4 (*Continued*)

in relation to the other four colleges in his area. This section is the clincher because these hard facts are most likely to convince the committee that the proposed plan is cost-effective. So as not to clutter his discussion, the writer adds an appendix containing a table of figures for the above comparisons.

CHAPTER SUMMARY

A proposal is an offer to do something or a suggestion that some action be taken. Three common types of proposals are the planning proposal, the research proposal, and the sales proposal. These can be internal or external, solicited or unsolicited. The planning proposal answers this central reader question:

> – What are the benefits of following your suggestions for change?

The research proposal answers these questions:

> – Why is this project worthwhile?
> – What, exactly, qualifies you to undertake this project?
> – What are its chances of success?

The sales proposal answers the question:

> – Why will your service or product serve our needs better than that of your competitors?

Any successful proposal will answer applicable questions about *what, why, how, when,* and *how much.*

To ensure a quality proposal, follow these guidelines:

1. Use the appropriate format and supplements.
2. Be sure that your subject is focused and your purpose worthwhile.
3. Identify all related problems.
4. Offer realistic methods.
5. Provide concrete and specific information.
6. Use visuals whenever possible.
7. Maintain the appropriate level of technicality.
8. Create a tone that connects with your readers.

As you plan and write your proposal, work from a detailed outline that has a distinct introduction, body, and conclusion:

1. In your introduction, answer the *what* and *why* and make the subject, background, and purpose of your proposal perfectly clear. Establish the need for and benefits of the project, along with your qualifications for doing the work. Identify your data sources, any limitations of your plan, and its scope.

2. In your body section, answer the *how, when,* and *how much.* Spell out your plan by enumerating methods, work schedules, materials and equipment, personnel, facilities, costs, expected results, and feasibility.

3. In your conclusion, summarize the key points and stimulate the readers toward taking action on your proposal.

REVISION CHECKLIST

As you revise your early drafts, use this list to check the quality of your proposal's format, content, arrangement, and style.

Format

1. Is the short, internal proposal cast as a memo, and is the short, external proposal written as a letter?

2. Is the long proposal cast as a formal document with adequate supplements to serve the different needs of different readers?

3. Is the format professional in appearance?

4. Are headings logical and adequate?

5. Does the title forecast clearly and accurately the proposal's subject and purpose?

Content

1. Is the subject original, concrete, and specific?

2. Is the purpose clear and worthwhile?

3. Does everything in the proposal support its stated purpose?

4. Does the proposal *show* as well as *tell* (opinions based on fact, assertions backed up by evidence, convincing reasons)?

5. Is the proposed plan, service, or product beneficial?

6. Are the proposed methods practical and realistic?

7. Are all related problems identified?

8. Is the proposal free from overstatement?

9. Are visuals used effectively and whenever possible?

10. Is the proposal's length appropriate to the complexity of the subject?

Arrangement

1. Are the sections arranged in a recognizable pattern: a distinct introduction, body, and conclusion?

2. Does each section contain the *relevant* headings from the general outline on page 447?

3. Does the introduction stimulate interest and answer the reader questions on page 447 that are applicable?

4. Does the body spell out and justify your plan by answering the applicable reader questions on page 450?

5. Does the conclusion encourage readers to accept the proposal?

6. Are there clear transitions between related ideas?

Style

1. Is the proposal text written at an appropriate level of technicality for its intended primary audience?

2. Are the supplements written at a level of technicality that follows the guidelines on pages 445–446?

3. Is the informative abstract written at a level of technicality that can be understood by all intended readers?

4. Does the tone connect with readers (confident, encouraging, and diplomatic)?

5. Is each sentence clear, concise, and fluent (pages 39–50)?

6. Is the language convincing and precise (pages 53–56)?

7. Is the proposal written in correct English (mechanics and usage — see Appendix A)?

EXERCISES

1. After consultation with instructors in your major or with people on the job, write a memo to your instructor identifying the kinds of proposals most often written in your field. Are such proposals usually short or long, internal or external, solicited or unsolicited? Who are the decision makers in your field? Provide at least one detailed scenario of a work situation that calls for a proposal. Identify both the primary and secondary audience before you begin writing. Include your sources of information in your memo.

2. Assume that the chairperson of your high school English Department has asked you, as a recent graduate, for suggestions about changing the English curriculum to better prepare students for later writing challenges. Write a proposal, based on your experience since high school, that will move readers to action. (Primary audience = the English Department chairperson

and English faculty; secondary audience = members of the school committee.) In your external, solicited proposal be sure to identify problems, needs, and benefits and to spell out a realistic plan. Review the general outline on page 447 before selecting specific headings for your proposal.

3. From a faculty member of your major department, obtain a copy of a recent proposal for changes in the department's course offerings or staffing, or for solving a departmental problem (e.g., instituting a minor, adding new courses, dropping certain courses, hiring new faculty, changing degree requirements for a major, or the like). Using the revision checklist, evaluate the proposal for effectiveness. Submit your findings in a memo to your instructor, or be prepared to discuss your evaluation in class, using an opaque projector.

4. After identifying your primary and secondary audience, compose a short planning proposal for improving an unsatisfactory situation in the classroom, on the job, or in your dorm or apartment (e.g., poor lighting, drab atmosphere, health hazards, poor seating arrangements). Choose a problem or situation whose resolution is more a matter of common sense and lucid observation than of intensive research. Be sure to (a) identify the problem clearly, give a brief background, and stimulate reader interest; (b) state clearly the methods proposed to solve the problem; and (c) conclude with a statement designed to gain reader support of your proposal.

5. Write a research proposal aimed at your instructor (or an interested third party) requesting approval for completing the final term project (most likely an analytical report or a formal proposal). Identify clearly the subject, background, purpose, and benefits of your planned inquiry, as well as the intended audience, scope of inquiry, data sources, methods of inquiry, and a task timetable. Be certain that sufficient resources are available for you to carry out the necessary primary and secondary research for your project. Convince your reader of the soundness and usefulness of the project.

6. As an alternate term project to the formal analytical report (see Chapter 17), develop a long proposal for solving a major problem, improving a key situation, or satisfying an urgent need in your school, community, or job. Choose a subject of sufficient complexity to justify a formal proposal, a topic requiring a good deal of research (mostly primary). Identify an audience (other than your instructor) who will use your proposal for a specific purpose. Compose an audience analysis, using the sample on page 453 as a model.

Here are some possible subjects for your proposal:

— improving living conditions in your dorm or fraternity/sorority
— creating a student-operated advertising agency on campus
— creating a day care center on campus
— creating a new business or expanding an existing business
— saving labor, materials, or money on the job
— improving working conditions
— supplying a product or service to clients or customers

– improving campus facilities for the handicapped
– a community waste-recycling program
– a bicycle safety program in your community
– increasing tourist trade in your town
– eliminating traffic hazards in your neighborhood
– developing plans to handle parking problems (e.g., plans to handle campus parking and traffic patterns for sports events)
– reducing energy expenditures on the job
– improving security in dorms or in the college library
– improving in-house training or job-orientation programs
– deciding how a church or synagogue might best use a bequest of money
– creating a jogging path through the campus
– creating a bike path or special bike lane on main streets
– creating a new student organization within the government (say, one to handle Chicano affairs)
– finding ways for an organization to raise money
– improving faculty advisement of students
– the purchase of a new piece of equipment
– improving the food service on campus
– the purchase of new uniforms for an athletic or musical organization
– easing freshmen students through the transition to college
– improving bomb-alert procedures in the student union building

17

Analyzing Data and Writing a Formal Report

DEFINITION

PURPOSE OF ANALYSIS

TYPICAL ANALYTICAL PROBLEMS
"Will *X* Work for a Specific Purpose?"
"Is *X* or *Y* Better for a Specific Purpose?"
"Why Does *X* Happen?"
"Is *X* Practical in a Given Situation?"
Combining Types of Analyses

ELEMENTS OF AN EFFECTIVE
ANALYSIS
Clearly Identified Problem or Question
A Report with No Bias
Accurate and Adequate Data
Fully Interpreted Data
Valid Conclusions and Recommendations
Clear and Careful Reasoning
Appropriate Length and Visuals

FINDING, EVALUATING, AND
INTERPRETING DATA
Choose the Most Reliable Sources
Distinguish Hard from Soft Evidence
Avoid Specious Reasoning

PLANNING AND WRITING THE FORMAL
REPORT
Maintain a Flexible Approach
Work from a Detailed Outline
Introduction
Body
Conclusion
Support the Text with Supplements

APPLYING THE STEPS

CHAPTER SUMMARY

REVISION CHECKLIST

EXERCISES

DEFINITION

Analytical reports are question-answering or problem-solving reports. Analysis is basic to practically all our thinking, and in this sense all the assignments we have written have involved analysis. A summary, for instance, requires an analysis of the original to reveal its significant points; expanded definition often includes an analysis of the parts of the item being defined; a description or a process explanation requires an analysis of the item or process. In each case, we divide the subject into its parts (partition) and group these parts within specific categories according to their similarities (classification).

The analysis of data is somewhat different from these earlier assignments: it is based not only on observation but also on investigation and research. The subjects of earlier assignments had observable structures: for example, the stethoscope described in Chapter 8 has distinct parts and the tree-felling instructions in Chapter 9 have distinct steps. Your subject was wholly before you, intact, waiting only to be discussed.

In an analysis of data, however, your subject is not an item that needs describing or a process that needs explaining. Instead it is a question that needs to be answered or a problem that needs to be solved. The results of your analysis might influence a major decision. Say, for example, you received the following assignment from your supervisor:

> Investigate the feasibility of opening a branch of our company in Falls City and select the best available location.

Clearly, to get this job done, you will have to do more than observe your subject in order to describe it clearly. Since decisions will depend on your findings, you must *seek out* and *interpret* all data that will help you make the best recommendations. This is where you apply the research activity discussed in Chapter 12, and it is where you face your greatest reporting challenge.

PURPOSE OF ANALYSIS

On the job, you apply analytical skills to problems, proposals, and planning. A structural engineer, for example, analyzes material and design in evaluating the plans for a suspension bridge or a skyscraper. The wise investor analyzes economic trends, treasurers' reports, past performance, dividend rates, and market conditions before buying stock in a corporation. A legal defense is built on the attorney's analysis and logical reconstruction of facts in a case. Medical diagnosis relies on the precise identification and classification of symptoms, along with the communication of findings, interpretations, and recommendations for treatment.

As a new employee you may be asked to evaluate a new assembly technique on the production line, or to locate and purchase the best machinery at the best price. Sometimes you will have to identify the cause of a monthly drop in sales, the reasons for low employee morale, the causes of an accident or fire, or the reasons for equipment failure. As a manager you might need to evaluate a proposal for company expansion or investment. The list is endless, but the process is basically the same: (1) making a plan, (2) finding the facts, (3) interpreting the findings, and (4) drawing conclusions and making recommendations.

TYPICAL ANALYTICAL PROBLEMS

The aim of an analytical report is to show how you arrived at your conclusions. Your plan of attack will depend on your subject, your purpose, and your reader's needs. Here are some typical analytical problems:

"Will X Work for a Specific Purpose?"

Analysis can answer practical questions. Suppose, for instance, that your employer is concerned about the damaging effects of psychological stress on personnel. He or she may ask you to investigate the claim that transcendental meditation has therapeutic benefits — with an eye toward instituting a TM program for employees. You would design your analysis to answer this question: "Does TM have therapeutic benefits?" (See the sample report later in this chapter.) The analysis would follow a *questions-answers-conclusions* structure. Because the report might lead to decisions and action, you would probably include recommendations based on your conclusions.

"Is *X* or *Y* Better for a Specific Purpose?"

Analysis is also used in comparisons of machines, processes, business locations, political candidates, or the like. Assume, for example, that you manage a ski lodge and you need to answer this question: Which of the two most popular types of ski binding is best for our rental skis? In a comparative analysis of the *Salamon 555* and the *Americana* bindings, you might assess the strengths and weaknesses of each in a point-by-point comparison: toe release, heel release, ease of adjustment, friction, weight, and cost. Or you might use an item-by-item comparison, discussing the first whole binding, then the next.

The comparative analysis follows a *questions-answers-conclusions* structure and is designed to help the reader make a choice. Examples are readily found in magazines such as *Consumer Report* and *Consumer's Digest*.

"Why Does *X* Happen?"

The problem-solving analysis is designed to answer questions like this: Why do independent television service businesses have a high failure rate? (See the sample report later in this chapter.) This kind of analysis follows a variation of the questions-answers-conclusions structure: namely, *problem-causes-solution*. Such an analysis has the following steps:

1. identifying the problem
2. examining all possible and probable causes and narrowing them to more definite ones
3. proposing solutions

An analysis of student disinterest in most campus activities would follow the same structure. The results of a problem-solving analysis can sometimes be condensed and classified in a troubleshooting chart, as in Figure 17-1.

Another kind of problem-solving analysis is done to predict an effect: "What are the consequences of my dropping out of school?" Here, the structure is *proposed action–probable effects–conclusions and recommendations*.

"Is *X* Practical in a Given Situation?"

The feasibility analysis assesses the practicality of a plan or project: Will the consumer tastes of Hicksville support a gourmet food and wine shop? In a variation of the question-answers-conclusions structure, this type of analysis

GENERAL TROUBLESHOOTING CHART

If the amplifier is otherwise operating satisfactorily the more common causes of trouble may generally be attributed to the following:

1. Incorrect connections or loose terminal contacts. Check the speakers, record player, tape deck, antenna and line cord.
2. Improper operation. Before operating any audio component, be sure to read the manufacturer's instructions.
3. Improper location of audio components. The proper positioning of components, such as speakers and turntable, is vital to stereo.
4. Defective audio components.

Following are some other common causes of malfunction and what to do about them.

PROGRAM	SYMPTOM	PROBABLE CAUSE	WHAT TO DO
AM, FM or MPX reception	a. Constant or intermittent noise heard at certain times or in a certain area.	* Discharge or oscillation caused by electrical appliances, such as fluorescent lamps, TV sets, D.C. motors, rectifier and oscillator	* Attach a noise limiter to the electrical appliance that causes the noise, or attach it to the power source of the amplifier.
		* Natural phenomena, such as atmospherics, static, and thunderbolt	* Install an outdoor antenna and ground the amplifier to raise the signal-to-noise ratio.
		* Insufficient antenna input due to reinforced concrete walls or long distance from the station	* Reverse the power cord plug-receptacle connections. * If the noise occurs at a certain frequency, attach a wave trap to the ANT. input.
		* Wave interference from other electrical appliances	* Place the set away from other electrical appliances.
	b. Needle of the tuning meter does not move sharply.	* Needle movement is not necessarily related to the sensitivity of the amplifier.	* Tune the set for maximum signal strength.
AM reception	a. Noise heard at a particular time of day, in a certain area or over part of the dial.	* Natural phenomenon	* Install an antenna for maximum antenna efficiency. See "ANTENNA" in the Operating Instructions. * In some cases, the noise can be eliminated by grounding the amplifier or reversing the power cord plug-receptacle connections.

Reprinted by permission from Sansui Electric Company, Ltd.

FIGURE 17-1 A Sample Troubleshooting Chart

uses *reasons for–reasons against,* with both sides supported by evidence. Business owners often use this type of analysis.

Combining Types of Analyses

You probably have noticed that these four major types of analytical problems overlap considerably. Any one study may in fact require the answer to two or more of these questions. The sample report on pages 519–538 is both a feasibility analysis and a comparative analysis. It is designed to answer these two questions: Will a boutique pay in town? and If so, what location is best?

Regardless of the question or problem, the analytical report differs from the research report discussed in Chapter 13. The research report is simply designed to *inform* readers — not to move them to action. Although the analytical report relies on the same research techniques, it is designed both to *inform* and to *advise* readers.

ELEMENTS OF AN EFFECTIVE ANALYSIS

The analytical report incorporates many writing strategies learned in previous assignments, along with the guidelines that follow.

Clearly Identified Problem or Question

To save time and energy, make sure you know what you're looking for. If your car's engine fails to turn over when you switch on your ignition, for instance, you would wisely check your battery and electrical system before dismantling the engine. Apply a similar focus to your report.

Earlier in this chapter, an employer posed this question: "Will transcendental meditation help reduce psychological stress among my employees? The question obviously requires the answers to three other questions: What are the therapeutic claims of TM? Are they valid? Will TM work in this situation? How transcendental meditation got established, how widespread it is, who practices it, whether it is taught in schools, and any other questions about transcendental meditation are not relevant to this problem (although some questions about background might be useful in the report's introduction). The point is that it is important to define the central questions and think through any subordinate questions they may imply. Only then can you determine the kinds of data or evidence you need.

With the central questions clearly identified, the writer of the report on TM can formulate her statement of purpose:

> This report examines some of the claims about therapeutic benefits made by practitioners of transcendental meditation and communicates its findings to the general reader.

The writer might have mistakenly begun instead with this statement:

> This report examines transcendental meditation and communicates its findings to the general reader.

Notice how her purpose statement effectively narrows her approach by expressing the *exact basis* of her analysis: not transcendental meditation (a huge topic), but the alleged *therapeutic benefits* of TM.

Define your purpose by condensing your approach to a basic question: Does TM have therapeutic benefits? or Why have our sales dropped steadily for the past three months? You can then restate the question as a declarative sentence in your purpose statement.

A Report with No Bias

Interpret evidence objectively. Stay on track by beginning with an unbiased title. Consider these two title versions:

> 1. The Accuracy of the Talmo Seismograph in Predicting Earth Tremors
> 2. An Analysis of the Accuracy of the Talmo Seismograph in Predicting Earth Tremors

The first version suggests that the report will discuss the Talmo Seismograph as an accurate predicting device. In contrast, the second version signals readers that the report will answer questions about the device's accuracy. Throughout your analysis, rely on your evidence; do not force personal viewpoints on your material.

Accurate and Adequate Data

Do not misrepresent data by excluding vital points found in the original. Assume, for example, that you are asked to recommend the best chainsaw for a tree-cutting company. Reviewing reports of various tests, you come across this information:

> The Bomarc chainsaw proved the easiest to operate of all six brands tested. However, it also had the fewest safety features of all brands.

If you cite these data, present both points, not simply the first — even though you may personally prefer the Bomarc brand.

Reserve personal comments or judgments for your conclusion. As space permits, include the full text of interviews or questionnaires in appendixes.

Fully Interpreted Data

Interpret the meaning of your data, emphasizing key facts and explaining their relationships. Interpretation is the heart of the analytical report. You might, for example, interpret the chainsaw data this way:

> Our cutting crews often work suspended by harness, forty to sixty feet above the ground. Furthermore, much of their work is done in highly remote areas. This means that safety features should be our first concern in selecting chainsaws. Despite its ease of operation, the Bomarc saw does not suit our purposes.

By saying "This means . . ." you are helping readers grasp the essential message suggested by your evidence. *Simply listing your findings is not enough.* Spell out the meaning.

Valid Conclusions and Recommendations

An effective conclusion appeals not to *emotion* ("You will love this device") but to *reason* ("This device will best serve your needs"). When analyzing a controversial subject, be especially careful to remain objective. Assume, for example, that you work in law enforcement administration and have been asked to study this question: Is the prisoner furlough system working in our state? Do justice to this topic by making sure that your data gathering is complete, that your interpretations are not colored by prior opinion, and that your conclusions and recommendations are based on the facts.

When you do reach definite conclusions, state them with assurance and authority. Avoid statements that hedge or seem noncommittal ("It would seem that . . ." or "It looks as if . . ."). Be direct ("The earthquake danger at our proposed plant site is high."). If, on the other hand, your analysis does not reach a definite conclusion, do not force a simplistic one on your material.

Clear and Careful Reasoning

Technical reporting is not simply a mechanical process of collecting and recording information. If it were, machines could be programmed for the job. Each step of your analysis requires decisions about what to record, what to exclude, and where to go next. Even the planning stage calls for creativity in selecting the best approach (problem-solving, feasibility, comparison, etc.). As you evaluate your data (Is this reliable and important?), interpret your evidence (What does it mean?), and make recommendations based on your conclusions (How should the reader act?), you might have to adjust your original plan. You cannot know what you will find until you have searched. Remain flexible enough to revise your plan in the light of new evidence.

Appropriate Length and Visuals

Depending on the problem or question, your analysis may range in length from a short memo to a long report, complete with supplements (see Chapter 10). Make it just long enough to show readers how you have arrived at your conclusions.

Use visuals generously (see Chapter 11). Graphs are especially useful in an analysis of trends (rising or falling sales, radiation levels, etc.). Tables, charts, photographs, and diagrams work well in comparative analyses.

FINDING, EVALUATING, AND INTERPRETING DATA

Information sources and search techniques (discussed in Chapter 12) are only part of your task. As you sort out the pieces, you must evaluate the quality of your evidence as well as interpreting it. Facts or statistics out of context can be interpreted in many ways, and you are ethically bound to find truthful answers and to communicate your findings without distortion. To do so, choose reliable sources, distinguish hard from soft evidence, and avoid specious reasoning.

Choose the Most Reliable Sources

Make sure that each source is reputable, objective, and authoritative. Assume, for example, that you are analyzing the question about the alleged benefits of transcendental meditation. You could expect claims made in a reputable professional journal, such as the *New England Journal of Medicine,* to have scientific reliability. Also, a reputable magazine written for the technically informed (not expert) reader, such as *Scientific American,* would be a dependable source. On the other hand, you would probably suspect claims in glamour or movie magazines. Even claims made in monthly "digests," which offer simplistic, reductive, and largely undocumented "wisdom" to mass reading audiences, should be taken lightly.

Interview only those people who have practiced TM for a long time. Anyone who has practiced for only a few weeks could hardly assess bodily changes reliably, and to the extent that you rely on personal experience reports, including those who have practiced TM for a long time, you need a representative sample. Even favorable reports from ten successful practitioners would be a small sample unless those reports were supported by formal laboratory and test data. On the other hand, with a hundred reports from people ranging from college students to judges and doctors, you might not have "proved" anything, but your evidence would be somewhat persuasive.

Your own experience, also, is not a valid base for generalizing. We cannot tell whether our personal experience is in fact representative, regardless of how long we might have practiced meditation. If you have practiced TM with success, interpret your experience within the broader context of your collected sources.

Some social, political, or economic issues (e.g., the prisoner furlough system or causes of inflation) are always controversial. Such issues will never be resolved. Although we can get verifiable factual data and we can reason fairly persuasively on some subjects, no amount of close reasoning by any expert and no supporting statistical analysis will, in the same sense, "prove" anything about a controversial subject. Somebody's thesis that the balance of payments is at the root of the inflation problem cannot be proved. Likewise, one could only *argue* (more or less effectively) that federal funds would or would not alleviate poverty or unemployment. In other words, some cause-and-effect problems are more resistant to a certain solution than others, no matter how reliable the sources.

Given these difficulties, you should resist the temptation to report easy but misleading answers. It is better to report no answers at all than incorrect ones. Take no claims for granted and cross-check your data through all available sources.

Distinguish Hard from Soft Evidence

Hard evidence consists of observable facts. It can stand up under testing because it is fact and is verifiable. Soft evidence consists of opinion. It may collapse under testing unless the opinion is substantiated and unbiased.

Base your conclusions on hard evidence whenever possible. Early in your analysis you might read an article that makes positive claims about the effects of TM, but the article provides no data from scientific measurements of pulse, blood pressure, or metabolic rates. Although your own experience and opinion might fully agree with the author's, you should not hastily conclude that TM is beneficial to everyone. So far, you have only two opinions — yours and the author's — without any scientific support (e.g., tests of a representative cross-section under controlled conditions). Any conclusions at this point would rest dangerously on soft evidence. Only after a full survey of reliable sources can you decide which conclusions are supported by the bulk of your evidence.

Until evidence proves itself hard, consider it soft. Suppose, for instance, that a local merchant claims that a certain vacant shop is the best business location in town, but does not support that assertion with facts. Before you rent the shop, cross-check with several other sources and observe customer traffic in adjoining businesses at first hand. On the other hand, if the previous owner gave you the same judgment, even though he did not support it with fact, you probably could take it very seriously — particularly if you knew that he was honest and that he had vacated the shop because he had purchased a $125,000 retirement home at the age of fifty-five.

A related difficulty in assessing evidence is the confusion of probable, possible, and definite causes in a problem-solving analysis. Sometimes a definite cause can be identified easily (e.g., "The engine's overheating is caused by a faulty radiator cap"), but usually a good deal of searching and thought are needed to isolate a specific cause. Suppose, for instance, that you set out to tackle this question: Why are there no children's day care facilities on our state college campus? A private brainstorming session (see Appendix B) yields this list of possible causes:

- lack of need among students
- lack of interest among students, faculty, and staff
- high cost of liability insurance
- lack of space and facilities on campus
- lack of trained personnel
- prohibition by state law
- lack of legislative funding for such a project

Assume that you proceed with interviews, questionnaires, and research into state laws, insurance rates, and local availability of qualified personnel. In the course of your inquiry, you rule out some items, but others emerge as probable causes. Specifically, you find a need among students, high campus interest, an abundance of well-qualified people for staffing, and no evidence of state laws prohibiting such a project. Three probable causes remain: lack of funding, high insurance rates, and lack of space. Further inquiry shows that lack of funding and high insurance rates *are* issues. These causes, however, can be eliminated by creating other sources of revenue: charging a fee for each child, holding fund-raising drives, diverting funds from other campus organizations, and so on. Finally, after careful examination of available campus space and after consultation with school officials, you arrive at one definite cause: lack of adequate space and facilities.[1]

Early in your analysis you might have based your conclusions hastily on soft evidence (e.g., an opinion — buttressed by a newspaper editorial — that the

---

[1] Of course, one could argue that the lack of adequate space and facilities is somehow related to the problem of funding. And the fact that the college is unable to find funds or space may be related to the fact that student need is not sufficiently acute or interest sufficiently high to exert real pressure. Lack of space and facilities, however, emerges as the *immediate* cause.

campus was apathetic). Now you can base your conclusions on solid, factual evidence. You have moved from a wide range of possible causes to a narrower range of probable causes, then to a definite cause. Because you have covered your ground well, your recommendations have credibility and authority.

Sometimes it is difficult to trace a problem to a single cause, but this reasoning process can be tailored to most problem-solving analyses. Although very few except the simplest effects will have one cause, usually one or more principal causes will emerge. By narrowing the field you can focus on the real issues.

Avoid Specious Reasoning

Reasoning that is specious seems correct at first glance, but is not so when scrutinized. Reasoning based on soft evidence is often specious. Sweeping generalities are often the product of specious reasoning. Conclusions that have been speciously derived fail to stand up under testing. Assume, for example, that you are an education consultant. A local community has asked you to analyze the accuracy of IQ testing as a measure of intelligence and as a predictor of students' performance. Reviewing your collected evidence, you find a positive correlation between low IQ scores and low achievers, and vice versa. You then verify your own statistics by examining a solid cross-section of reliable sources. At this point you might feel justified in concluding that IQ tests do measure intelligence and do predict performance accurately. This conclusion, however, would be specious unless you could show that:

1. Neither parents, teachers, nor the children tested had seen individual test scores and had thus been able to develop biased attitudes.

2. Children testing in all IQ categories had later been exposed to an identical curriculum at an identical pace. In other words, they were not channeled into tracked programs on the basis of their test scores.

Your total data could be validly interpreted only within the context of these two variables (items having different values under different conditions).

Even hard evidence can support specious reasoning unless it is interpreted precisely, objectively, and within a context that accounts for all variables.

PLANNING AND WRITING THE FORMAL REPORT

Maintain a Flexible Approach

Your analysis will develop its own direction as it proceeds, depending on what you find at each point in your search. Because you will be writing and revising while searching, you will need clear points of reference to remain on track.

Pose and answer the following questions at various points in your search and in your writing.

1. What am I looking for?
2. How should I structure my inquiry to obtain this information?
3. How will I best communicate my process of inquiry and my findings?

These are ongoing questions whose answers can constantly change as you work your way through the research and the writing. Initially, question 1 will be answered in your statement of purpose; question 2 will be answered in your tentative or working outline; and question 3 will be answered in your formal topic or sentence outline — the blueprint for your actual report. As you search and write you might find these answers changing, so be flexible enough to modify your approach in case the unexpected happens. Here are a few possibilities:

1. Near the end of your research you uncover issues additional to those described in your statement of purpose (e.g., you discover that TM has no effect on certain people). Thus you have to modify your original statement of purpose.

2. You think of new major topics that are not in your working outline or find that information on one of your original topics is unavailable. As a result, your working outline might need additions, deletions, or shuffling of parts.

3. As you write your first draft, you find that your report is choppy and disorganized. Therefore, you must rearrange the sections in your formal outline before writing another draft.

Your finished report will be the product of countless decisions and revisions. Review your approach often to maintain control. Revise and reshuffle as often as necessary.

Work from a Detailed Outline

An effective report grows from a good outline. The following model outline can be adapted to most analytical reports:

I. INTRODUCTION
    A. Definition, Description, and Background of the Question, Issue, Problem, or Item
    B. Purpose of the Report and Intended Audience
    C. Sources of Information
    D. Working Definitions (here or in a glossary)
    E. Limitations of the Study
    F. Scope of the Inquiry (with topics listed in ascending or descending order of importance)

II. COLLECTED DATA
    A. First Topic for Investigation
       1. Definition
       2. Findings
       3. Evaluation of findings
       4. Interpretation of findings
    B. Second Topic for Investigation
       1. First subtopic
          a. Definition
          b. Findings
          c. Evaluation of findings
          d. Interpretation of findings
       2. Second subtopic
          etc.

III. CONCLUSION
    A. Summary of Findings
    B. Comprehensive Interpretation of Findings
    C. Recommendations and Proposals (as needed)

This outline is only tentative. Revise it if necessary.

The three sample reports that follow are built on the framework of this model outline. The first report, "Survival Problems of Television Service Businesses," is not reproduced in its entirety; however, each major section (introduction, body, and conclusion) is illustrated and preceded by a brief explanation. The second report, "Analytical Report on the Therapeutic Benefits of Transcendental Meditation," is entirely reproduced. The third paper, "An Analysis to Determine the Feasibility and the Best Location for Opening a Boutique in Pelham, Massachusetts," is included for further study and discussion.

Each report responds to a slightly different question or problem. The first tackles the question, Why does X happen? The second, Will X work for a specific purpose? The third tackles two questions: Is X practical in a given situation? and Is X or Y better for a specific purpose? At least one of these reports should serve as a model for your own analysis.

Introduction

The introduction identifies the subject of the report, describes and defines the question or problem, and provides relevant background. Identify your intended audience and discuss briefly your sources of data, along with reasons for omitting certain data (e.g., key person not available for interview). List working definitions, unless you have so many that you need a glossary. If you do use a glossary and appendixes, refer to them at this point. Finally, define the scope of your report by listing all major topics discussed in the body.

SURVIVAL PROBLEMS OF TELEVISION
SERVICE BUSINESSES

INTRODUCTION

Definition, Description, and Background

A television service business specializes in repairing selected home entertainment products. The modern household contains TV sets, master antenna systems, and related electronic equipment, which require periodic service.

The "TV repairman" (or electronics technician) has become a household necessity. Like the family plumber, electrician, physician, and lawyer, the electronics technician is considered one of the professionals who keep the American home functioning.

Because of their recent arrival (TV has been in general use for just over twenty-eight years) the professional personality of technicians escapes public analysis. Service costs regarded as excessive cause public distrust. Many feel that the TV service field is lucrative and that the technicians are sometimes dishonest. Technicians are aware of this public image.

Coupled with the public relations problem is the history of a technology that constantly changes. Service technicians perennially struggle against obsolescence. "The state of the art" is the phrase constantly ringing in their ears, diverting attention from the need for sound business management. Poor public relations, efforts to keep up with an everchanging technology, and poor financial management all contribute to the high rate of TV service business failure.

Purpose of Report, and Intended Audience

This analysis has been prompted by the fact that TV service businesses rank second only to auto service stations in bankruptcies. The purpose of the report is to increase public understanding of this important element of today's service technology, but it is written primarily to direct the attention of technician–shop owners to potentially fatal problems that may plague their businesses.

Sources of Information

The main source of information for this report is a study of a small business familiar to this writer for several years. An interview with the present owner is attached in Appendix A. A second interview with another local shop owner is attached in Appendix B. The Department of Labor Statistics and several trade magazines and journals round out the sources. Published material on this subject is understandably scarce.

Limitations of Study

This study is limited by the narrow sampling possibilities within our geographical area. Moreover, it is limited by the reluctance of many small business owners to "tell it like it is." Owners of TV service shops, specifically, are most hesitant to disclosed financial records. The owner of a very large TV and appliance center that does its own servicing refused an interview. Unfortunately, it was the only large facility in our area.

Working Definitions

Television service shop: The conventional TV service shop is essentially a service center for home entertainment products. Items routinely serviced include TV sets, radios, stereo equipment, and antenna systems. Service businesses exist with or without sales. Some businesses refuse to sell equipment because of the added bookkeeping and inventory problems.

Service technician: An electronics service technician is a person trained to service home entertainment products. Outside technicians make service calls, and inside technicians do the bench work. According to the Department of Labor, in 1965 there were 115,000 people engaged in radio and television service. One-third were self-employed. Of the remainder, two-thirds worked in TV shops, and the others were employed by manufacturers.[1] In 1972, the number of technicians was reported at 140,000 with roughly the same breakdown as in 1965.[2] In 1974, technicians numbered 135,000 with the same breakdown.[3] This last figure is striking because it shows a marked decrease in a field considered by the Department of Labor to have a scarcity of personnel. The rapid drop also occurred in a decade when the population and the number of items to be serviced increased greatly.

Scope

The topics of investigation in this report are, in ascending order of importance, the history of television service businesses, the problem of public relations, the demand for growing technical competence, and the problems of financial management.

Body

The body divides a complex subject into related topics and subtopics, ranked in order of their importance (ascending or descending).

Your purpose is to achieve a rational and useful partition of the subject into its main parts and those, in turn, into their appropriate subparts. Carry your division as far as you can to make sense of the topic. In the sample that follows, the major topic, "The Problems of Financial Management," is divided into two subtopics: "Technician versus Manager" and "Cash Flow." This last subtopic, in turn, is divided into three sub-subtopics: "Sales," "Collection for

Service," and "Credit Arrangements." These divisions and subdivisions keep the writer on track and help readers follow the inquiry. Be sure that readers can draw conclusions identical to your own on the basis of your evidence. The logic of a problem-solving analysis requires that you discuss all possible causes while narrowing your focus to probable and then definite causes. Sift, evaluate, and interpret clues to reach a valid conclusion. The process might be diagrammed like this:

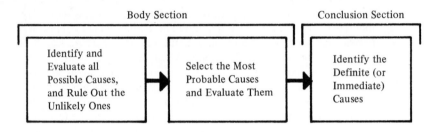

One major topic of the television service report follows. To save space, the other major topics are omitted.

COLLECTED DATA

The Problems of Financial Management

Technician versus Manager

Technician-owners cannot function productively as both full-time technicians and full-time business managers. Without improving their technical proficiency, they cannot keep the quality of their service up to professional standards. On the other hand, they must maintain the detailed financial records that are the backbone of any successful business. Shop owners usually attend evening courses and leave their workbenches for a part of the day to attend to the management of their business. A shop with several technicians can easily permit this arrangement, but, in a one- or two-man shop these multiple commitments can spell disaster.

Some small shops have turned to outside bookkeeping and accounting agencies. Computerized methods have reduced the cost of these services, to the relief of the harried shop owner. Many shops showing only a marginal profit, however, cannot afford outside help. Their only recourse is to request assistance from the Small Business Administration, which provides advice, financial analysis, and small loans, but these services are offered only on a limited, first-come, first-served basis.

Cash Flow

The cash flow of a small service business is affected by sales, collection for service, and credit arrangements.

Sales. Some service shops also sell TV sets, stereos, and other home-entertainment products. In many cases this policy has resulted in their downfall because of the owner's lack of financial discipline. Most manufacturers permit "floor plans." TV sets, for example, are consigned to a seller for a small down payment (usually 10 percent of retail). The seller is required to pay for each item *as it is sold,* or within specified time limits, but many shop owners use this money to finance their operation and it is just a matter of time before the day of reckoning arrives. Full payment is due and the money has been spent. This too-familiar situation has led to countless bankruptcies.

Collection for Service. Most shops generate their own capital. Therefore, strict collection for services rendered is the common practice. Signs on the walls of most shops ask for cash only — no charges and no checks. When the customer is unable or unwilling to pay, however, the technician must hold the set for collateral. This results in a shop cluttered with sets that are generating no capital.

Credit Arrangements. Most service work is done on a cash-on-delivery basis. Only warranty work (paid for by manufacturers) is done on credit. Manufacturers are dependable and usually pay for warranty work monthly. Checks totaling between $500 and $1,000 per month are quite welcome and generally cover most of the fixed monthly costs (rent, utilities, etc.) of a small service business.

Evaluation and Interpretation

Some owners use cash that they owe to manufacturers. Others have their shops cluttered with items held as collateral. The overriding problem in financial management, however, stems from inadequate record-keeping. This is where technician-owners were found to be weakest. They simply do not have time to leave the bench work to do "paperwork." And without good records they are unable to keep track of and manage their cash flow.

Conclusion

The conclusion culminates your report. In the body, you divided your subject into its components for investigation. Here you pull together your findings. Your conclusion is likely to be of most interest to readers because it answers the questions that sparked the analysis in the first place.

Here you summarize, interpret, and recommend. Although you have interpreted evidence at each step of your analysis, your conclusion pulls the strands together in a broader interpretation and suggests a plan of action, where appropriate. Because of its importance, this final section must be consistent in three ways:

1. Your summary must reflect accurately the body of your report.
2. Your overall interpretation must be consistent with the findings in your summary.
3. Your recommendations must be consistent with the purpose of the report, the evidence presented, and the interpretations expressed.

Your summary and interpretations should lead logically to your recommendations. Here is the conclusion section of the television service report:

CONCLUSION

Summary of Findings

The History of Television Service Businesses

Many survival problems in today's service business derive from early habits and traditions. Virtually all early television technicians were military veterans trained as "radio mechanics." They moved into shops and went right to work with little business training. Their management expertise was confined to a working knowledge of a very active cash register.

Public Relations

Price gouging, which occasionally occurred in the early years, has been eliminated by the profession's development of ethical standards. Suspicions linger, however, largely because customers have little understanding of the electronic repairs needed for their sets. As a result, many customers turn to service departments in larger stores. Among the businesses studied, there seemed to be an embarrassing lack of public relations effort. The reasons given for this deficiency were lack of time and lack of advertising funds.

The Demand for Technical Competence

The shop owners studied were all technicians of the highest caliber, who continually upgrade their skills. Realistically, a shop cannot endure without strong professional standards. Customers simply would not return. In shops with several technicians, individuals have begun to specialize. One shop has a TV department, an auto radio and stereo department, and a commercial electronic department that services such items as microwave ovens and motel master systems for signal distribution. All technicians surveyed feel that specialization will improve individual competence.

Financial Management

Shop owners who insist on doing the bench work themselves and who neglect the daily management of the business are unable to maintain proper financial records or to stabilize their cash flow. With few exceptions, unsound financial management leads to business failure.

Comprehensive Interpretation of Findings

The areas of public relations and financial management pose threats to the welfare of independent TV service businesses.

Public Relations

Official and consumer attitude toward this profession still reflects the "radio mechanic" days of the postwar years. Technicians are still viewed as "TV repairmen" who have easily learned a simple trade and who keep customers uninformed so they can overcharge them. Consumers' lack of familiarity with the product and servicing techniques is compounded by technicians' unwillingness or inability to better inform their customers.

Financial Management

Unsound financial management is the largest cause of TV service business failures. When the books are not balanced no business can survive for long.

Recommendations

Service technicians show a puzzling lack of initiative to erase the "radio mechanic" and "TV repairman" image. This attitude and the owner-technician's hesitancy to become a real "businessman" are problems requiring positive action.

Collective action by shop owners is the best way to strengthen public relations. Local technicians' guilds should design and sponsor advertising to inform consumers and to improve the technician's image. Even though work may be plentiful, technicians owe it to themselves to convince customers that they are competent, reliable, and trustworthy.

Service shop owners must adopt more efficient practices for operating their businesses. They should enroll at local colleges in courses such as primary management, accounting, and business methods. Properly trained, they will understand such important items as balance sheets and income statements, as well as their responsibilities in the tax accounting of their business. Professionalism in management will show immediate results. Any problems with financing or cash flow will become obvious, thus enabling owners to take remedial action.

Support the Text with Supplements

Submit your completed report with these supporting documents, in order:

- title page
- letter of transmittal
- table of contents

– table of illustrations
– abstract
– **report text (introduction-body-conclusion)**
– glossary (as needed)
– appendixes (as needed)
– note page ⎱
– bibliography ⎰ (or alphabetical or numbered list of references)

Each of these supplements is discussed in Chapter 10. Footnotes and other forms of documentation are discussed in Chapter 13.

APPLYING THE STEPS

The two analytical reports in Figures 17-2 and 17-3 are written for general audiences. Each is patterned after our model outline. The first report is designed to answer a practical question. The second combines a feasibility analysis with a comparative analysis. To illustrate an alternative to the author-year documentation system, the second report uses a numbered list of references.

AUDIENCE ANALYSIS

One purpose of this report is to satisfy a requirement in a technical writing course, so one copy is accompanied by a letter of transmittal (not shown) to the instructor. The author writes, however, for another audience as well: her part-time employer at a large day care center.

As a result of a community program to hire the elderly, many of the center's staff are qualified professionals well beyond retirement age. The center's director has expressed concern about the stress placed on elderly employees who have to cope with energetic preschoolers all day long. Thus, Bonnie Kelly decides to research a topic that will have practical benefits for her colleagues, in addition to satisfying a course requirement. In a letter of transmittal to her employer, the author recommends the use of transcendental meditation (TM) for reducing stress on the staff.

Both Bonnie's professor and her employer (and interested colleagues) can be considered laypersons. They want to know *if* TM will work to lower stress, but before they can be persuaded by primary and secondary research evidence, they have to understand *how* TM works. Therefore, the first two parts of the body section describe the functions of the *mantra* and *dive*. (A glossary spells out the meaning of these two key terms.)

To increase audience interest, the author includes, in her introduction, a brief history of TM and a description of the technique.

Like any worthwhile analysis, this report will enable readers to make an *educated* decision, based on facts.

ANALYTICAL REPORT ON THE

THERAPEUTIC BENEFITS

OF

TRANSCENDENTAL MEDITATION

Prepared for

Dr. John Laws
Technical Writing Instructor
Cape Cod Community College
West Barnstable, Massachusetts

by

Bonnie Kelly

December 29, 1981

FIGURE 17-2   An Analytical Report

16 Slocum Road
Arlington, Massachusetts 03146
December 29, 1981

Barbara Hawkes, Director
Amherst Day Care Center
56 Branch Street
Arlington, Massachusetts 03146

Dear Ms. Hawkes:

At our September staff meeting, several employees complained
of the heavy stress that results from looking after a room
full of active youngsters. I wondered whether transcendental
meditation might be a way to decrease stress, so I decided to
research the possibility.

As you will see in the enclosed report, medical evidence
strongly suggests that TM does reduce the effects of stress.
In addition to its measurable physiological effects on blood
pressure and metabolism, TM may have immeasurable emotional
benefits as well. My interviews with two practitioners of
TM provides further evidence of its pronounced benefits.

On the basis of these findings, you may wish to consider a
TM program for the day care staff -- for those willing to
participate in such an experiment.

Please share this report with any interested staff members,
and call me anytime you have questions.

Sincerely yours,

Bonnie Kelly

Bonnie Kelly

FIGURE 17-2 (*Continued*)

iii

TABLE OF CONTENTS

Page

LETTER OF TRANSMITTAL   ii

TABLE OF FIGURES   iv

INFORMATIVE ABSTRACT   v

INTRODUCTION   1
  Definition   1
  Background   1
  Description   1
  Purpose   2
  Data Sources   2
  Working Definitions   3
  Scope   3

COLLECTED DATA   3
  Function of the Mantra   3
  Function of the Dive   5
  Physiological Effects of TM   6
    Lowering of Blood Pressure   6
    Relief from Bronchial Asthma   7
    Decrease in Metabolism   7
  First-Person Testimonials   9
  Continuing Research   9

CONCLUSION   10
  Summary of Findings   10
  Comprehensive Interpretation of
              Findings   11
  Recommendation   12

GLOSSARY   13

APPENDIX   14

REFERENCES   15

FIGURE 17-2   (*Continued*)

iv

TABLE OF FIGURES

                                                                Page

1.   The Mantra's Effect on the Nervous System   4

2.   Changes in Oxygen Consumption During TM   8

FIGURE 17-2   (*Continued*)

v

INFORMATIVE ABSTRACT

The technique of transcendental meditation, as taught by Maharishi Mahesh Yogi, has given rise to therapeutic claims by many of its practitioners. Scientific research on the physiological changes that accompany meditation supports these therapeutic claims, especially in the areas of hypertension, bronchial asthma, and metabolism. Research and interview data suggest that transcendental meditation may have important future applications in medicine. Those claims by meditators that TM lowers blood pressure and helps the meditator to achieve a more relaxed and productive lifestyle should generally be believed. Anyone wishing to relieve physiological stress should consider practicing TM.

FIGURE 17-2   (*Continued*)

INTRODUCTION

Definition

Transcendental meditation is an act of intense contemplation during which one's consciousness rises above common thoughts or ideas.

Background

The person behind the spread of transcendental meditation is a Hindu monk with a long, gray beard and a relaxed, friendly manner, known as the Maharishi Mahesh Yogi; his style is an interesting combination of Eastern spirituality and Western paganism and incorporates a revival of some of the ancient Vedic tradition of India (based on the oldest, most sacred Hindu writings), which he absorbed from his teacher, Guru Dev (Time, 1975). It is generally believed that TM has caught on so quickly because the Maharishi has been able to present Eastern culture in a way that is compatible with Western lifestyles and values.

Description

TM is simple to learn and convenient to practice, and it seems to work even if one does not believe it will. The TM technique is as simple as getting into a comfortable

FIGURE 17-2 (*Continued*)

2

upright position, closing the eyes, and relaxing for twenty

minutes each morning and evening. While relaxing, the medi-

tator repeats an assigned mantra.* The technique is simple,

yet meditators universally agree that the benefits are

enormous. The physical benefits, in particular, have

attracted the interest of the scientific world. Accumulating

scientific evidence of the benefits of meditation seems to be

convincing skeptics that TM may indeed hold the answer to a

number of physiological and related psychological problems

(Schultz, 1972:24).

Purpose

     This report examines some of the claims for therapeutic

benefits expressed by practitioners of TM and communicates

its findings to the general reader.

Data Sources

     Information for this report was compiled from a broad

sample of reputable publications and from personal interviews

with two long-time practitioners of TM.

*See Glossary.

FIGURE 17-2   (*Continued*)

3

Working Definitions

Working definitions are included in the Glossary.

Scope

The major topics in this analysis are as follows: function of the mantra, function of the dive,* and the physiological effects of TM.

COLLECTED DATA

Function of the Mantra

The mantra is a sacred formula -- a collection of syllables that parallels the individual's specific rhythm of pulse and respiration, when silently repeated over and over. The Maharishi argues that the mantra is vital to the whole technique of TM. He contends that the vibrations of the mantra arise just from thinking the sound, and thereby affect the nervous system by dissolving stress and freeing the mind to pursue more "subtle levels of awareness" (Campbell, 1973:56). This process of stress reduction is represented in Figure 1. One could compare the mantra's effect

*See Glossary

FIGURE 17-2 (*Continued*)

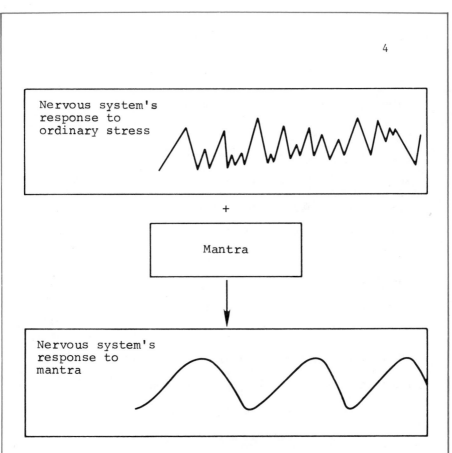

4

Nervous system's response to ordinary stress

+

Mantra

Nervous system's response to mantra

<u>Figure 1.</u>   <u>The Mantra's Effect on the Nervous System</u>

to the vibratory regularity created by rocking the cradle to soothe a baby, or contrast it to the vibratory shock of some modern music.

Resonance frequency, according to Dr. B. Glueck,

FIGURE 17-2   (*Continued*)

5

director of a Connecticut psychiatric hospital, can have a
quieting and stabilizing effect on the nervous system
(Fiske, 1972). Dr. Herbert Benson, a leading researcher of
TM, suggests that the mantra acts to shut out distracting
sounds, thereby blocking the individual's related responses
(Fiske, 1972). To date, scientific evidence seems to support
the Maharishi's assertion that the mantra exerts an important
calming effect.

Function of the Dive

While relaxing comfortably and thinking the mantra sound
over and over, the meditator experiences a dive. It is as
though a swimmer is paddling along on the surface of a pool
and then plunges into the deepest part, only to emerge again
on the surface a few minutes later, refreshed and relaxed.

While the meditator is experiencing this dive -- equivalent, researchers say, to several hours of deep sleep --
several physiological changes occur in the body, including
lowered oxygen consumption and lowered blood pressure
(Wallace and Benson, 1972:87). These biochemical changes,
in many cases, lead to better mental and physical health.

FIGURE 17-2 (*Continued*)

6

Researchers, most prominently Dr. Herbert Benson of Harvard
and Dr. R. K. Wallace of UCLA, have published articles in
scholarly, scientific, and medical journals to document the
physiological changes in meditators. Maharishi International
University, recently established in Iowa, has collected
these scientific findings into a pamphlet of charts and
graphs with accompanying documentation (1975). The collected evidence strongly suggests that the mantra and the
dive are real events with profound physiological effects.

Physiological Effects of TM

Although it is not within the scope of this report to
analyze all medical findings, interpretations are included
that discuss the effects of TM on hypertension, asthma, and
metabolism.

Lowering of Blood Pressure

The blood pressure of twenty-two hypertensive patients
was recorded 1,119 times before and after they learned TM.
The patients' decrease in blood pressure after practicing
TM was statistically significant (Benson and Wallace, 1972).
TM apparently has clinical value in helping hypertensive
patients control their blood pressure.

FIGURE 17-2 (*Continued*)

7

Relief from Bronchial Asthma

Ninety-four percent of bronchial asthmatic patients who learned TM showed improvement as determined by the physiological measurement of airway resistance. Sixty-one percent showed improvement as reported by their physicians, and the same percentage made personal reports of their own improvement (Honsberger and Wilson, 1973). These findings indicate that TM has a positive effect on bronchial asthmatic patients.

Decrease in Metabolism

Research has shown that oxygen consumption and metabolic rate markedly decrease during meditation, indicating a state of deep rest. Furthermore, Wallace and Benson (1972) report that the partial pressure of oxygen and carbon dioxide in the blood remain essentially constant during meditation; thus, the decrease in total oxygen consumption during TM is not caused by a manipulation in breathing pattern or a forced deprivation of oxygen. The decreased consumption of oxygen is a normal physiological change caused by a lowered requirement for oxygen by the cells during this effortless process. Figure 2 shows some of the changes in oxygen consumption

FIGURE 17-2   (*Continued*)

8

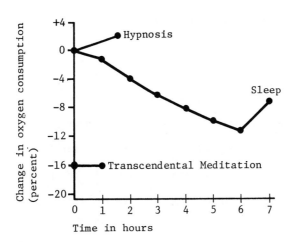

Figure 2.   Changes in Oxygen Consumption During TM

Source: R. K. Wallace and H. Benson, "The Physiology of Meditation," Scientific American,
226, No. 2 (February 1972), p. 90. Copyright ⓒ 1972 by Scientific American, Inc.
All rights reserved.

that occur while a person practices TM in comparison to

undergoing hypnosis or sleeping. The graph shows that oxygen

consumption increases slightly under hypnosis, whereas it

decreases by 8 percent after five hours' sleep. During TM,

however, oxygen consumption is reduced by 16 percent within

FIGURE 17-2   (*Continued*)

9

a few short moments. Again, this evidence supports the
theory that TM exerts a major relaxing influence on the mind
and body.

First-Person Testimonials

Two practitioners personally contacted insist that TM
is of great assistance in controlling their particular ailments. Jane Smith, age thirty-three, insists that her controlled diabetic condition is managed by the practice of TM
(1980). Ms. Smith has practiced TM for three years. Martha
Corey, a middle-aged woman who was seriously troubled by high
blood pressure and circulatory difficulty thirteen months
ago, has lowered her blood pressure and lost the sense of
numbness in her hands and in one leg as a result, she claims,
of her meditation sessions (1980). I have noticed that this
woman is becoming more alert and seems to have more energy
than she had two years ago. The full text of the interview
with Ms. Corey is contained in the Appendix, and again reinforces the positive claims made for TM.

Continuing Research

Thousands of businesspeople, scientists, teachers, and

FIGURE 17-2   (*Continued*)

10

homemakers have taken up TM, reporting such beneficial effects

as freedom from tension, mental well-being, heightened energy,

and increased creativity (Livingstone, 1973:18). Doctors and

researchers at leading hospitals are investigating TM as a

treatment for certain diseases and as a means of therapy for

drug addiction (Williams, 1972:74). Yet, despite the Wallace

and Benson research and that of other scientists, no one is

quite sure how TM works. It has been demonstrated that

definite physiological changes do take place during medita-

tion, but so far no studies have explained how TM works on

specific emotional problems. To answer this question, re-

searchers at the Institute for the Living in Hartford, Con-

necticut, have begun a three-year study of the specific

effects of TM on emotional disturbances (Graham, 1972). This

continuing professional interest indicates that psychologists

have been sufficiently impressed by the physiological findings

to make further studies on TM's influence on the emotions.

CONCLUSION

Summary of Findings

This report has analyzed some of the claims for

FIGURE 17-2 (*Continued*)

11

therapeutic benefits of transcendental meditation. Findings

indicate scientific support for the theory that the mantra

and dive have measurable and beneficial effects in three

physiological areas: TM lowers blood pressure in hyper-

tensive patients, decreases airway resistance in asthmatic

patients, and slows metabolism by reducing oxygen consumption.

These scientific findings have been further validated by in-

terviews with two TM practitioners and by my personal obser-

vations. Continuing scientific research suggests that TM

may have further application in the treatment of emotional

disorders.

<u>Comprehensive Interpretation of Findings</u>

TM is not a physical and emotional panacea, but evidence

suggests that TM does produce definite and measurable physi-

cal changes in meditators. The uses to which TM may be put

to combat some of our modern illnesses are as yet unknown.

We do know, however, that the claims of TM practitioners are

being seriously considered by scientists, and we will watch

closely for the results of their research and the subsequent

application of TM to various disorders. In general, TM seems

FIGURE 17-2 (*Continued*)

12

to offer the possibility of a more relaxed and productive

lifestyle for its practitioners.

Recommendation

    Claims made about the therapeutic benefits of transcen-

dental meditation are convincing. Anyone wishing to relieve

physiological stress should seriously consider practicing TM.

FIGURE 17-2   (*Continued*)

13

GLOSSARY

Dive: an experience in transcendental meditation when the
      attention of the meditator "passes from the super-
      ficial to deeper levels of thought and then frequently
      beyond, into the state of pure awareness." (Campbell,
      p. 56.)

Mantra: a word or phrase, usually Hindi, meaningless to the
        user, which is assigned by the teacher to the medi-
        tator to be used in his meditation sessions as a
        vehicle by which to obtain a more subtle level of
        thought. The choice of a specific mantra is con-
        sidered most important and is based on the per-
        sonality of the individual meditator. Once assigned,
        the mantra is not to be disclosed for fear it will
        be used by individuals for whom it was not intended.

FIGURE 17-2 (*Continued*)

14

APPENDIX

Personal interview with M. C. of Radford, Massachusetts, who has been practicing Transcendental Meditation for two and one-half years, with brief intervals of nonmeditation.

Question: Did you begin practicing TM with the idea of losing weight or for any other health reason?
Answer: No, I actually started because the instructor had been a regular house guest for several months. When he started a TM group in town, I decided to enroll -- but I had no specific expectations.

Question: Have you noticed any change in your health since you have been practicing TM?
Answer: Yes, my blood pressure has remained within normal range. For years earlier, it had been difficult to control and highly unpredictable. Also I now find it much easier to relax.

Question: Are you aware of any other changes?
Answer: Yes, I feel much more alert and energetic than I had before becoming a meditator.

Question: Would you say that TM is effective medical therapy for a person who tends to be nervous and hypertensive?
Answer: It's as effective as a lot of pills and shots that doctors are so quick to give out, and it's much more pleasant.

FIGURE 17-2 (*Continued*)

15

REFERENCES

Benson, H. and Wallace, R. K. 1972. Decreased blood pressure in hypertensive subjects who practiced meditation. Circulation 45: 10.

Campbell, A. 1973. Transcendental meditation: seven states of consciousness. London: Victor Gollancz.

Corey, M. 1980. Interview about the effects of TM, December 9.

Editors. Fundamentals of progress. 1975. Fundamentals of progress: scientific research on the transcendental meditation program. Ames, Iowa: Maharishi International University Press.

Fiske, E. B. 1972. Thousands finding meditation eases stress. The New York Times (11 December):13.

Graham, E. 1972. Transcendent trend: meditation technique, once haven of young, gains wider following. The Wall Street Journal (31 August):3.

Honsberger, R. and Wilson, A. F. 1973. The effects of transcendental meditation upon bronchial asthma. Clinical Research 22, no. 2:121–125.

Livingstone, R. 1973. Business tries meditating. The New Englander 20, no. 2:15–18.

Schultz, T. 1972. What science is discovering about the beneficial effects of meditation. Today's Health 15, no. 4:20–25.

Smith, J. 1980. Interview about the effects of TM, December 10.

FIGURE 17-2 (*Continued*)

16

Time. 1975. The TM craze: four minutes to bliss. Time
    (13 October):71.

Wallace, R. K. and Benson, H. 1972. The physiology of medi-
    tation. Scientific American 226, no. 2:84-90.

Williams, G. 1972. Transcendental meditation: can it fight
    drug abuse? Science Digest 71, no. 2:74-79.

FIGURE 17-2 (*Continued*)

AUDIENCE ANALYSIS

This report is addressed to the author's professor, as a course requirement, but we might well say that its primary audience is the writer *herself*.

About to graduate as a business major, Paula Sweetman hopes to open her own retail clothing store. As a first step, she has to answer two questions: Is it feasible? and, if so, Where is the best location? The answers, in this case, will not be found in library books. Instead, Paula will rely mostly on primary research information (interviews, observation, review of census data, etc.).

For other interested readers (a general audience), Paula opens her report with an expanded definition of *boutique*, along with a description of her proposed shop. This information will be especially useful if Paula applies for a loan to finance her venture. (The loan officers at local banks are not likely to know much about the character and function of boutiques, and, of course, her research data would demonstrate to the banks that Paula is making an *educated* decision.)[2]

CHAPTER SUMMARY

Analytical reports are question-answering or problem-solving reports. An analysis of data requires that you collect evidence from various sources and use this evidence to draw specific conclusions and make specific recommendations. As you plan your report consider which of these questions or combination of questions your analysis is intended to answer:

1. Will X work for a specific purpose?
2. Is X or Y better for a specific purpose?
3. Why does X happen?
4. Is X practical in a given situation.

Condense your approach to a basic question and then restate it as a declarative sentence in your statement of purpose. Sometimes you will work with a combination of these approaches.

After identifying the problem or question, interpret all data objectively and fully in order to reach valid conclusions. At each step of your analysis, make careful decisions about what to record, what to exclude, and where to go next. Make the report itself long and detailed enough to show readers how you have arrived at your conclusions. Use visual aids generously and, except for a memo report, use a formal report format with supplements.

[2] For the sake of simplicity, no front matter (title page, letter of transmittal, table of contents, table of illustrations, and informative abstract) is reproduced here, nor are the appendixes containing the full text of three interviews, illustrations of various boutique displays, and a map of the town.

AN ANALYSIS TO DETERMINE THE FEASIBILITY AND

THE BEST LOCATION FOR A BOUTIQUE

IN PELHAM, MASSACHUSETTS

INTRODUCTION

Definition, Background, and Description of a Typical Boutique

A boutique is a small, informal specialty shop that
specializes in wearing apparel and accessories for women.

The boutique originated in France in the early 1960s.
Parisian designers created these shops as places for displaying their ready-to-wear apparel. Drenched in perfume
and prestige, the designer-owned boutique is the place to
shop for clothing in France. Most boutiques are decorated
from ceiling to floor with striking merchandise displays
and exotic artifacts. Shopping here is an experience in
itself. An alluring element of these French boutiques is
that the clothes are precisely fit to the customer. The
boutiques offer one or two free fittings, which is one of
the reasons why the prices are so high (1:30).

On the simpler side of the international boutique scene
are the Italian shops, which seem to concentrate more on

FIGURE 17-3   An Analytical Report

2

merchandise and less on decor. Their approach to merchan-

dising is simple, practical, and relatively inexpensive.

Impact display is confined to the shop windows. Brand-name

labels are boldly printed on shop doors and windows to

attract attention to the merchandise before the customer

enters the shop. Often the circle display is used. Here,

merchandise complements other merchandise in several distinct

circles of clothing, each making a different fashion state-

ment. Although not as elaborate as the French boutique, the

Italian boutique provides a pleasant, relaxing atmosphere in

which to shop (1:31-32).

The boutique boom hit the United States during the early

1960s. In New York City, specialty shops of every imaginable

shape, size, and style opened. In essence, the American

adaptation of the boutique is a shop that offers highly

specialized, ready-to-wear or made-to-order apparel in

unusual styles. Also offered are exotic artifacts, a wide

variety of accessories, and the widest possible range of

prices (2:27).

Description of Proposed Boutique

P. J. Sweetman's is an informal, one-room boutique that

FIGURE 17-3   (*Continued*)

3

specializes in clothing for young to middle-aged women.
Figure 1 illustrates a free-flow layout (without parallel
aisles). The ideal shop is 28 feet wide and 32 feet long,
for a total of 896 square feet. Selling space occupies 754
square feet of the shop. This includes the entry way, all
displays, racks, cases, counters, dressing rooms, and the
customer sitting area. The remaining 142 square feet consists of the employees' bathroom and a small storage room,
both of which are nonselling space. The storage area will
be used mainly to store gift boxes, hangers, bags, etc.

P. J. Sweetman's carries a low volume of high-quality
merchandise, which falls into three categories: career
sportswear, evening wear, and accessories.

Purpose of Report and Intended Audience

This report has two purposes: to determine the feasibility of opening a boutique in Pelham and to identify the
best location for such a shop. The areas considered are
Main Street in Pelham and the Pelham Mall, two miles from
downtown. The report is written for the general reader and
should prove helpful to anyone planning to open a shop in
either of these areas.

FIGURE 17-3 (*Continued*)

4

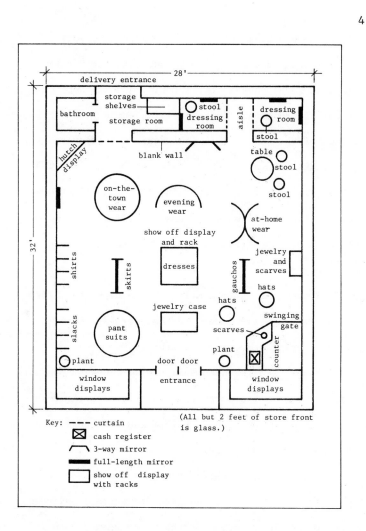

Figure 1. The Proposed Free-Flow Layout of P. J. Sweetman's

FIGURE 17-3 (*Continued*)

5

Data Sources

Data sources include magazines, books, demographic charts, personal observation, and interviews and discussions with local businesspeople.

Limitations of Study

Limitations of this study include a few uncooperative interviewees and unavailable population and income statistics for the recent year. Also, firsthand observation was limited to the winter/spring season.

Working Definitions

1. Anchor stores are the major department stores that draw the bulk of mall shoppers because of wide variety and moderate prices.

2. Target market is the group of customers, classified according to age, sex, or income bracket, whom the retailer wishes to serve.

Scope

Following a discussion of marketing opportunities for Sweetman's in Pelham, this report evaluates the site possibilities at Pelham Mall and on Main Street in terms of type

FIGURE 17-3 (*Continued*)

6

and number of competing businesses and availability and cost
of desirable space.

COLLECTED DATA

Market Opportunities in Pelham, Massachusetts

Definition

P. J. Sweetman's potential target market consists of the
women between the ages of 21 and 45 who live in Bailey County
year round. Their number would have to approach 10,000 in
order to make this proposed boutique feasible.

Findings

Recent census figures indicate that an estimated 12,700
women between the ages of 21 and 45 live in Bailey County
year round (3:7). During summer months, tourism swells this
number to over 50,000 (4:206). Table 1 breaks down P. J.
Sweetman's year-round target market by area and town.

Evaluation and Interpretation of Findings

Because these census figures are recent, they can be
considered a reliable indicator of the distribution by town
and area of P. J. Sweetman's target population. Table 1
indicates that most women between the ages of 21 and 45 live

FIGURE 17-3 (*Continued*)

7

Table 1. P. J. Sweetman's Potential Target Market Population
(year-round)

LOCATION	POPULATION
Upper Cape Area	5,225 total
Barnard[a]	2,066
Fitchfield	2,325
Monroe	199
Shoreville	635
Mid-Cape Area	4,814 total
Pelham	2,578
Deary	746
Youngville	1,490
Lower Cape Area 1	1,920 total
Brewsville	207
Colrain	496
Elwood	268
Hashby	594
Orono	355
Lower Cape Area 2	782 total
Princeton	408
Thames	148
Warren	226

[a]Excluding Orange Air Force Base

Source: Cape Hope Planning and Economic Development Commission, Census Population and Housing, 1976.

FIGURE 17-3 (*Continued*)

8

in the Upper and Mid-Cape areas. Both these areas are within twenty miles (a half hour's driving time) of Pelham. Only Lower Cape area 2, with the smallest target population, is over thirty miles away. Because mild winters make the roads on Cape Hope quite navigable, the bulk of the target population would have quick and easy access to P. J. Sweetman's year-round.

Number of Competing Businesses

Competing businesses are those retail operations that offer merchandise similar to that offered by P. J. Sweetman's. Direct competition consists of stores specifically designed as boutiques and catering to the same customer tastes and age bracket. Indirect competition consists of larger stores with one or two departments offering some styles and prices roughly similar to P. J. Sweetman's.

Pelham Mall

Direct Competition. One retail operation in the Pelham Mall that would be in direct competition with P. J. Sweetman's is El Tarantula, a small boutique chain that specializes in extravagant clothing and accessories for women. El Tarantula began on Main Street in Pelham during the 1960s.

FIGURE 17-3 (*Continued*)

9

Following the success of this shop, a second shop was opened

in the Pelham Mall. Owned and operated by Barbara and

Irving Jackson, El Tarantula caters to the top 5 percent of

the consumer income range. Ms. Jackson does the buying,

window displays, and shop layouts. She buys three times

yearly -- twice in Europe and once in New York. According

to Mr. Jackson, 1½ percent of their merchandise is designer

merchandise. The remainder are just "high quality goods."

Aside from an occasional radio ad, this boutique does little

advertising (5).

Indirect Competition. Two retail stores in the Pelham

Mall would be competing indirectly with P. J. Sweetman's.

The first, Mason's, a federated department store, moved to

the Pelham Mall from Main Street in 1970. Because it is a

department store, Mason's will not be a major competitor.

Mason's career sportswear department may be in competition

because it carries similar merchandise and price lines. The

second indirect competitor would be Smith and Turner, a

department store specializing in apparel for young to middle-

aged women. Smith and Turner will not be a major competitor

FIGURE 17-3 (*Continued*)

10

because it carries such a variety of styles. In a few sportswear lines, however, it may compete indirectly.

Interpretation of Findings. Competition at the Pelham Mall is not a major concern for P. J. Sweetman's. The only direct competitor, El Tarantula, prices its merchandise two to three times higher than P. J. Sweetman's and therefore caters to a higher-income market. The indirect competitors are both department stores. Because of their "department store" images, they do not draw the specialty-shop clientele.

Main Street

Direct Competition. There are two ladies' apparel boutiques on Main Street directly competing with P. J. Sweetman's. The first is El Tarantula at 352 Main Street. It is identical to its sister shop in the Pelham Mall. The second is Entropy, a small, ladies' apparel boutique at 604 Main Street. Entropy specializes in casual sportswear for young women. Its merchandise and prices are similar to P. J. Sweetman's. Entropy, however, concentrates on the eighteen- to twenty-five-year-old market instead of the twenty-one to forty-five market.

Indirect Competition. Some indirect competition might

FIGURE 17-3 (*Continued*)

11

stem from two larger stores. First is Drake's, located at
585 Main Street. Drake's carries a few lines similar to
those that will be carried by P. J. Sweetman's, but this
store cannot be regarded as a major competitor because it
caters to an older, higher-income market. The second is
Hanley's, located at 326 Main Street. Hanley's carries a
few similar lines, but not enough to be a direct competitor.
This store caters primarily to a lower-income market in a
variety of areas such as little girls' wear, nightwear,
etc. (6).

Interpretation of Findings. Competition on Main Street
is not a primary concern for P. J. Sweetman's. The first
direct competitor, El Tarantula, sells high-priced merchandise. The second, Entropy, has merchandise and prices
similar to P. J. Sweetman's but our target markets barely
overlap. The two indirect competitors cater either to
different age groups or to different income markets.

Availability and Cost of Desirable Space

As mentioned earlier, the proposed layout for P. J.
Sweetman's requires roughly 900 square feet of space.

FIGURE 17-3 (*Continued*)

12

The Pelham Mall

The Pelham Mall provides for one-stop shopping in a pleasant environment. It is clean, well lit, climate controlled, and tastefully decorated. These positive features make retail space scarce and expensive.

Findings. There is presently no retail space available in the Pelham Mall, but within the calendar year, construction will be completed on an addition to the mall. The exact size and number of shops have not yet been determined.

Rental fees at the Pelham Mall now range from $7.00 to $14.00 per square foot per year, depending on the location, but according to Mr. Talbot, future tenants of the new wing will most likely be paying $15.00 per square foot regardless of location. Increased construction costs and competition for space are cited as the reasons for this increase (7).

Interpretation of Findings. Retail space in the Pelham Mall should become available within two months before the proposed opening date of P. J. Sweetman's. The high rental costs, however, would have to be passed along to the consumers in higher merchandise costs, and higher prices could drastically affect sales in an unestablished business.

FIGURE 17-3 (*Continued*)

13

Main Street

Findings. The following locations are presently available for leasing on Main Street:

1. At 564 Main Street, on the corner of Main and Brown
    Streets, is a two-family house that has been con-
    verted into two individual 1000-square-foot shop
    areas. One shop is occupied by a delicatessen and
    the other is vacant. Situated in the center of the
    business district where the traffic flow is heaviest,
    this building has an excellent location. The rent
    is $250.00 monthly, excluding utilities. The site
    has several disadvantages, however:

    a. The building exterior is in very poor condition.

    b. The adjacent delicatessen would not complement
        P. J. Sweetman's.

    c. Because the building is set back twenty feet
        from the street, foot traffic would be limited.

2. At 638 Main Street, in the West End, is an 800-
    square-foot vacant shop situated between a pizza
    parlor and a laundromat. The building appears rela-
    tively new both inside and out. The rent is $175.00

FIGURE 17-3 (*Continued*)

14

monthly, excluding utilities. This location also
has several disadvantages:

a. The building is approximately thirty feet from
the sidewalk, making access inconvenient.

b. The location itself is far west of the central
business district.

c. The adjacent businesses do not complement P. J.
Sweetman's.

d. The adjacent businesses do not generate much
traffic flow.

3. At 598 Main Street is a large, two-family house that
has been converted into two 850-square-foot shops.
Both the interior and exterior of the building are
in good condition. One of the shops is occupied by
a waterbed store and the other is vacant. Both the
size and condition of this shop are ideal for P. J.
Sweetman's needs. The rent is $200.00 monthly,
excluding utilities. Because of its location, however, the site has two disadvantages:

a. The building is far west of the central business
area.

FIGURE 17-3 (*Continued*)

15

b. The building is set back about twenty feet from
the sidewalk.

4. At 502 Main Street is the Mini Mall. This is an old
department store that has been remodeled inside and
sectioned into small shop areas. The building is in
good condition both inside and out. It is clean,
well lit, heated, air conditioned, and in the
process of being carpeted. Individual shops are
rented on a seasonal or yearly basis. The largest
shop available is 550 square feet, rented on a
yearly basis at $175.00 monthly, including utilities.
The business location of this shop is good but there
are certain disadvantages:

a. The vacant shop is not large enough to serve
P. J. Sweetman's needs.

b. The handmade kit shop next to the vacant shop
would not complement P. J. Sweetman's.

c. There is already an apparel shop in this
building, Doomsday Designs, specializing in
wraparound skirts.

d. Most of the shops in this building seem

FIGURE 17-3 (*Continued*)

16

designed to cater to tourists by offering

inexpensive goods and souvenir items.

Interpretation of Findings. The yearly rental for each

of these sites on Main Street is substantially lower than

that for a shop of equivalent size in the Pelham Mall. None

of these possibilities, however, offers the combination of

good location, easy foot-traffic access, adequate floor

space, and overall desirability needed by P. J. Sweetman's.

CONCLUSIONS AND RECOMMENDATIONS

Summary of Findings

Market Opportunities

An estimated 12,700 women between the ages of 21 and

45 live in Bailey County year round, with the figure swelling

to over 50,000 in the summer. The majority of the year-round

target population lives within twenty miles of Pelham.

Site Possibilities at the Pelham Mall

The one direct competitor with P. J. Sweetman's in the

Pelham Mall caters to a higher-income market. The two

indirect competitors are department stores that do not

generally draw specialty-shop clientele. With the

FIGURE 17-3 (*Continued*)

17

completion of a new wing, retail space will soon be available
at an approximate cost of $15.00 per square foot per year
(cost for 900 square feet = $13,500 per year).

Site Possibilities on Main Street, Pelham

The two direct competitors with P. J. Sweetman's on Main
Street either cater to a higher-income market or to a
generally younger age group. The two indirect competitors
are larger stores that cater to different age groups and
income markets. Several retail sites are available with
yearly rentals ranging from $2100 (with utilities) to $3600
(without utilities), but each of these has disadvantages in
regard to location, foot-traffic access, floor space, or
overall desirability.

Comprehensive Interpretation of Findings

Market Opportunities

The present target market in the Cape Hope area is
clearly large enough to support the operation of this proposed boutique.

Site Possibilities at the Pelham Mall

The Pelham Mall is not presently the best location for

FIGURE 17-3 (*Continued*)

18

P. J. Sweetman's. Most consumers come to the mall to shop in the anchor stores, not to browse through the small shops. In order to afford the high rent, the small shops are forced to charge higher prices. Thus consumers tend to avoid the boutiques and other specialty shops. Instead, they shop in the department stores where the prices are lower and the styles less exotic. P. J. Sweetman's, as a new, unknown business, would not have an established clientele for some time. By beginning its operation in the mall, P. J. Sweetman's might never develop a following of established customers.

Site Possibilities on Main Street, Pelham

Main Street, Pelham, is the best location for P. J. Sweetman's. Consumers come to Main Street primarily to browse through the boutiques and specialty shops. Shop rentals on Main Street are not nearly as high as those in the mall. Therefore, the Main Street shops need not charge such high prices for comparable goods, but there are presently no suitable sites available.

FIGURE 17-3 (*Continued*)

19

Recommendations

1. Because P. J. Sweetman's has an adequate target market and is clearly a feasible venture, we should proceed from the proposal stage to the planning stage by carefully surveying current fashion trends and by contacting the appropriate designers, importers, and wholesale distributors.

2. We should ask all local rental agents to alert us immediately of sites that become available for rent within the central business district of Main Street. Periodic advertisements describing our site needs should also be placed in the local newspaper.

3. If, following the success of P. J. Sweetman's on Main Street, we have built a substantial clientele, we should reevaluate the mall possibilities and perhaps open another shop at the mall. We must, however, establish a predictable customer flow before taking this step.

FIGURE 17-3   (*Continued*)

20

REFERENCES

1. Sloane, C. W. The little boutiques of Roma and Firenze.
   Fashion Retailing (December 1976):30-32.

2. Carey, S. Boutiques bustling all over town. The New
   York Times, 2 July 1966:27.

3. Haskel Associates Estimates. U.S. census of population
   PC (1)-B23 (1970).

4. Census of population -- general, social, and economic
   characteristics, 1970. Cape Hope Planning and Economic
   Development Commission.

5. Jackson, I. F., Owner of El Tarantula Boutique, in an
   interview about retailing practices, 21 February 1977.

6. Talbot, E., Manager of Pelham Mall, in an interview
   about retail space, 4 May 1977.

7. Howard, R. P., President of the Downtown Pelham Association, in an interview about competition among local
   fashion retail stores, 6 May 1977.

FIGURE 17-3 (*Continued*)

As you sift through data and write your report, choose the most reliable sources, distinguish hard from soft evidence, and avoid specious reasoning.

Follow these steps in planning and writing your analysis:

1. Pose and answer these questions at various points in your search and in your writing:
   a. What am I looking for?
   b. How should I structure my inquiry to obtain this information?
   c. How will I best communicate my process of inquiry and my findings to my reader?

   Remain flexible enough to modify your approach as you go along.
2. Make a detailed outline and develop your report from it:
   a. In your introduction, make clear the subject of the report, describe and define the problem or question, and explain whatever background is necessary and relevant. Identify your intended reader and discuss briefly sources of data, along with any reasons for omitting certain data. List working definitions or place them in a glossary. If you do use a glossary or appendixes, mention them. Finally, list all major topics to be discussed in your body.
   b. In your body section, divide your subject into major topics and related subtopics. Carry the division as far as you can to make sense of the topic. At each level of division, define the topic, discuss your related findings, and evaluate and interpret the findings.
   c. In your conclusion, summarize the major findings in the body, explain the overall meaning of your findings, and make recommendations based on your overall interpretation.

Revise your report as needed and submit your final draft with all necessary supplements.

REVISION CHECKLIST

Use this list to refine the content, arrangement, and style of your report.

Content

1. Does the report grow from a clear statement of purpose (answering a practical question, examining quality, solving a problem, measuring feasibility)?
2. Is the title unbiased and accurate?
3. Is the report's length appropriate to its subject (e.g., long enough to show how you arrived at your conclusions)?

4. Are all limitations of the report spelled out?
5. Is each topic defined before it is discussed?
6. Are visuals used effectively and whenever possible?
7. Is the analysis based on hard evidence (no unsupported opinions)?
8. Is it free from specious reasoning?
9. Are all data sources credible?
10. Are all data accurate?
11. Are all data unbiased?
12. Are all data complete?
13. Are all data fully interpreted?
14. Is the documentation adequate and correct?
15. Are recommendations based on objective interpretations of data?

Arrangement

1. Does the report text have a fully developed introduction, body, and conclusion?
2. Are headings appropriate and adequate?
3. Are there enough transitions between related ideas?
4. Is the report supplemented by all needed front matter (title page, letter of transmittal, table of contents, table of illustrations, informative abstract)?
5. Is the report supplemented by all needed end matter (glossary, list of references, appendix)?

Style

1. Is the level of technicality appropriate for the stated audience?
2. Are front-matter pages numbered in small roman numerals?
3. Are all sentences clear, concise, and fluent?
4. Is the language convincing and precise?
5. Is the report written in correct English (see Appendix A)?

Now list those elements of the report that need improvement.

EXERCISES

1. List five or six college courses you have taken. Explain briefly the kinds of analytical processes you practiced in each course. *Hint:* Course syllabi may provide an outline for each discussion.

2. *Self-help analysis:* Choose a problem in your life (low grades, poor love life, insufficient time for relaxation, depression, anxiety, etc.). Define the problem clearly in several sentences. In a private brainstorming session (see Appendix B), compile a list of possible causes. After carefully evaluating each possible cause, compile a list of probable causes, thereby narrowing your original list. Now, evaluate each probable cause in an attempt to emerge with one or more definite causes. Summarize your findings; draw conclusions based on your evidence and make recommendations for solving the problem.

3. *In class:* Divide into groups of about eight. Choose a subject for group analysis — preferably, a campus issue — and partition the topic through a group brainstorming session. Next, select major topics from your list and classify as many items as possible under each major topic. Finally, draw up a working outline that could be used for an analytical report on this subject.

4. Prepare a questionnaire based on your work in exercise 3 and administer it to members of your campus community. List the findings of your questionnaire and your conclusions in clear and logical form.

5. Explain, in detail, how analysis forms the basis of each technical writing technique studied in this course.

6. In the periodical section of your library, find examples of reports or articles that use each of the following types of analysis:

    a. answering a practical question (Will X work for a specific purpose?)
    b. comparing two or more items (Is X or Y better for a specific purpose?)
    c. solving a problem (Why does X happen?)
    d. assessing feasibility (Is X practical in a given situation?)

Provide full bibliographical information, along with a *descriptive* abstract (see page 79) of each article.

7. In the periodical and newspaper section of your library, compile a list of sources, by title, ranked in general order of reliability: (a) List five highly reliable sources. (b) List five sources that are less reliable. Briefly explain the reason for each choice by discussing your criteria for judgment.

8. The statements below are followed by false or improbable conclusions. In order to prevent specious generalizations, what specific supporting data or evidence would be needed to justify each conclusion?

    a. Eighty percent of black voters in Mississippi voted for John Jones as governor. Therefore, he is not a racist.
    b. Fifty percent of last year's college graduates did not find desirable jobs. Therefore, college is a waste of time and money.
    c. Only 60 percent of incoming freshmen eventually graduate from this college. Therefore, the college is not doing its job.
    d. He never sees a doctor. Therefore, he is healthy.
    e. This house is expensive. Therefore, it must be well built.

9. Write an informative abstract of the report in Figure 17-3. (You may wish to review Chapter 4.)

10. Write an analytical report using the following guidelines:
    a. Choose a subject for analysis from the list at the end of this exercise, from your major, or from an area of interest.
    b. Identify the problem or question so you will know exactly what you are looking for.
    c. Restate the main question as a declarative sentence in your statement of purpose.
    d. Identify an audience — other than your instructor — who will use your information for a specific purpose.
    e. Hold a private brainstorming session (see Appendix B) to generate major topics and subtopics.
    f. Use the topics to make an outline based on the model outline in this chapter. Divide as far as necessary to identify all points of discussion.
    g. Make a list of all sources (primary and secondary) that you will investigate in your analysis.
    h. Write a proposal memo to your instructor (pages 436–438), describing the problem or question and your plan for analysis. Attach a working bibliography to your memo.
    i. Use your working outline as a guide to research and observation. Evaluate sources and evidence, and interpret all evidence fully. Modify your outline as needed.
    j. Submit a progress report to your instructor (pages 406–408), describing work completed, problems encountered, and work remaining.
    k. Write an analysis of your audience's needs and level of technical knowledge. (Use the samples on pages 498 and 518 as models.)
    l. Write the report for your stated audience and include topic headings and transitions between topics. Work from a clear statement of purpose and be sure that your reasoning process is shown clearly. Verify that your evidence, conclusions, and recommendations are consistent.
    m. After writing your first draft, make any needed changes in the outline and revise your report according to the revision checklist. Include all necessary supplements.
    n. Exchange reports with another class member for further suggestions for revision.
    Here are some possible subjects for your analysis:

    – the effects of a vegetarian diet on physiological energy
    – the causes of student disinterest in campus activities
    – the student transportation problem to and from your college
    – two or more brands of tools or equipment from your field
    – noise pollution from nearby airport traffic
    – the effects of toxic chemical dumping in your community
    – the advisability of earning a graduate degree in your field
    – the adequacy of veterans' benefits

- the adequacy of security on your campus or in your dorm
- the adequacy of your college's remediation program
- the feasibility of opening a particular business
- the best location for a new business
- causes of the high drop-out rate of students in your college
- the pros and cons of condominium ownership
- the feasibility of moving to a certain area of the country
- job opportunities in your career field
- effects of the 200-mile limit on the fishing industry in your coastal area
- the effects of budget cuts on public higher education in your state
- the best nonprescription cold remedy
- the adequacy of zoning laws in your town
- radio and TV interference caused by electrical devices through 115 VAC (60 hz) power lines
- effect of population increase on your local water supply
- adequacy of your student group health insurance policy
- the feasibility of using biological pest control as an alternative to pesticides
- the feasibility of large-scale desalination of sea water as a fresh water source
- effective water conservation measures that can be employed in your area
- the effects of legalizing gambling in your state
- effective measures for relieving the property-tax burden in your town
- the causes of low employee morale in the company where you work part time
- causes of poor television reception in your area
- the effects of thermal pollution from a local nuclear power plant on marine life
- effective measures for the reclamation of strip-mined land
- the feasibility of using wood as an energy source
- two or more brands of a wood-burning stove
- the feasibility of converting your home to solar heating
- the adequacy of police protection in your town
- effective measures for improving the fire safety of your home
- the best energy-efficient, low-cost housing design for your area
- effective measures for improving productivity in your place of employment
- reasons for the success of a specific restaurant (or other business) in your area
- the feasibility of operating a campus food co-op
- the best investment (real estate, savings certificates, stocks and bonds, precious metals and stones, etc.) over the past ten years
- the problem of water supply and waste disposal for new subdivisions in your town
- the advisability of home birth (as opposed to hospital delivery)

18

Oral Reporting

DEFINITION

PURPOSE OF ORAL REPORTS

IDENTIFYING THE BEST TYPE OF
    FORMAL REPORT
  The Impromptu Delivery
  The Memorized Delivery
  The Reading Delivery
  The Extemporaneous Delivery

PREPARING THE EXTEMPORANEOUS
    DELIVERY
  Know Your Subject
  Identify Your Audience
  Plan Your Report
    Statement of Purpose
    Homework
    Introduction-Body-Conclusion
      Sentence Outline
    Visual Aids
    Delivery Time
  Practice Your Delivery
    Feedback
    Organization
    Tone
  Anticipate Audience Questions

DELIVERING THE EXTEMPORANEOUS
    REPORT
  Use Natural Body Movements
    and Posture
  Speak with Confidence, Conviction,
    and Authority
  Moderate Voice Volume, Tone,
    Pronunciation, and Speed
  Maintain Eye Contact
  Read Audience Feedback
  Be Concise
  Summarize Effectively
  Leave Time for Questions and Answers

CHAPTER SUMMARY

REVISION CHECKLIST

EXERCISES

DEFINITION

An oral report is any spoken factual statement requiring preparation and fore-thought. Thus, oral reports can be anything from a brief, informal discussion to prepared formal speeches or lectures. This chapter will discuss the more formal reporting situations, which require planning and preparation.

Like the written report, the oral report must be clear, informative, and tech-nically appropriate for its audience, but compared with its written equivalent, the oral report has both advantages and disadvantages. An oral report is more personal, may be more memorable, is a time saver, and elicits direct audience response; in contrast, a written report is easier to refine and organize, can be more complex, can be studied at the reader's own pace, and can be easily reviewed.

PURPOSE OF ORAL REPORTS

Your own oral reports will vary in style, range, complexity, and formality in various situations. Your most formal level of oral reporting may include con-vention speeches, reports at national sales meetings, reports to officers and other company personnel about a new proposal or project, speeches to civic groups in your community, and so on.

These formal talks may be designed to *inform* (e.g., to describe a new pro-cedure for handling customer complaints), to *persuade* (e.g., to induce com-pany officers to vote a pay raise for employees), or to do both. In any case, the higher your status, on the job or in the community, the more you will have to give formal talks.

IDENTIFYING THE BEST TYPE OF FORMAL REPORT

The techniques for delivering an oral report determine its effectiveness. Here are the possibilities:

The Impromptu Delivery

The impromptu ("off-the-cuff") delivery is ineffective only for formal reports. It is difficult to be unified, coherent, fluent, and informative without an outline, notes, and a rehearsal or two. Don't comfort yourself by saying "It's all in my head." Get your plan down on paper.

The Memorized Delivery

In the memorized delivery, you write out your report in full and then memorize every detail. You might sound like a parrot, and if you forget a word, phrase, or line, disaster strikes! Because your personality and body language influence audience interest, you can hardly expect to be a natural speaker while mechanically reciting your lines.

The Reading Delivery

Simply reading your written report aloud can be boring. Without maintaining eye contact with your audience, and varying your gestures and tone of voice at the right times, you might look like a robot. If you *do* plan to read the report, study it well and practice aloud to decrease reliance on the text. Otherwise, you may slur or mispronounce words or keep your nose stuck to the text throughout your talk, in fear of losing your place.

The Extemporaneous Delivery

An extemporaneous delivery is carefully planned, practiced, and based on notes that keep you on track. This is the most widely used and effective speaking technique. By following a set of notes in sentence-outline form, you maintain control of your material. Also, you speak in a natural, conversational style, with only brief glances at your notes. Because an extemporaneous delivery is based on ideas (represented by topic sentences to jog your memory) rather than fully developed paragraphs to be read or memorized, you can adjust your pace and diction as needed. In other words, instead of coming across as a droning, mechanical voice, you are seen as a distinct personality speaking to

other personalities. Thus, you won't stare at the ceiling as you recite memorized lines, or at your report as you read it word for word.

PREPARING THE EXTEMPORANEOUS DELIVERY

Plan your report step by step to stay in full control and to build your confidence. Some suggestions follow.

Know Your Subject

A sure way to appear ridiculous is to speak on a subject you don't fully understand or know too little about. Do your homework, *exhaustively.* Be prepared
to explain and defend each assertion and statement of opinion with fact. When
you know your subject, you can avoid tentative and equivocating statements
that begin with "I feel," "I guess," "I suppose," "I imagine," and so on. Your
audience has come to hear a knowledgeable speaker. Don't disappoint them.

Identify Your Audience

Everyone has sat through lectures that were either boring or confusing. Surely
you don't want to impose this kind of agony on your own audience. Just as you
expect another speaker to be interesting, informative, and clear, live up to
these same expectations for your audience. At one extreme, speakers who over-
simplify and belabor obvious or trivial points are boring; at the other extreme,
those who speak in generalities or fail to explain complex information are
confusing. Adjust the amount of detail and the level of technicality to your
particular audience.

Many audiences contain people with varied levels of technical understanding. Therefore, unless you know the background of each person, prepare a
report that speaks to a general audience (see Chapter 2), as in a typical classroom of mixed majors.

As you plan your report, answer these questions about your audience:

1. What major points do I want to make?
2. How can I develop each of these points to ensure interest and understanding?

If your subject is controversial, consider carefully the average age, political
views, educational level, and socioeconomic status of most audience members.
Then decide how to speak effectively without offending anyone.

Plan Your Report

Statement of Purpose

Formulate, on paper, a statement of purpose, no more than two or three sentences long. Why, specifically, are you speaking on this subject? To whom are you speaking? What purpose do you hope to achieve?

Homework

Begin gathering data well ahead of time. Use the summarizing techniques discussed in Chapter 4 to identify and organize major points.

If the oral report is simply a spoken version of your written report, you'll need substantially less preparation. Simply expand your outline for the written report into a full-sentence outline. For our limited purposes, we will assume that your oral report is based on a written report.

Introduction-Body-Conclusion Sentence Outline

Figure 18-1 shows a typical sentence outline for an oral report. It is based on a written report about twenty pages long. Each sentence in this outline is a topic sentence for a paragraph the speaker will develop in detail. (Review outlining techniques in Chapter 7.) Review each part of your outline to identify areas where visual aids might help (see Chapter 11). Flip charts and drawings on overhead projectors are especially effective for small (classroom-size) audiences.

Before practicing your delivery you might transfer your speech outline to 3 × 5 notecards, which you can hold in one hand and shuffle as needed to stay on track. Otherwise, insert the outline pages in a looseleaf binder for easy flipping. In either case, type or print clearly, leaving substantial white space between statements so that you can locate material at a glance.

Visual Aids

Well-chosen visuals can increase audience interest tremendously. Remember, however, that visuals are not a substitute for your report, but only a supplement. Select visuals that will clarify and enhance your talk — without making you fade into the background.

Use visuals whenever possible to emphasize an important point, when *showing* will be more effective than simply *telling* (review Chapter 11). The map

```
                                              Arnold Borthwick

                    ORAL REPORT OUTLINE


Purpose:  By informing Cape Cod residents about the dangers
          to the Cape's fresh water supply posed by rapid
          population growth, this report is intended to
          increase local interest in the problem.

  I.  INTRODUCTION

      A. Do you know what you are drinking when you turn on

         the tap and fill a glass?

      B. The quality of our water is high, but not guaranteed

         to last forever.

      C. The somewhat unique natural storage facility for our

         water supply creates a dangerous situation.

      D. The Cape's rapidly increasing population could

         easily pollute our water.

      E. In fact, pollution in some towns has already begun.

 II.  BODY

      A. The groundwater is collected and held in an aquifer.

         1. This water-bearing rock formation forms a broad,

            continuous arch under the entire Cape (Visual

            Aid 1).

         2. The lighter fresh water flows on top of the

            heavier salt water.

      B. With increasing population, vast amounts of sewage
```

FIGURE 18-1   Sentence Outline for an Oral Report

and solid waste from landfill dumps invade the
aquifer.

   1. As wastes flow naturally toward the sea, they
      sometimes invade the drawing radii of various
      town wells (Visual Aid 2).

   2. The Cape's sandy soil causes rapid seepage of
      wastes into the groundwater stored in the
      aquifer.

C. Increased population also causes an overdraw on
   some town wells, resulting in salt water intrusion,
   which contaminates many wells.

D. Salt and calcium used in snow removal add to the
   problem by entering the aquifer from surface runoff.

E. The effects of continuing pollution of the Cape's
   water table will be far-reaching.

   1. Drinking water will have to be piped in over a
      hundred miles from the Quabbin reservoir, at
      great expense.

   2. The Cape's beautiful fresh water ponds will be
      unfit for swimming.

   3. All aquatic and aviary marsh life will be
      threatened.

FIGURE 18-1   (*Continued*)

4. The sensitive ecological balance of Cape Cod's
   environment will be destroyed.

F. Such damage would, in turn, lead to economic disaster for Cape Cod's major industry -- tourism.

III. CONCLUSION

A. In summary, from year to year, this problem becomes
   more real than theoretical.

B. The conclusion is obvious: If the Cape is to survive ecologically and financially, immediate steps
   must be taken to preserve the quality of our only
   water supply.

C. The following recommendations offer a starting point
   for effective action:

1. Restrict population density in all Cape towns by
   creating larger lot requirements for private
   home building.

2. Keep strict watch on proposed high-density
   apartment and condominium projects.

3. Institute a committee in each town to educate its
   residents about the importance of conserving
   fresh water, thereby reducing the draw on town
   wells.

FIGURE 18-1   (*Continued*)

4. Prohibit the use of salt, calcium, and other
   additives in the sand spread on snow-covered
   winter roads.

5. Identify alternatives to land-fill dumps for
   solid waste disposal.

D. This crucial issue deserves the immediate attention
   of every Cape Cod resident.

E. Conduct a question-and-answer session.

FIGURE 18-1   (*Continued*)

in Figure 18-2 ($3' \times 5'$ as used in the actual talk) is designed to reinforce Arnold Borthwick's explanation of how solid wastes can contaminate wells. In this case, the *telling* is clarified greatly by the *showing*.

As you plan visuals for your report, ask yourself these questions:

1. What points in my report do I want to highlight? What specific type of visual will provide the best emphasis in each case?

2. How large should the visual(s) be in order to be seen clearly by the entire audience?

3. What hardware is available (slide projector, opaque projector, overhead projector, film, videotape, blackboard)? Which is the best for my purpose and audience? How far in advance do I have to request this equipment?

4. Can I make (or have made) drawings, sketches, charts, graphs, or maps as needed? Can transparencies (for an overhead projector) be made up or slides collected? Do handouts have to be typed and reproduced?

Follow the suggestions below for using visuals in your oral report:

1. Keep the visuals simple. Your audience will not have time to study each one in great detail.

2. Limit their number (no more than four in a twenty-minute talk). Otherwise, your presentation may seem like a media event. Be selective in your choices of what to highlight with visuals.

3. Interpret each visual as you present it.

4. Stand aside when discussing a visual so that everyone can see it.

5. After discussing the visual, remove it to refocus audience attention on you.

6. If you plan to do drawings on a blackboard, do them beforehand (in multicolored chalk). Otherwise, your audience will be twiddling their thumbs while you draw away.

7. If the audience has to remember key information, you might prepare handouts for distribution *after* your talk.

8. Check out the room beforehand to make sure that you have the necessary space, electrical outlets, furniture, etc., for your visual equipment. If you will be addressing a large audience by microphone and are planning to point to certain features on your visuals, be sure that the microphone is movable. Don't forget a pointer if you need one.

Delivery Time

Because your oral report is an extended summary of your written version, aim for a maximum delivery time of twenty minutes. Longer talks may cause your audience to lose attention (the outline in Figure 18-1 is so designed). Time yourself in your practice sessions and trim as needed.

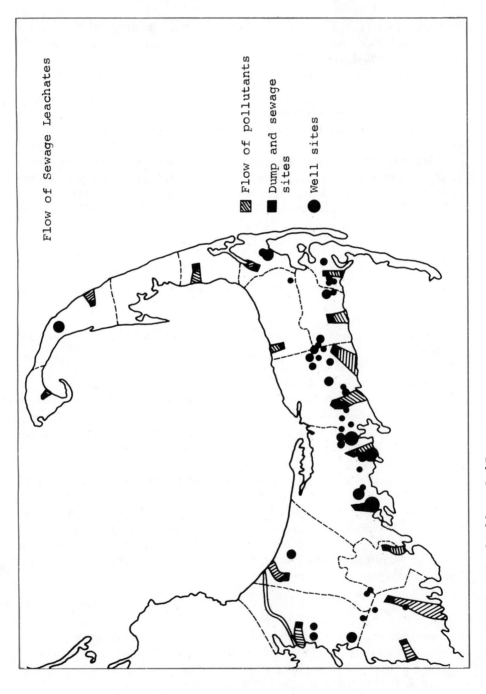

FIGURE 18-2   A Visual Aid for an Oral Report

Practice Your Delivery

Hold several practice sessions to learn the geography of your report. Then you won't fumble your actual delivery.

Feedback

Try to practice at least once before friends. Otherwise, use a full-length mirror and a tape recorder. Assess your rate of speaking from your friends' comments or from your taped voice (which, by the way, will sound high to you) and adjust your pace if necessary. Revise any parts that your friends find unclear. Have them evaluate your organization and tone.

Organization

Is your delivery unified, coherent, and logically developed? How well does it hang together? Will the audience be able to follow your chain of reasoning? Do you use enough transitional statements to reinforce the logical connection between related ideas?

Tone

Maintain a conversational tone. Speaking as you do in the classroom should be appropriate.

Anticipate Audience Questions

Consider the parts of your report that might elicit questions and challenges from your audience. "You might need to clarify or justify information that is new, controversial, disappointing, or in any way surprising. Be prepared to give factual support for all conclusions, recommendations, assertions, and statements of opinion. If you doubt the validity of any statement, leave it unsaid. Try to predict your audience's responses so you'll be ready to field questions confidently and effectively.

DELIVERING THE EXTEMPORANEOUS REPORT

Capitalize on your good preparation by using the following guidelines for your delivery.

Use Natural Body Movements and Posture

If you move and gesture as you normally would in a conversation, your audience will be more relaxed. Nothing seems more pretentious than a speaker who works through a series of rehearsed moves and artificial gestures. Also, maintain a reasonable posture.

Speak with Confidence, Conviction, and Authority

Show your audience that you believe in what you say. Avoid qualifiers ("I suppose," "I'm not sure," "but . . . ," "maybe," etc.). Also, clean up verbal tics ("er," "ah," "uuh," "mmm," etc.), which do a poor job of filling in the blank spaces between statements. If you seem to be apologizing for your existence, you won't be impressive. Speaking with authority, however, is not the same as speaking like an authoritarian.

Moderate Voice Volume, Tone, Pronunciation, and Speed

When using a microphone, people often speak too loudly. Conversely, without a microphone, they may speak too softly. Be sure that you can be heard clearly without shattering eardrums. You might ask your audience about the sound and speed of your delivery after speaking a few sentences. Your tone should be confident, sincere, friendly, and conversational.

Because nervousness can cause too-rapid speech and unclear or slurred pronunciation, pay close attention to your pace and pronunciation. Usually, the rate you feel is a bit slow will be just about right for your audience.

Maintain Eye Contact

Eye contact is a key in relating to your audience. Do not stare at the ceiling, your notes, the rear wall, or your feet. Instead, look directly into your listeners' eyes to hold their interest. With a small audience, your eye contact is one of your best connectors. As you speak, establish eye contact with as many members of your audience as possible. In addressing a large group, maintain eye contact with those in the first few rows.

Read Audience Feedback

Addressing a live audience gives you the advantage of receiving immediate feedback on your delivery. Assess your audience's responses continually and make adjustments as needed. If, for example, you are laboring through a long list of facts, figures, examples, or statistical data, and you notice that people are dozing or moving restlessly, you might summarize the point you are making. Likewise, if frowns, raised eyebrows, or questioning looks indicate confusion, skepticism, or indignation, you can backtrack a bit with a specific example or a more detailed explanation. By tuning in to your audience's reactions, you can avoid leaving them confused, hostile, or simply bored.

Be Concise

Say what you came to say, then summarize and close — politely and on time. Don't punctuate your speech with "clever" digressions that pop into your head. Unless a specific anecdote was part of your original plan to clarify a point or increase interest, avoid excursions. Remember that each of us often finds what we have to say much more engaging and entertaining than our listeners do!

Summarize Effectively

Before ending, take a moment to summarize the major points and to reemphasize anything of special importance. As you conclude, thank your listeners.

Leave Time for Questions and Answers

As you begin, inform your audience that a question-and-answer period will follow. Announce a specific time limit (such as ten minutes) so you will not become embroiled in lengthy public debates over certain points. Then you can conclude the session gracefully, without making anyone feel that he or she has been cut off abruptly or arbitrarily excluded from the discussion. Answer each question as fully as you can and don't be afraid to admit ignorance. If you can't answer a question, say so, and move on to the next question. Terminate the session by saying "We have time for one more question," or the like.

Planning and practice make oral reports effective. As in writing, *control* is central. On a basic level, just filling a page with words can be called *writing*. Similarly, the mere utterance of intelligible sounds can be called *speaking*. The effective speaker, however, communicates with confidence, sophistication, and purpose. As with all skills, practice makes perfect. Therefore, instead of avoiding public speaking opportunities, seek them out.

CHAPTER SUMMARY

In general comparison with its written equivalent, an oral report elicits direct audience response, is more personalized, may have a greater effect, and saves time. In contrast, a written report is easier to refine and organize, can be more complex, can be studied by readers at leisure, and can be easily reviewed.

When preparing a formal report (in this case, based on a written equivalent), don't try to memorize the report, don't expect to read the report to your audience, and don't try to cook it up on stage because you think "it's all in your head." Your typical situation will be an extemporaneous one in which the report is by no means written out, but for which you have ample time for preparation and in which you have notes to rely on.

Follow these steps in preparing your delivery:

1. Know your subject matter so you can speak with authority.
2. Identify the technical level of your audience and plan accordingly.
3. Plan your report by writing a statement of purpose, doing your homework, constructing an introduction-body-conclusion sentence outline, and working out a reasonable delivery time.
4. Include well-chosen visual aids.
5. Practice your delivery before friends or using a mirror and tape recorder.
6. Try to anticipate the kinds of questions your audience will ask.

Follow these guidelines in delivering your report:

1. Move and gesture as you normally would in a conversation; keep a good, natural posture.
2. Speak with confidence, conviction, and authority — as though you know what you are talking about.
3. Keep tabs on your voice volume, tone, pronunciation, and speed.
4. Maintain eye contact, talking to your listeners, not at them.
5. Read audience feedback for any adjustments you might need to make.
6. Be concise; say what you came to say, and close.
7. Near closing, summarize your main points.
8. Conclude gracefully, leaving time for questions and answers.

REVISION CHECKLIST

Use this list to refine the content, arrangement, and style of your delivery.

Content

1. Is the report content suited to the makeup and needs of my audience?
2. Do I begin with a clear statement of purpose?

3. Does the report itself achieve my stated purpose (deliver what title and purpose statement promise)?

4. Do I know what kinds of questions to expect from my audience?

5. Do I have adequate and appropriate visuals to ensure audience interest and understanding?

6. Do I have enough concrete facts to support my assertions?

Arrangement

1. Are the introduction-body-conclusion sections of my report clearly differentiated and fully developed?

2. Can I follow my outline with only brief glances?

3. Is my report coherent?

4. Do I stick to my purpose without digressing?

5. Do I summarize effectively, pulling everything together before concluding?

Style

1. Is my delivery relaxed and personable?

2. Am I comfortable with body movements and posture?

3. Do I speak with confidence and authority but without sounding pretentious?

4. Do I pronounce all words distinctly?

5. Are the volume, tone, and speed of my delivery effective?

6. Do I maintain good eye contact with my audience/mirror reflection?

7. Is my delivery clear?

Now list those elements of your delivery that need improvement.

EXERCISES

1. In a one- or two-paragraph memo to your instructor, identify and discuss the kinds of oral reporting duties you expect to encounter in your career.

2. In a two-paragraph memo to your instructor, identify your biggest fear about oral reporting. Discuss the reasons for this anxiety, describing it in detail. Finally, propose solutions to your problem.

3. Design an oral report for presentation to your technical writing class. (The report may be based on one of your written reports.) Outline your

presentation in sentence form and include at least two visual aids. Practice at home with a tape recorder and mirror or a friend as audience. Evaluate your delivery according to the revision checklist.

4. Observe a lecture or speech at your college and evaluate it according to the revision checklist. Write a memo to your instructor (without naming the speaker) identifying any weak areas and making recommendations for improving the delivery. Identify the strong areas as well.

Appendix A

Review of Grammar, Usage, and Mechanics

No matter how vital and informative a message may be, its credibility can be damaged by basic errors. Any of these errors — an illogical, fragmented, or run-on sentence; faulty punctuation; or a poorly chosen word — stands out and mars otherwise good writing. Not only do such errors confuse and annoy the reader, but they also speak badly for the writer's attention to detail and precision. Your career will make the same demands for good writing that your English classes do. The difference is that evaluation (grades) in professional situations usually shows in promotions, reputation, and salary.

None of the material here should be new to you. Everyone studies the ground rules of our language from the earliest grades onward. For reasons that are not yet fully understood, however, many writers continue to have trouble with "the basics." Although the material in this appendix offers no overnight solutions to long-standing writing problems, it does provide a simple and practical guide for making basic repairs. Spend an occasional few minutes reviewing sections that fit your needs, and you may discover that some troublesome writing problems are easier to solve than you had realized.

Table A-1 contains the standard correction symbols along with their interpretation and page references. When your instructor or proofreader marks a symbol on your paper, turn to the appropriate section (here or in Chapter 3) for explanations and examples that will help you make corrections quickly and easily. This appendix is for your reference; use it when you need it, as you would a dictionary or thesaurus.

SENTENCE PARTS AND TYPES

Sentence Parts

A sentence is a statement that contains a subject and a verb and expresses a complete idea. Much more important than this textbook definition of sentence,

TABLE A-1 Correction Symbols

Symbol	Meaning	Page	Symbol	Meaning	Page
ab	abbreviation	603–605	*, /*	comma	584, 587–593
agr p	pronoun/referent agreement	576	*-- /*	dashes	597–598
agr sv	subject/verb agreement	574–576	*··· /*	ellipses	596
			! /	exclamation point	585–586
appr	inappropriate diction	51–52	*- /*	hyphen	598
			ital	italics	596
bias	biased tone	57–58	*( ) /*	parentheses	597
ca	pronoun case	578–579	*. /*	period	584, 585
cap	capitalization	605–606	*? /*	question mark	585
chop	choppy sentences	48–49, 572	*" / "*	quotation	591, 595–596
cl	clutter word	47	*; /*	semicolon	584, 586
coh	paragraph coherence	33–34	*qual*	needless qualifier	47
cont	contraction	594–595	*red*	redundancy	44
coord	coordination	572–573	*rep*	needless repetition	45
cs	comma splice	569–571	*ref*	faulty reference	576–577
dgl	dangling modifier	579–580	*ro*	run-on sentence	571
euph	euphemism	54	*seq*	sequence of development in a paragraph	34–39
exact	inexact word	55–56	*shift*	sentence shift	582–583
frag	sentence fragment	567–569	*sp*	spelling	607
gen	generalization	54	*st mod*	stacked modifiers	40–41
jarg	needless jargon	52–53	*str*	paragraph structure	29–33
len	paragraph length	29	*sub*	subordination	573–574
lev	level of technicality	13–18	*th op*	"th" sentence openers	45
mng	meaning ambiguous	40	*trans*	transition	599–602
mod	misplaced modifier	580–581	*trite*	triteness	53
noun ad	noun addiction	46	*un*	paragraph unity	33
om	omitted word	39–40	*v*	voice	41–43
over	overstatement	54	*var*	sentence variety	49–50
par	parallelism	581–582	*w*	wordiness	43–47
pct	punctuation		*wo*	word order	41
ap /	apostrophe	593–595	*ww*	wrong word	50–58
[ ] /	brackets	597	*#*	numbers	606–607
: /	colon	584, 587	*¶*	begin new paragraph	28–29
			no ¶	no paragraph	28–32

however, is our innate understanding of how groups of words function as sentences. As an illustration, consider this nonsense statement:

> In the cronk, the crat midingly pleted the mook smurg.

Although the only words we recognize in the example are "in" and "the," we can say that this is a sentence. Why? Because in some place, something did something to something else. There is a subject, "crat," which did the doing; there is a verb, "pleted," which is in the past tense; there is an adverb, "midingly," which modifies the verb, telling us how the crat pleted; there is an adjective, "mook," which modifies "smurg"; there are three nouns, "cronk," "crat," and "smurg"; "cronk" is the object of the preposition "in," and "smurg" is the object of the verb "pleted." So, without understanding the words, we can see that we all know something about language — how words work to make sentences. We don't know what the idea is but we do know that it is complete.

Let's look at these sentence parts, and others, in more detail.

Subject

The subject is the actor of the sentence — the noun or pronoun that usually precedes the predicate (the verb and other words that explain it) and about which we say something or ask a question.

> The **technician** works too hard.
> Why does the **technician** work too hard?
> The **new, young, ambitious technician in the materials testing department** works too hard.

In this last sentence, the simple subject is "technician," and the complete subject (with all the words that explain the subject) is in boldface.

Predicate

The predicate is made up of the verb and any words that modify and explain it. The predicate usually is what is said about the subject's action or being.

> The technician **works**.
> Who **is** the technician?
> The technician **works until she can no longer stand up.**

In this last sentence, the simple verb is "works," and the complete predicate (with all the words that explain the simple predicate) is in boldface.

Object

An object is something acted on directly or indirectly by a verb, or governed by a preposition.

Direct Object.   A direct object is a noun or noun substitute that receives or is otherwise affected by the predicate's action.

> The technician delivered **the blueprints.**
> Why did the technician refuse **the job offer?**
> I don't know **where the technician is.**
> **What** did the technician say?

Indirect Object.   An indirect object is a noun or noun substitute that states to whom or for whom (or what) the predicate acts.

> I sent **the technician** a message.
> She wrote **the employer** a letter of recommendation for the technician.

Object of a Preposition.   The object of the preposition is a noun or noun substitute joined to another part of the sentence by a preposition (*across, after, between, by, for, in, near, up, with,* and other "relationship" words). "Technician" in the last example above is an object of the preposition.

> The exhausted technician collapsed on the **floor.**

Objective Complement

An objective complement is a word or group of words that further explains the subject's action on the direct object.

> The union elected Sarah **president.**
> I consider the technician **an asset.**

Subjective Complement

A subjective complement is a word or group of words that further explains the subject.

> The technician is **talented.**
> All our technicians appear **ambitious.**

Phrase

A phrase is a group of words lacking a subject or a predicate, or both. There are several kinds of phrases.

> Infinitive Phrase   Sam likes **to be challenged** by a job. (*functions as direct object*)
>
> Prepositional Phrase   Sarah is not content to remain **in a technician's role.** (*functions as adverb*)

Verbal Phrase	Sam **will be working** until his dying day. (*functions as verb*)
Gerund Phrase	**Working constantly** can be damaging. (*functions as noun*)
Participial Phrase	**Hoping for a promotion,** Sarah worked overtime every week. (*functions as adjective*)

Clause

A clause is a group of words that contains a subject and a predicate. It may be independent (main) or dependent (subordinate). An independent clause can stand alone as a sentence.

> The technician works too hard.

A dependent clause cannot stand alone as a sentence; it can serve as a noun, an adjective, or an adverb. A dependent clause always needs an independent clause to complete its meaning.

Noun Clause	Sam hopes **that he can work forever.** (*as direct object*) **Whomever Sam works with** is inspired by his talent. (*as subject*)
Adjective Clause	Sally, **who works constantly,** is the success story of our firm. (*modifies "Sally"*)
Adverb Clause	Sally is exhausted **because she works too hard.** (*modifies the subjective complement "exhausted"*)

Below are the types of sentences that can be made by combining these parts.

Sentence Types

Simple Sentences

A simple sentence contains one clause with one predicate.

> Sam works.
> **Sam works** too hard.
> On any given day, **Sam works** too hard.
> On any given day, **Sam,** the new chemistry technician, **works** too hard for any employee.

Each of these is a simple sentence. Although the subject and verb gradually are expanded, and an object, adjectives, adverbs, and prepositional phrases are added, the kernel sentence remains "Sam works."

Compound Sentences

A compound sentence contains two or more main clauses, each with a subject and a predicate. The clauses usually are joined by coordinating conjunctions ("and," "but," "or," "nor," "for") or by a semicolon or colon.

> Sam works constantly, **and** he gets thinner.
> Sam works constantly; he gets thinner.

Ideas in a compound sentence are roughly equal in importance; therefore, they are expressed in equal (parallel) grammatical form.

> Sam **works** all day, he **studies** all evening, and he **travels** all weekend.

Complex Sentences

A complex sentence has two or more clauses that are not equal in importance. Instead, it has a dependent clause and an independent clause; the former depends on the latter for completion of its meaning.

> Because Sam works too hard, **he is thin.** (*The second clause is independent.*)
> **Have you heard from the technician** who is assigned to this project? (*The first clause is independent.*)

Because one clause depends on the other, they should not be separated by anything stronger than a comma. Words such as "who," "which," "although," "after," "when," and "because" (subordinating conjunctions), placed at the beginning of an independent clause, will make it dependent.

A complex sentence can have more than one dependent clause — as long as it has an independent clause.

> After Sam works all day and studies all evening, **he is ready to face another day,** even though he traveled all weekend.

Compound-Complex Sentences

A compound-complex sentence has at least two independent clauses and one dependent clause.

> Even though Sam's job is exhausting, **he finds time to study, and he hopes to get his master's degree within two years.** (*The second and third clauses are independent.*)

Now let's look at some of the things that can go wrong in sentences, along with ways to avoid common errors.

COMMON SENTENCE ERRORS

Any piece of writing is only as good as each of its sentences. Sometimes, in haste to complete an assignment, you might overlook weak spots. Here are the most common sentence errors, along with suggestions for easy repairs.

Sentence Fragment

As we said earlier, a sentence can be defined as the expression of a logically complete idea. Any complete idea must contain a subject and a verb and must not depend on another complete idea in order to make sense. Your sentence might contain several complete ideas, but it must contain at least one!

> Although he was nervous, he grabbed the line, and he saved the sailboat.
>   (*incomplete idea*)   (*complete idea*)   (*complete idea*)

However long or short your sentence is, it should make sense to your reader. If the idea is not complete — if your reader is left wondering what you mean — you probably have left out an essential element (the subject, the verb, or another complete idea). Such a piece of a sentence is called a *fragment*.

> Grabbed the line. (*a fragment because it lacks a subject*)

> Although he was nervous. (*a fragment because — although it has a subject and a verb — it needs to be joined with a complete idea to make sense*)

The only exception to the rule for sentences is when we give a command (Run!) in which the subject (you) is understood. Because this is a logically complete statement, it can properly be called a sentence. So can this one:

> Sam is an electronics technician.

Again the idea is logically complete. Although your readers may have some questions about Sam, they cannot fail to understand your meaning: somewhere there is a person; the person's name is Sam; the person is an electronics technician.

Suppose instead we write:

> Sam an electronics technician.

This statement is not logically complete, therefore not a sentence. The reader is left asking, "What about Sam the electronics technician?" The verb — the word that makes things happen — is missing. By adding a verb we can easily change this fragment to a complete sentence.

> Simple Verb   Sam **is** an electronics technician.

> Verb plus Adverb   Sam, an electronics technician, **works hard.**

Dependent Clause,   **Although he is well paid,** Sam, an electronics tech-
Verb, and Subjective   nician, **is not happy.**
Complement

Do not, however, mistake the following statement — which seems to contain
a verb — for a complete sentence:

Sam being an electronics technician.

Such "ing" forms do not function as verbs unless accompanied by such other
verbs as *is, was,* and *will be.* Again, readers are left in a fog unless you complete your idea with an independent clause.

Sam, being an electronics technician, **was responsible for checking the
circuitry.**

Likewise, remember that the "to + verb" form does not function as a verb.

To become an electronics technician.

The meaning is unclear unless you complete the thought.

To become an electronics technician, **Sam had to complete a two-year
apprenticeship.**

Sometimes we can inadvertently create fragments by adding certain words
— *because, since, if, although, while, unless, until, when, where,* and others —
to an already complete sentence. We then change our independent clause
(complete sentence) to a dependent clause.

Although Sam is an electronics technician.

Such words subordinate the words that follow them so that an additional idea
becomes necessary to make the first statement complete. That is, they make
the statement dependent on an additional idea, which must itself contain a
subject and a verb and be a complete sentence. (See "Complex Sentences" and
"Faulty Subordination.") Now we have to round off the statement with a complete idea (an independent clause).

Although Sam is an electronics technician, **he hopes to become an electrical engineer.**

Note: Be careful not to use a semicolon or a period, instead of a comma, to
separate elements in the previous sentence. Because the incomplete idea (dependent clause) depends on the complete idea (independent clause) for its
meaning, you need only a *pause* (symbolized by a comma), not a *break* (symbolized by a semicolon), between these ideas. In fact, many fragments are
created when the writer uses too strong a mark of punctuation (period or

semicolon) between a dependent and an independent clause, thereby severing the needed connection. (See our later discussion of punctuation.)

Here are some fragments from student reports. Each is repaired in several ways. Can you think of any other ways of making these statements complete?

Fragment   She spent her first week on the job as a researcher. **Selecting and compiling technical information from digests and journals.**

Correct   She spent her first week on the job as a researcher, selecting and compiling technical information from digests and journals.

She spent her first week on the job as a researcher. She selected and compiled technical information from digests and journals.

She spent her first week on the job as a researcher by selecting and compiling technical information from digests and encyclopedias.

Fragment   **Because the operator was careless.** The new computer was damaged.

Correct   Because the operator was careless, the new computer was damaged.

The operator's carelessness resulted in damage to the new computer.

The operator was careless; as a result, the new computer was damaged.

Fragment   **When each spool is in place.** Advance your film.

Correct   When each spool is in place, advance your film.

Be sure that each spool is in place before advancing your film.

Comma Splice

In a sentence fragment, an incomplete statement is isolated from items on which it depends for its completion by too strong a punctuation mark: a period or semicolon is mistakenly used in place of a comma. In a comma splice, on the other hand, two complete ideas (independent clauses), which should be separated by a period or a semicolon, are incorrectly joined by a comma, as follows:

Sarah did a great job, she was promoted.

There are several possibilities for correcting this error:

1. Substitute a period followed by a capital letter:

   Sarah did a great job. She was promoted.

2. Substitute a semicolon to signal a relationship between the two items:

   Sarah did a great job; she was promoted.

3. Use a semicolon with a connecting adverb (a transitional word):

   Sarah did a great job; **consequently** she was promoted.

4. Use a subordinating word to make the less important sentence incomplete, thereby dependent on the other:

   Because Sarah did a great job, she was promoted.

5. Add a connecting word after the comma:

   Sarah did a great job, **and** she was promoted.

Your choice of construction will depend, of course, on the exact meaning or tone you wish to convey. Each of the following comma splices can be repaired in the ways mentioned above.

Comma Splice
This is a fairly new technique, therefore, some people don't trust it.

Correct
This is a fairly new technique. Some people don't trust it.

This is a fairly new technique; therefore, some people don't trust it.

Because this is a fairly new technique, some people don't trust it.

This is a fairly new technique, **so** some people don't trust it.

Comma Splice
Ms. Jones was a strict supervisor, she was well liked by her employees.

Correct
Ms. Jones was a strict supervisor. She was well liked by her employees.

Ms. Jones was a strict supervisor; **however,** she was well liked by her employees.

Although Ms. Jones was a strict supervisor, she was well liked by her employees.

Ms. Jones was a strict supervisor, **but** she was well liked by her employees.

Ms. Jones was a strict supervisor; she was well liked by her employees.

Comma Splice A current is placed on the wires entering and leaving the meter, the magnetic field generated by the current moves the coil.

Correct A current is placed on the wires entering and leaving the meter. The magnetic field generated by the current moves the coil.

A current is placed on the wires entering and leaving the meter; the magnetic field generated by the current moves the coil.

A current is placed on the wires entering and leaving the meter; **consequently** the magnetic field generated by the current moves the coil.

When a current is placed on the wires entering and leaving the meter, the magnetic field generated by the current moves the coil.

A current is placed on the wires entering and leaving the meter, **and** the magnetic field generated by the current moves the coil.

Run-on Sentence

The run-on sentence, a cousin to the comma splice, crams too many ideas together without providing needed breaks or pauses between thoughts.

Run-on The hourglass is more accurate than the waterclock for the water in a water clock must always be of the same temperature in order to flow with the same speed since water evaporates it must be replenished at regular intervals thus not being as effective in measuring time as the hourglass.

Like a runaway train, this statement is out of control. Here is a corrected version:

Revised The hourglass is more accurate than the waterclock because water in a water clock must always be of the same temperature to flow at the same speed. Also, water evaporates and must be replenished at regular intervals. These temperature and volume problems make the waterclock less effective than the hourglass in measuring time.

Choppy Sentences

It is possible to write grammatically correct sentences that, nonetheless, read like Dick-and-Jane sentences in a third-grade reader. Short, choppy sentences cause tedious reading for your audience and bad publicity for you.

Choppy   Brass-plated prongs are not desirable. They do not always make a good contact in the outlet. They also rust or corrode. Some of the cheaper plugs also have no terminal screws. The conductors in the cord are soldered to the prongs. Sometimes they are just wrapped around them. These types often come as original equipment on small lamps and appliances. They are not worth repairing when a wire comes loose.

Correct this problem by combining related ideas within single sentences and by using transitions to increase coherence.

Revised   Brass-plated prongs are not desirable because they do not always make a good contact in the outlet and they rust and corrode. Furthermore, some of the cheaper plugs, which often come as original equipment on small lamps and appliances, have no terminal screws: the conductors in the cord are either soldered to the prongs or just wrapped around them. Therefore, these plugs are not worth repairing when a wire comes loose.

Notice that the original eight sentences have been replaced by three. (See also pages 48–50.)

Faulty Coordination

Give ideas of equal importance equal emphasis by joining them with coordinating conjunctions: *and, but, or, no, for, so,* and *yet.*

This course is difficult **but** worthwhile.
My horse is old **and** gray.
We must decide to support **or** reject the dean's proposal.

Do not coordinate excessively.

Excessive   The climax in jogging comes after a few miles **and** I
Coordination   can no longer feel stride after stride **and** it seems as if I am floating **and** jogging becomes almost a reflex **and** my arms **and** legs continue to move **and** my mind no longer has to control their actions.

Revised	The climax in jogging comes after a few miles when I can no longer feel stride after stride. By then I am jogging almost by reflex, nearly floating, my arms and legs still moving, my mind no longer having to control their actions.

Avoid coordinating ideas that can't be sensibly connected:

Faulty	John had a weight problem and he dropped out of school.
Revised	John's weight problem made him so depressed that he couldn't study, so he quit school.
Faulty	I was late for work and wrecked my car.
Revised	Late for work, I backed out of the driveway too quickly, hit a truck, and wrecked my car.

Don't use "try and." Use "try to" instead.

Faulty	I will try and help you.
Revised	I will try to help you.
Faulty	Bill promised to try and be on time.
Revised	Bill promised to try to be on time.

Faulty Subordination

When you subordinate, you make the less important ideas dependent on the most important idea. Through subordination you can combine related short sentences and also emphasize the most important idea. Consider, for instance, these two ideas:

Joe studies hard. He has a learning disability.

Because these ideas are each expressed as a simple sentence, they appear to be coordinate (equal in importance). But if you wanted to express your opinion about Joe's chances of succeeding, you would need a third sentence: "His handicap probably will make success impossible"; or "His will power will help him succeed." An easier and more concise way to inject your opinion is by combining ideas and subordinating the one that deserves less emphasis:

Despite his learning disability (*subordinate idea*), Joe studies hard (*independent idea*).

This first version suggests that Joe will succeed. Below, subordination is used to suggest the opposite:

Despite his diligent studying (*subordinate idea*), Joe has a learning disability (*independent idea*).

Be sure to place the idea you want emphasized in the independent clause; don't write

> Although Alfred is receiving excellent medical treatment, he is seriously ill.

if you mean to suggest that Alfred has a good chance of recovering.

Don't coordinate when you should subordinate:

> Weak   Television viewers can relate to an athlete they idolize and they feel obligated to buy the product endorsed by their hero.

Of the two ideas in the above sentence, one is the cause, the other the effect. Emphasize this relationship through subordination:

> Revised   Because television viewers can relate to an athlete they idolize, they feel obligated to buy the product endorsed by their hero.

When combining several ideas within a sentence, decide which is most important, and make the other ideas subordinate to it — don't simply coordinate:

> Faulty   This employee is often late for work, and he writes illogical reports, and he is a poor manager, and he should be fired.

> Revised   Because this employee is often late for work, writes illogical reports, and has poor management skills, **he should be fired.** (*The last clause is independent.*)

Don't overstuff sentences by subordinating excessively:

> Overstuffed   This job, which I took when I graduated from college, while I waited for a better one to come along, which is boring, where I've gained no useful experience, makes me anxious to quit.

> Revised   Upon college graduation, I took this job while waiting for a better one to come along. Because I find it boring, and have gained no useful experience, I am anxious to quit.

Faulty Agreement — Subject and Verb

Failure to make the subject of a sentence agree in number with the verb is a common error. Happily, it's an error easily corrected and avoided. We are not likely to use faulty agreement in short sentences, where subject and verb are not far apart. Thus we are not likely to say "Jack eat too much" instead of "Jack eats too much," but in more complicated sentences — in which the subject is separated from its verb by other words — we sometimes lose track of the subject-verb relationship.

> Faulty   The lion's **share** of diesels **are** sold in Europe.

Although "diesels" is closest to the verb, the subject is "share," a singular subject that must agree with a singular verb.

> Correct   The lion's **share** of diesels **is** sold in Europe.

Agreement errors are easy to correct when the subject and verb are identified.

> Faulty   There **is** an estimated 29,000 **women** living in our county.

> Correct   There **are** an estimated 29,000 **women** living in our county.

> Faulty   A **system** of lines **extend** horizontally to form a grid.

> Correct   A **system** of lines **extends** horizontally to form a grid.

A second situation that causes trouble in subject-verb agreement occurs when we use indefinite pronouns such as *each, everyone, anybody,* and *somebody.* They function as subjects and usually take a singular verb.

> Faulty   **Each** of the crew members **were** injured.

> Correct   **Each** of the crew members **was** injured.

> Faulty   **Everyone** in the group **have** practiced long hours.

> Correct   **Everyone** in the group **has** practiced long hours.

Sometimes agreement problems can be caused by collective nouns such as *herd, family, union, group, army, team, committee,* and *board.* They can call for a singular or plural verb — depending on your intended meaning. When denoting the group as a whole, use a singular verb.

> Correct   The **committee meets** weekly to discuss new business.
>
> The editorial **board** of this magazine **has** high standards.

To denote individual members of the group, however, use a plural verb.

> Correct   The **committee have** voted unanimously to hire Jim.
>
> The editorial **board are** all published authors.

Yet another problem occurs when two subjects are joined by *either ... or* or *neither ... nor.* Here, the verb is singular if both subjects are singular, and plural if both subjects are plural. If one subject is plural and one is singular, the verb agrees with the one that is closest.

Correct   Neither **John** nor **Bill works** regularly.

Either **apples** or **oranges are** good vitamin sources.

Either Felix or his **friends are** crazy.

Neither the boys nor their **father likes** the home team.

If, on the other hand, two subjects (singular, plural, or mixed) are joined by *both . . . and*, the verb will be plural. Whereas *or* suggests "one or the other," *and* suggests a combination of the two subjects, thereby requiring a plural verb.

Correct   **Both** Joe and Bill **are** resigning.
The **book and** the **briefcase appear** expensive.

Faulty Agreement — Pronoun and Referent

A pronoun can make sense only if it refers to a specific noun (its referent or antecedent), with which it must agree in gender and number. It is easy enough to make most pronouns agree with their respective referents.

Correct   **Joe** lost **his** blueprints.
The **workers** complained that **they** were treated unfairly.

Some cases, however, are not so obvious. When, for example, an indefinite pronoun like *each, everyone, anybody, someone,* and *none* serves as the pronoun referent, the pronoun itself is singular.

Correct   **Anyone** can get **his** degree from that college.
Anyone can get **his or her** degree from that college.
Each candidate described **her** plans in detail.

Faulty Pronoun Reference

Whenever a pronoun is used, it must refer to one clearly identified referent; otherwise, your message will be vague and confusing.

Ambiguous   **Sally** told **Sarah** that **she** was obsessed with her job.

Does "she" refer to Sally or Sarah? Several interpretations are possible:

1. Sally is obsessed with her job.
2. Sally thinks that Sarah is obsessed with her (Sally's) job. (Sarah is envious.)
3. Sally thinks that Sarah is obsessed with her own job.
4. Sally is obsessed with Sarah's job.
5. Sally thinks that someone else is obsessed with her (Sally's) job.

6. Sally thinks that someone else is obsessed with Sarah's job.
7. Sally thinks that someone else is obsessed with some other person's job.
8. Sally thinks that someone else is obsessed with her own job.

Correct   Sally told Sarah, "I'm obsessed with my job."
Sally told Sarah, "I'm obsessed with your job."
Sally told Sarah, "You're obsessed with [your, my] job."
Sally told Sarah, "She's obsessed with [her, my, your] job."

Avoid using *this, that,* or *it* — especially to begin a sentence — unless the pronoun refers to a specific antecedent (referent).

Vague   As he drove away from his menial **job,** boring **lifestyle,** and damp **apartment,** he was happy to be leaving **it** behind.

Correct   As he drove away, he was happy to be leaving his menial job, boring lifestyle, and damp apartment behind.

Vague   The problem with our **defective machinery** is only compounded by the new **operator's incompetence. That** makes me angry!

Correct   I am angered by the problem with our defective machinery as well as by the new operator's incompetence.

Vague   Water boils at 212°F and freezes at 32°F, which makes it usable as a coolant in most parts of the country. Antifreeze is required to prevent **this.** Some manufacturers recommend **this** on cars with air conditioning because of the possibility of the heater core freezing.

Notice that the first "this" has no specific referent, and the second seems to refer to "antifreeze" but is placed too far from its referent. The meaning of the message is obscured by such vague construction.

Inaccurate   Increased blood pressure is caused by the narrowing of the blood vessels, making the pressure higher as **it** attempts to flow through the blood vessels.

Here, "it" seems to refer to "pressure," which is absurd.

Correct   Increased blood pressure is caused by the narrowing of the blood vessels, making the pressure higher as the blood attempts to flow through the vessels.

Faulty Pronoun Case

The case of a pronoun — nominative, objective, or possessive — is determined by the role it plays in the sentence: as a subject, an object, or an indicator of possession.

If the pronoun serves as the subject of a sentence (*I, we, you, she, he, it, they, who*), its case is *nominative.*

> **She** completed her graduate program in record time.
> **Who** broke the chair?

When a pronoun follows a version of the verb *to be* (a linking verb), it further explains (complements) the subject, and thus its case is nominative.

> It was **she.**
> The chemist who perfected our new distillation process is **he.**

If the pronoun serves as the object of a verb or a preposition (*me, us, you, her, him, it, them, whom*), its case is *objective.*

Object of the Verb	The employees gave **her** a parting gift.
Object of the Preposition	Several colleagues left with **him.** To **whom** do you wish to complain?

If a pronoun indicates possession (*my, mine, our, ours, your, yours, his, her, hers, its, their, theirs, whose*), its case is *possessive.*

> The brown briefcase is **mine.**
> **Her** offer was accepted.
> **Whose** opinion do you value most?

Here are some of the most frequent errors made in pronoun case:

Faulty	**Whom** is responsible to **who**? (*The subject should be nominative and the object should be objective.*)
Correct	**Who** is responsible to **whom**?
Faulty	The debate was between Marsha and **I**. (*As object of the preposition, the pronoun should be objective.*)
Correct	The debate was between Marsha and **me.**
Faulty	**Us** board members are accountable for our decisions. (*The pronoun accompanies the subject, "board members," and thus should be nominative.*)
Correct	**We** board members are accountable for our decisions.

Faulty   A group of **we** managers will fly to the convention. (*The pronoun accompanies the object of the preposition, "managers," and thus should be objective.*)

Correct   A group of **us** managers will fly to the convention.

Hint: By deleting the accompanying noun from the two latter examples, we can identify easily the correct pronoun case ("We . . . are accountable . . ."; "A group of us . . . will fly . . .").

Faulty Modification

The word order (syntax) of a sentence determines its effectiveness and meaning. Certain words or groups of words are modified (i.e., explained or defined) by other words or groups of words. Prepositional phrases, for example, usually define or limit adjacent words:

> the foundation **with the cracked wall**
> the repair job **on the old Ford**
> the journey **to the moon**
> the party **for our manager**

As do phrases with "ing" verb forms:

> the student **painting the portrait**
> **Opening the door,** we entered quietly.

Or phrases with "to + verb" form:

> **To succeed,** one must work hard.

Or certain clauses:

> the man **who came to dinner**
> the job **that I recently accepted**

When using modifying phrases to begin sentences, we can get into trouble unless we read our sentences carefully.

Dangling   **Answering the telephone,** the cat ran out the open
Modifier   door.

Here, the introductory phrase signals the reader that the noun beginning the main clause (its subject) is what or who is answering the telephone. The absurd message occurs because the opening phrase has no proper subject to modify; it *dangles.*

Correct   As Mary answered the telephone, the cat ran out the open door.

A dangling modifier can also obscure the meaning of your message.

> Dangling Modifier — **After completing the student financial aid application form,** the Financial Aid Office will forward it to the appropriate state agency.

Who completes the form — the student or the financial aid office? Here are some other dangling modifiers that make the message confusing, inaccurate, or downright absurd:

> Dangling Modifier — **While walking,** a cold chill ran through my body.
>
> Correct — While **I** walked, a cold chill ran through my body.
>
> Dangling Modifier — **After a night of worry,** the lights came on.
>
> Correct — **After we had worried all night,** the lights came on.
>
> Dangling Modifier — Impurities have entered our bodies **by eating chemically processed foods.**
>
> Correct — Impurities have entered our bodies by **our** eating chemically processed foods.
>
> Dangling Modifier — **By planting different varieties of crops,** the pests were unable to adapt.
>
> Correct — By planting different varieties of crops, **farmers** prevented the pests from adapting.

The word order of adjectives and adverbs in a sentence is as important as the order of modifying phrases and clauses. Notice how changing word order affects the meaning of these sentences:

> I **often** remind myself of the need to balance my checkbook.
> I remind myself of the need to balance my checkbook **often.**

Be sure that modifiers and the words they modify follow a word order that reflects your meaning.

> Misplaced Modifier — Harry typed another memo on our new electric typewriter **that was useless.** (*Was the typewriter or the memo useless?*)
>
> Correct — Harry typed another useless memo on our new electric typewriter.
>
> *or*
>
> Harry typed another memo on our new, useless electric typewriter.

Misplaced | He read a report on the use of nonchemical pesti-
Modifier | cides **in our conference room.** (*Are the pesticides to be used in the conference room?*)

Correct | In our conference room, he read a report on the use of nonchemical pesticides.

Misplaced | She volunteered **immediately** to deliver the radioac-
Modifier | tive shipment. (*Volunteering immediately, or delivering immediately?*)

Correct | She immediately volunteered to deliver....

or

She volunteered to immediately deliver....

Faulty Parallelism

Express items of the same importance in the same grammatical form:

Correct | We here highly resolve ... that government **of the people, by the people, for the people** shall not perish from the earth.

The above statement describes the government by using three modifiers of equal importance. Because the first modifier is a prepositional phrase, the others must be also. Otherwise, the message would be garbled, like this:

Faulty | We here highly resolve ... that government **of the people, which the people created and maintain, serving the people** shall not perish from the earth.

If you begin the series with a noun, use nouns throughout the series; likewise for adjectives, adverbs, and specific types of clauses and phrases.

Faulty | The new apprentice is **enthusiastic, skilled,** and **you can depend on her.**

Correct | The new apprentice is **enthusiastic, skilled,** and **dependable.** (*all subjective complements*)

Faulty | In his new job, he felt **lonely** and **without a friend.**

Correct | In his new job, he felt **lonely** and **friendless.** (*both adjectives*)

Faulty | She plans **to study** all this month and **on scoring well** in her licensing examination.

Correct | She plans **to study** all this month and **to score** well in her licensing examination. (*both infinitive phrases*)

Faulty   She **sleeps** well and **jogs** daily, **as well as eating** high-protein foods.

Correct   She **sleeps** well, **jogs** daily, and **eats** high-protein foods. (*all verbs*)

To improve coherence in long sentences, repeat words that introduce parallel expressions:

Faulty   Before buying this property, you should decide whether you plan to settle down and raise a family, travel for a few years, or pursue a graduate degree.

Correct   Before buying this property, you should decide whether you plan **to settle** down and raise a family, **to travel** for a few years, or **to pursue** a graduate degree.

Make all headings parallel in your table of contents, your formal outline, and your report.

Faulty   A. Picking the Fruit
         1. When to pick
         2. Packing
         3. Suitable temperature
         4. Transport with care

The logical connection between steps in that sequence is obscured because each heading is phrased in a different grammatical form.

Correct   A. Picking the Fruit
         1. Choose the best time
         2. Pack the fruit loosely
         3. Store at a suitable temperature
         4. Transport with care

Other forms of phrasing would also be correct here, as long as each item is expressed in a form parallel to all other items in the series.

Sentence Shifts

Shifts in point of view damage coherence. If you begin a sentence or paragraph with one subject or person, don't shift courses.

Shift in Person   When **you** finish such a great book, **one** will have a sense of achievement.

Correct   When **you** finish such a great book, **you** will have a sense of achievement.

Shift in Number **One** should sift the flour before **they** make the pie.

Correct **One** should sift the flour before **one** makes the pie. (**Or** better: Sift the flour before making the pie.)

Don't begin a sentence in the active voice and then shift to the passive voice.

Shift in Voice **He delivered** the plans for the apartment complex, and the building site **was also inspected by him.**

Correct **He delivered** the plans for the apartment complex and also **inspected** the building site.

Don't shift tenses without good reason.

Shift in Tense She **delivered** the blueprints, **inspected** the foundation, **wrote** her report, and **takes** the afternoon off.

Correct She **delivered** the blueprints, **inspected** the foundation, **wrote** her report, and **took** the afternoon off.

Don't shift from one mood to another (e.g., from imperative to indicative mood in a set of instructions).

Shift in Mood **Unscrew** the valve and then steel wool **should be used** to clean the fitting.

Correct **Unscrew** the valve and then **use** steel wool to clean the fitting.

Don't shift from indirect to direct discourse within the same sentence.

Shift in Discourse Jim wonders **if he will get the job** and will he like it?

Correct Jim wonders **if he will get the job** and **if he will like it.**

Will Jim get the job, and will he like it?

EFFECTIVE PUNCTUATION

Punctuation marks are like road signs and traffic signals. They govern reading speed and provide clues for navigation through a network of ideas; they mark intersections, detours, and road repairs; they draw attention to points of interest along the route; and they mark geographic boundaries. In short, punctuation marks provide us with a practical and simple way of making ourselves understood. They take up the slack created when the spoken message (made clear by the speaker's tone, pitch, volume, speaking rate, pauses, body movements, and facial expressions) is transposed into a written message (silent, static words on a page). In fact, effective punctuation can often make the written message clearer than its spoken equivalent.

As an experiment, copy a paragraph — without the punctuation — from any book, and try to read it clearly.

Before we discuss individual punctuation marks in detail, a common sense review of the relationship among the four used most often (period, semicolon, colon, and comma) might help. These marks can be ranked in order of their relative strengths.

1. *Period.* The strongest mark. A period signals a complete stop at the end of an independent idea (independent clause). The first word in the idea following the period begins with a capital letter.

> Jack is a fat cat. His friends urge him to diet.

2. *Semicolon.* Weaker than a period but stronger than a comma. A semicolon signals a brief stop after an independent idea but does not end the sentence; instead, it provides advance notice that the independent idea that follows is *closely related* to the previous idea.

> Jack is a fat cat; he eats too much.

3. *Colon.* Weaker than a period but stronger than a comma. A colon usually follows an independent idea and, like the semicolon, signals a brief stop but does not end the sentence. The colon and semicolon, however, are never interchangeable. A colon provides an important cue: it symbolizes "explanation to follow." Information after the colon (which need not be an independent idea) explains or clarifies the idea expressed before the colon.

> Jack is a fat cat: he weighs forty pounds. (*The information after the colon answers "How fat?"*)
>
> *or*
>
> Jack is a fat cat: forty pounds worth! (*The second clause is not independent.*)

Note: As long as any two adjacent ideas are independent they may correctly be separated by a period. Sometimes, a colon or a semicolon may be more appropriate for illustrating the logical relationship between two given ideas. When in doubt, however, use a period.

4. *Comma.* The weakest of these marks. A comma does not signal a stop at the end of an independent idea but only a pause within or between ideas in the sentence. A comma often indicates that the word, phrase, or clause set off from the independent idea cannot stand alone but must rely on the independent idea for its meaning.

> Jack, **a fat cat**, is jolly. (*In that sentence, the phrase within commas depends on the independent idea for its meaning.*)

> **Although he diets often,** Jack is a fat cat. (*Because the first clause depends on the second, any stronger mark would create a fragment.*)

A comma is rarely appropriate between two independent clauses unless there is a conjunction.

> Comma Splice   Jack is a fat cat, he eats too much.

So we see that punctuation marks, like words, convey specific meanings to the reader. These meanings are further discussed in the sections that follow.

End Punctuation

The three marks of end punctuation — period, question mark, and exclamation point — work like a red traffic light by signaling a complete stop.

Period

A period ends a sentence. Periods end some abbreviations.

> Ms.   Assn.   N.Y.
> M.D.   Inc.   B.A.

Periods serve as decimal points for figures.

> $15.95
> 21.4%

Question Mark

A question mark ends a sentence asking a direct question.

> Where is the balance sheet?

Do not use a question mark to end an indirect question.

> Faulty   He asked if all students had failed the test?
>
> Correct   He asked if all students had failed the test.
> *or*
> He asked, "Did all students fail the test?"

Exclamation Point

Because exclamation points mean that you are excited or adamant, don't overuse them. Otherwise you might seem hysterical or insincere.

> Correct   Oh, no!
> Pay up!
> My pants are missing!

Use an exclamation point only when the expression of strong feeling is appropriate.

Semicolon

A semicolon usually works like a blinking red traffic light at a deserted intersection by signaling a brief but definite stop.

Semicolon Separating Independent Clauses

Most commonly, semicolons separate independent clauses (logically complete ideas) whose contents are closely related.

> The project was finally completed; we had done a good week's work.

The semicolon can replace the conjunction-comma combination that joins two independent ideas.

> The project was finally completed, and we were elated.
> The project was finally completed; we were elated.

The second version emphasizes the sense of elation.

Semicolons Used with Adverbs as Conjunctions and Other Transitional Expressions

Semicolons must accompany adverbs and other expressions that connect related independent ideas (*besides, otherwise, still, however, furthermore, moreover, consequently, therefore, on the other hand, in contrast, in fact,* etc.).

> The job is filled; however, we will keep your résumé on file.

> Your background is impressive; in fact, it is the best among our applicants.

Semicolons Separating Items in a Series

When items in a series contain internal commas, semicolons provide clear separations between items.

> We are opening branch offices in the following cities: Santa Fe, New Mexico; Albany, New York; Montgomery, Alabama; and Moscow, Idaho.

> Members of the survey crew were John Jones, a geologist; Hector Lightweight, a draftsman; and Mary Shelley, a graduate student.

Colon

A colon works like a flare in the middle of the road. It signals you to stop and then proceed paying close attention to the situation ahead, the details of which will be revealed as you move ahead. Usually, a colon follows an introductory statement that requires a follow-up explanation.

> We need the following equipment immediately: a voltmeter, a portable generator, and three pairs of insulated gloves.

> She is an ideal colleague: honest, reliable, and competent.

> Two candidates are clearly superior: John and Marsha.

In most cases — with the exception of *Dear Sir:* and other salutations in formal correspondence — colons follow independent statements (logically and grammatically complete). Because colons, like end punctuation and semicolons, signal a full stop, they are never used to fragment a complete statement.

> Faulty   My plans include: finishing college, traveling for two years, and settling down in Boston.

No punctuation should follow "include."
Colons can introduce quotations.

> The supervisor's message was clear enough: "You're fired."

As shown on page 584, a colon normally replaces a semicolon in separating two related, complete statements, when the second statement directly explains or amplifies the first.

> His reason for accepting the lowest-paying job offer was simple: he had always wanted to live in the Northwest.

The statement following the colon explains the "reason" mentioned in the statement preceding the colon.

Comma

The comma is the most frequently used — and abused — punctuation mark. Unlike the period, semicolon, and colon, which signal a full stop, the comma signals a *brief pause*. Thus, the comma works like a blinking green traffic light for which you slow down without coming to a dead stop. As we said earlier, a comma should never be used to signal a *break* between independent ideas; it is not strong enough.

Comma as a Pause Between Complete Ideas

In a compound sentence where a coordinating conjunction (*and, or, nor, for, but*) connects equal (independent) statements, a comma is usually placed immediately before the conjunction.

> This is a high-paying job, but the stress is high.

> This vacant shop is just large enough for our boutique, and the location is excellent for walk-in customer traffic.

Without the conjunction, each of the above statements would suffer from a comma splice, unless the comma were changed to a semicolon or a period.

Comma as a Pause Between an Incomplete and a Complete Idea

A comma is usually placed between a complete and an incomplete statement in a complex sentence to show that the incomplete statement depends for its meaning on the complete statement (i.e., the incomplete statement cannot stand alone, separated by a break symbol such as a semicolon, colon, or period).

> **Because he is a fat cat,** Jack diets often.
> **When he eats too much,** Jack gains weight.

Above, the first idea is made incomplete by a subordinating conjunction (*since, when, because, although, where, while, if, until,* etc.), which here connects a dependent with an independent statement. The first (incomplete) idea depends on the second (complete) for wholeness. When the order is reversed (complete idea followed by incomplete), the comma can usually be omitted.

> Jack diets often **because he is a fat cat.**
> Jack gains weight **when he eats too much.**

Because commas take the place of speech signals, reading a sentence aloud should tell you whether or not to pause (and use a comma).

Commas Separating Items (Words, Phrases, or Clauses) in a Series

Use a comma to separate items in a series.

> **Sam, Joe, Marsha,** and **John** are joining us on the hydroelectric project.

> The office was **yellow, orange,** and **red.**

> He works hard **at home, on the job,** and even **during his vacation.**

> The new employee complained **that the hours were long, that the pay was low, that the work was boring,** and **that the foreman was paranoid.**

> **She came, she saw,** and **she conquered.**

Do not use commas when *or* or *and* is used between all items in the series.

>She is willing to work in San Francisco or Seattle or even in Anchorage.

Add a comma when *or* or *and* is used only before the final item in the series.

>Our luncheon special for Thursday will be coffee, rolls, steak, beans, and ice cream.

Without the comma, that sentence might cause the reader to conclude that beans and ice cream is an exotic new dessert.

Comma Setting off Introductory Phrases

Infinitive, prepositional, or verbal phrases introducing a sentence are usually set off by commas.

Infinitive Phrase	**To be or not to be,** that is the question.
Prepositional Phrase	**In Rome,** do as the Romans do. **In other words,** you're fired. **In fact,** the finish work was superb.
Verbal Phrase	**Being fat,** Jack was a slow runner. **Moving quickly,** the army surrounded the enemy.

When an interjection introduces a sentence, it is set off by a comma.

>**Oh,** is that the final verdict?

When a direct address introduces a sentence, it is set off by a comma.

>**Mary,** you've done a great job.

Comma Used to Avoid Ambiguity

By signaling a pause that you would make in speaking, a comma can increase the clarity of your statement.

Ambiguous	Outside the office building was colorful. (*Was the exterior of the building or the area surrounding the building colorful?*)
Clear	Outside, the office building was colorful.
Ambiguous	For Bill Smith's advice was a lifesaver. (*Was Bill Smith's advice a lifesaver, or was Smith's advice to Bill a lifesaver?*)
Clear	For Bill, Smith's advice was a lifesaver.

Read your sentences aloud to reveal any such ambiguities in your own writing.

Commas Setting off Nonrestrictive Elements

A restrictive phrase or clause describes, defines, or limits its subject in such a way that it could not be deleted without changing the meaning of the sentence.

> All candidates **who have work experience** will receive preference.

The clause, "who have work experience," defines "candidates" and is essential to the meaning of the sentence. Without this clause, the meaning would be entirely different.

> All candidates will receive preference.

The following sentence also contains a restriction.

> All candidates **with work experience** will receive preference.

The phrase, "with work experience," defines "candidates" and thus specifies the meaning of the sentence. Because these elements *restrict* the subject by limiting the category, "candidates," each forms an integral part of the sentence and is thus not separated from the rest of the sentence by commas.

In contrast, a nonrestrictive phrase or clause does not limit or define the subject; such an element is optional because it could be deleted without changing the essential meaning of the sentence.

> Our draftsperson, **who has only six weeks' experience,** is highly competent.

> This house, **riddled with carpenter ants,** is falling apart.

In each of those sentences, the modifying phrase or clause does not restrict the subject; each could be deleted as follows:

> Our new draftsperson is highly competent.
> This house is falling apart.

Unlike a restrictive modifier, the nonrestrictive modifier does not supply the essential meaning to the sentence; therefore, it is set off by commas from the rest of the sentence.

To appreciate how the use of simple commas can affect meaning, consider the following statements:

> Restrictive   Office workers **who drink martinis with lunch** have slow afternoons.

Because the restrictive clause limits the subject, "office workers," we interpret that statement as follows: some office workers drink martinis with lunch, and these have slow afternoons. In contrast, we could write

> Nonrestrictive   Office workers, **who drink martinis with lunch,** have slow afternoons.

Here the subject, "office workers," is not limited or defined. Thus we interpret the statement as follows: all office workers drink martinis with lunch and have slow afternoons.

Commas Setting off Parenthetical Elements

Items that interrupt the flow of a sentence are called parenthetical and are enclosed by commas. Expressions such as *of course, as a result, as I recall,* and *however* are parenthetical and may denote emphasis, afterthought, clarification, or transition.

Emphasis	This deluxe model, **of course,** is more expensive.
Afterthought	Your report format, **by the way,** was impeccable.
Clarification	The loss of my job was, **in a way,** a blessing.
Transition	Our warranty, **however,** does not cover tire damage.

So is a direct address.

> Listen, **my children,** and you shall hear . . .

A parenthetical expression at the beginning or the end of a sentence is set off with a comma.

> **Naturally,** we will expect a full guarantee.
> **My friends,** I think we have a problem.
> You've done a good job, **Jim.**
> **Yes,** you may use my name in your advertisement.

Commas Setting off Quoted Material

Quoted items included within a sentence are often set off by commas.

> The customer said, **"I'll take it,"** as soon as he laid eyes on our new model.

Commas Setting off Appositives

An appositive, a word or words explaining a noun and placed immediately after it, is set off by commas.

> Martha Jones, **our new president,** is overhauling all personnel policies.

> The new Mercedes, **my dream car,** is priced far beyond my budget.

> Alpha waves, **the most prominent of the brain waves,** are typically recorded in a waking subject whose eyes are closed.

> Please make all checks payable to Sam Sawbuck, **company treasurer.**

> Sarah, **my colleague,** has arrived.

Notice that the commas clarify the meaning of the last sentence. Without commas, the sentence would be ambiguous.

> Sarah my colleague has arrived.

Are you telling Sarah that your colleague has arrived?

> **Sarah,** my colleague has arrived.

Or are you saying that your colleague, Sarah, has arrived (as in the first version)?

Commas Used in Common Practice

Commas are used to set off the day of the month from the year, in a date.

> May 10, 1984

They are used to set off numbers in three-digit intervals.

> 11,215
> 6,463,657

They are also used to set off street, city, and state in an address.

> The bill was sent to John Smith, 184 Sea Street, Albany, New York 01642.

When the address is written vertically, however, the commas that are omitted are those that would otherwise occur at the end of each address line.

> John Smith
> 184 Sea Street
> Albany, New York 01642

If we put "Albany" and "New York" on separate lines, we wouldn't have a comma after "Albany" either.

Use commas to set off an address or date in a sentence.

> Room 3C, Margate Complex, is the site of our newest office.
> December 15, 1977, is my retirement date.

Use them to set off degrees and titles from proper names.

> Roger P. Cayer, M.D.
> Gordon Browne, Jr.
> Marsha Mello, Ph.D.

Commas Used Erroneously

Don't be a comma philanthropist. Avoid sprinkling commas where they are not needed or simply do not belong. In fact, you are probably safer using too few commas than using too many. Again, the overuse of commas generally can be avoided if you read your sentences aloud.

Faulty As I opened the door, he told me, that I was late. (*separates the indirect from the direct object*)

The most universal symptom of the suicide impulse, is depression. (*separates the subject from its verb*)

This has been a long, difficult, project. (*separates the final adjective from its noun*)

John, Bill, and Sally, are joining us on the design phase of this project. (*separates the final subject from its verb*)

An employee, who expects rapid promotion, must quickly prove his worth. (*separates a modifier that should be restrictive*)

I spoke in a conference call with John, and Marsha. (*separates two words linked by a coordinating conjunction*)

The room was, eighteen feet long. (*separates the linking verb from the subjective complement*)

We painted the room, red. (*separates the object from its complement*)

Apostrophe

Apostrophes are used for three purposes: to indicate the possessive, to indicate a contraction, and to indicate the plural of numbers, letters, and figures.

Apostrophe Indicating the Possessive

At the end of a singular word, or of a plural word that does not end in *s*, add an apostrophe plus an *s* to indicate the possessive.

The people's candidate won.
The chainsaw was Bill's.
The men's locker room burned.
The car's paint job was ruined by the hail storm.
I borrowed Doris's book.
Have you heard Ray Charles's song?

In some cases convention will require that you not add an *s*.

Correct Moses' death
for conscience' sake

Do not use an apostrophe to indicate the possessive form of either singular or plural pronouns:

> Correct   The book was hers.
> Ours is the best sales record.
> The fault was theirs.

At the end of a plural word that ends in *s,* add an apostrophe only.

> Correct   the cows' water supply
> the Jacksons' wine cellar

At the end of a compound noun, add an apostrophe plus an *s.*

> Correct   my father-in-law's false teeth

At the end of the last word in nouns of joint possession, add an apostrophe plus *s* if both own one item.

> Correct   Joe and Sam's lakefront cottage

and an apostrophe plus *s* to both nouns if each owns specific items.

> Correct   Joe's and Sam's passports

Apostrophe Indicating a Contraction

An apostrophe shows that you have omitted one or more letters in a phrase that is usually a combination of a pronoun and a verb.

> Correct   I'm   they're
> he's   you'd
> we'll   who's
> you're   who'll

Don't confuse *they're* with *their* or *there.*

> Faulty   there books
> their now leaving
> living their

> Correct   their books
> they're now leaving
> living there

Remember the distinction this way:

> Correct   Their boss knows they're there.

Don't confuse *it's* and *its. It's* means "it is." *Its* is the possessive.

> Correct   It's watching its reflection in the pond.

Don't confuse *who's* and *whose*. *Who's* means "who is," whereas *whose* indicates the possessive.

> Correct   Who's interrupting whose work?

Other contractions are formed from the verb and the negative.

> Correct   isn't   can't
>   don't   haven't
>   won't   wasn't

Apostrophe Indicating the Plural of Numbers, Letters, and Figures

There are no buts about it: The 6's on this new typewriter look like smudged G's, 9's are illegible, and the %'s are unclear.

Quotation Marks

Quotation marks set off the exact words borrowed from another speaker or writer. At the end of a quotation the period or comma is placed within the quotation marks.

> Correct   "Hurry up," he whispered.
>   She told me, "I'm depressed."

The colon or semicolon is always placed outside the quotation marks:

> Correct   Our contract clearly defines "middle-management personnel"; however, it does not clearly state the promotional procedures for this group.
>
>   You know what to expect when Honest John offers you a "bargain": a piece of junk.

Sometimes a question mark is used within a quotation that is part of a larger sentence.

> Correct   "Can we stop the flooding?" inquired the foreman.

When the question mark or exclamation point is part of the quote, it is placed within the quotation marks, thereby replacing the comma or period.

> Correct   "Help!" he screamed.
>   He asked John, "Can't we agree about anything?"

If, however, the question mark or exclamation point is meant to denote the attitude of the quoter instead of the quotee, it is placed outside.

> Correct   Why did he wink and tell me, "It's a big secret"?
>   He actually accused me of being an "elitist"!

When quoting a passage of fifty words or longer, indent the entire passage five spaces and single space between the lines of the passage to set it off from the text. Do not enclose the passage in quotation marks.

Use quotation marks around titles of articles, paintings, book chapters, and poems.

> Correct   The enclosed article, "The Job Market for College Graduates," should provide some helpful insights.

The title of a bound volume — book, journal, or newspaper — should be underlined to represent italics.

Finally, use quotation marks to indicate your ironic use of a word.

> Correct   He is some "friend"!

Ellipses

Use three dots in a row (. . .) to indicate that you have left some material out of a quotation. If the omitted words come at the end of the original sentence, a fourth dot indicates the period. Use several dots centered in a line to indicate that a paragraph or more has been left out. Ellipses help you save time and zero in on the important material within a quote.

> Correct   "Three dots . . . indicate that you have left some material out. . . . A fourth dot indicates the period. Several dots centered in a line . . . indicate . . . a paragraph or more. . . . Ellipses help you . . . zero in on the important material. . . ."

Italics

In typing or longhand writing, indicate italics by <u>underlining</u>. Use italics for titles of books, periodicals, films, newspapers, and plays; for the names of ships; for foreign words or scientific names; for emphasizing a word (used sparingly); for indicating the special use of a word.

> *The Oxford English Dictionary* is a handy reference tool.

> The *Lusitania* sank rapidly.

> He reads *The Boston Globe* often.

> My only advice is *caveat emptor.*

> *Bacillus anthracis* is a highly virulent organism.

> *Do not* inhale these spores, under any circumstances!

> Our contract defines a *full-time employee* as one who works a minimum of thirty-five hours weekly.

Parentheses

Use commas normally to set off parenthetical elements, dashes to give some emphasis to the material that is set off, and parentheses to enclose material that defines or explains the preceding statements.

> An anaerobic (airless) environment must be maintained for the cultivation of this organism.

> The cost of manufacturing our Beta II transistors has increased by 10 percent in one year (see Appendix A for full cost breakdown).

> This new three-colored model (made by Ilco Corporation) is selling well.

Notice that material between parentheses, like all other parenthetical material discussed earlier, can be deleted without harming the logical and grammatical structure of the sentence.

Also, use parentheses to enclose numbers or letters that segment items of information in a series.

> There are three basic steps to this process: (1) ... , (2) ... , and (3)....

Brackets

Use brackets within a quotation to add material that was not in the original quote but that is needed for clarification. Sometimes a bracketed word will provide an antecedent (or referent) for a pronoun.

> "She [Jones] was the outstanding candidate for the job."

Brackets can enclose information taken from some other location within the context of the quotation.

> "It was in early spring [April 2, to be exact] that the tornado hit."

Use brackets to correct a quotation.

> "His report was [full] of mistakes."

Use *sic* ("thus so") when quoting a mistake in spelling, usage, or logic.

> His secretary's comment was clear: "He don't [sic] want any of these."

Dashes

Dashes are effective — as long as they are not overused. Make dashes on your typewriter by placing two hyphens side by side. While parentheses deemphasize the enclosed material, dashes strongly emphasize it.

Used selectively, dashes can provide dramatic emphasis for a statement, but they should not be used flagrantly as a substitute for all other forms of punctuation. In other words, when in doubt, do not use a dash!

Dashes can be used to denote an afterthought

> Have a good vacation — but don't get sunstroke.

or to enclose an interruption in the middle of a sentence.

> The designer of this building — I think it was Wright — was, above all, an artist.

> Our new team — Jones, Smith, and Brown — is already compiling outstanding statistics.

Although they can often be used interchangeably with commas, dashes dramatize a parenthetical statement more than commas do.

> Mary, a true friend, spent hours helping me rehearse for my interview.
> Mary — a true friend — spent hours helping me rehearse for my interview.

Notice the added emphasis in the second version.

Hyphen

Use a hyphen to divide a word at your right-hand margin. Consult your dictionary for the correct syllable breakdown:

> com-puter
> comput-er

Actually, it is best to avoid altogether this practice of word division at the ends of lines in a typewritten text.

Use a hyphen to join compound modifiers (two or more words preceding the noun as a single adjective).

> the rough-hewn wood
> the well-written report
> the all-too-human error

Do not hyphenate these same words if they *follow* the noun.

> The wood was rough hewn.
> The report is well written.
> The error was all too human.

Hyphenate an adverb-participle compound preceding a noun.

> the high-flying glider

Do not hyphenate compound modifiers if the adverb ends in *ly*.

> the finely tuned engine

Hyphenate all words that begin with the prefix *self.*

> self-reliance
> self-discipline
> self-actualizing

Hyphenate to avoid ambiguity.

> re-creation (*a new creation*)
> recreation (*leisure activity*)

Hyphenate words that begin with *ex* only if *ex* means "past."

> ex-foreman
> expectant

Hyphenate all fractions, along with ratios that are used as adjectives and precede the noun.

> a two-thirds majority
> In a four-to-one vote they defeated the proposal.

Do not hyphenate ratios if they do not immediately precede the noun.

> The proposal was voted down four to one.

Hyphenate compound numbers from twenty-one through ninety-nine.

> Thirty-eight windows were broken.

Hyphenate a series of compound adjectives preceding a noun.

> The subjects for the motivation experiment were fourteen-, fifteen-, and sixteen-year-old students.

TRANSITIONS AND OTHER CONNECTORS

Transitions help us make our meaning clear — to signal the reader that we are in a certain time frame or a certain place, that we are giving an example, showing a contrast, shifting gears, concluding our discussion, or the like. Here is a list of the most commonly used transitions and the relationships they signal:

Addition	I am majoring in naval architecture; **furthermore,** I spent three years crewing on a racing yawl.

> moreover   and
> in addition   again
> also   as well as

Place	Here is the switch that turns on the stage lights. **To the right** is the switch that dims them.

> beyond   to the left

over	nearby
under	adjacent to
opposite to	next to
beneath	where
inside	

Time

The crew will mow the ball field this morning; **immediately afterward** we will clean the dugouts.

first	the next day
next	in the meantime
second	in turn
then	subsequently
meanwhile	while
at length	since
later	before
now	after

Comparison

Our reservoir is drying up because of the drought; **similarly,** water supplies in neighboring towns are dangerously low.

likewise
in the same way
in comparison

Contrast or Alternative

Felix worked hard; **however,** he received poor grades.
Although Jack worked hard, he was never promoted.

however	but
nevertheless	on the other hand
yet	to the contrary
still	notwithstanding
in contrast	conversely
otherwise	

Results

Jack fooled around; **consequently** he was shot by a jealous husband.
Because he was fat, he was jolly.

thus	thereupon
hence	as a result
therefore	so
accordingly	as a consequence

Example

Competition for part-time jobs is fierce; **for example,** eighty students applied for the clerk's job at Sears.

for instance	namely
to illustrate	specifically

Explanation | She had a terrible semester; **that is,** she flunked four courses.

in other words   in fact
simply stated   put another way

Summary or Conclusion | Our credit is destroyed, our bank account is overdrawn, and our debts are piling up; **in short,** we are bankrupt.

in closing   to sum up
to conclude   all in all
to summarize   on the whole
in brief   in retrospect
in summary   in conclusion

Pronouns serve as connectors, because a pronoun refers back to a noun that you have used in a preceding clause or sentence.

As the **crew** neared the end of the project, **they** were all willing to work overtime to get the job done.

Low employee morale is damaging our productivity. **This** problem needs immediate attention.

Synonyms (words meaning the same as other words) can also be effective connectors. In the whaling intelligence paragraph on page 34, for example, "huge and impressive mammals" (a synonym for "whales") helps tie the ideas together.

Repetition of key words or phrases is another good connecting device — as long as it is not overdone. Here is another example of effective repetition:

Overuse and drought conditions have depleted our water supply to a critical level. Because of our **depleted water supply,** we will need to enforce strict **water-**conservation measures.

Here, the repetition also emphasizes a critical problem.

The following paragraph lacks adequate transitions, and the sentences seem choppy and awkward:

Choppy | Technical writing is a difficult but important skill to master. It requires long hours of work and concentration. This time and effort are well spent. Writing is indispensable for success. Good writers derive great pride and satisfaction from their effort. A highly disciplined writing course should be a part of every student's curriculum.

Here is the same paragraph rewritten to improve coherence:

Revised   Technical writing is a difficult but important skill to master. **Thus** it requires long hours of work and concentration. This time and effort, **however,** are well spent **because** writing is an indispensable tool for success. **Moreover,** good writers derive great pride and satisfaction from their effort. A highly disciplined writing course, **therefore,** should be a part of every student's curriculum.

Besides increasing coherence *within* a paragraph, transitions and other connectors can underscore relationships *between* paragraphs by linking related groups of ideas. Here are two transitional sentences that could serve as the concluding sentences for certain paragraphs or as topic sentences for paragraphs that would follow; or they could stand alone for emphasis as single-sentence paragraphs:

Because the A-12 filter has decreased overall engine wear by 15 percent, it should be included as a standard item in all our new models.

With the camera activated and the watertight cover sealed, the diving bell is ready to be submerged.

These sentences look both ahead and back, thereby providing a clear direction for continuing discussion.

Topic headings, like those used in this book, are another device for increasing coherence. A topic heading is both a link and a separation between related, yet distinct, groups of ideas.

Finally, a whole paragraph can serve as a connector between major sections of your report. Assume, for instance, that you have just completed a section of a report on the advantages of a new oil filter and are now moving to a section on selling the idea to the buying public. Here is a paragraph you might write to link the two sections:

Because the A-12 filter has decreased overall engine wear by 15 percent, it should be included as a standard item in all our new models. Because tooling and installation adjustments will add roughly $100 to the list price of each model, however, we have to explain the filter's long-range advantages to the customers. Let's look at ways of explaining these advantages.

The effective use of transitional devices in your revisions is one way of transforming a piece of writing from adequate to excellent.

EFFECTIVE MECHANICS

Correctness in abbreviation, capitalization, use of numbers, and spelling is an important sign of your attention to detail. Don't ignore these conventions. (See Chapter 10 for format conventions.)

Abbreviations

(For correct abbreviations in documentation, see pages 324–331.)

In using abbreviations consider your audience; never use one that might confuse your reader. Often, abbreviations are not appropriate in formal writing. When in doubt, write the word out in full.

Abbreviate certain words and titles when they precede or immediately follow a proper name.

Correct	Mr. Jones	Raymond Dumont, Jr.
	Dr. Jekyll	Warren Weary, Ph.D.
	St. Simeon	

However, do not write abbreviations such as the following:

Faulty	He is a Dr.
	Pray, and you might become a St.

In general, do not abbreviate military, religious, and political titles.

Correct	Reverend Ormsby
	Captain Hook
	President Reagan

Abbreviate time designations only when they are used with actual times.

Correct	400 B.C.
	5:15 A.M.

Do not abbreviate these designations when they are used alone.

Faulty	Plato lived sometime in the B.C. period.
	She arrived in the A.M.

In formal writing, don't abbreviate days of the week, individual months, words such as *street* and *road* or names of disciplines such as *English*. Avoid abbreviating states, such as *Me.* for *Maine;* countries, such as *U.S.* for *United States;* and book parts such as *Chap.* for *Chapter, pg.* for *page,* and *fig.* for *figure.*

Use *no.* for *number* only when the actual number is given.

Correct	Check switch No. 3.

Abbreviate units of measurement only when they are used often in your report and are written out in full on first use. Use only those abbreviations that you are sure your reader will understand. Abbreviate items in a visual aid only if you need to save space.

Here is a list of common technical abbreviations for units of measurement:

ac	alternating current		kw	kilowatt
amp	ampere		kwh	kilowatt hour
A	angstrom		l	liter
az	azimuth		lat	latitude
bbl	barrel		lb	pound
BTU	British Thermal Unit		lin	linear
C	Centigrade		long	longitude
Cal	calorie		log	logarithm
cc	cubic centimeter		m	meter
circ	circumference		max	maximum
cm	centimeter		mg	milligram
cps	cycles per second		min	minute
cu ft	cubic foot		ml	milliliter
db	decibel		mm	millimeter
dc	direct current		mo	month
dm	decimeter		mph	miles per hour
doz	dozen		no	number
dp	dewpoint		oct	octane
F	Farenheit		oz	ounce
f	farad		psf	pounds per square foot
fbm	foot board measure		psi	pounds per square inch
fl oz	fluid ounce		qt	quart
FM	frequency modulation		r	roentgen
fp	foot pound		rpm	revolutions per minute
fpm	feet per minute		rps	revolutions per second
freq	frequency		sec	second
ft	foot		sp gr	specific gravity
g	gram		sq	square
gal	gallon		t	ton
gpm	gallons per minute		temp	temperature
gr	gram		tol	tolerance
hp	horsepower		ts	tensile strength
hr	hour		v	volt
in	inch		va	volt ampere
iu	international unit		w	watt
j	joule		wk	week
ke	kinetic energy		wl	wavelength
kg	kilogram		yd	yard
km	kilometer		yr	year

Here are some common abbreviations for reference in manuscripts:

anon.	anonymous	fig.	figure
app.	appendix	i.e.	that is
b.	born	illus.	illustrated
bull.	bulletin	jour.	journal
©	copyright	l., ll.	line(s)
c., ca.	about (c. 1963)	ms., mss.	manuscript(s)
cf.	compare	n.	note
ch.	chapter	no.	number
col.	column	p., pp.	page(s)
d.	died	pt., pts.	part(s)
diss.	dissertation	rev.	revised or review
ed.	editor	rep.	reprint
e.g.	for example	sec.	section
esp.	especially	sic	thus, so (to cite an
et al.	and others		error in the quote)
etc.	and so forth	trans.	translation
ex.	example	viz.	namely
f. or ff.	the following page or pages	vol.	volume

For abbreviations of other words, consult your dictionary. Most dictionaries have a list of abbreviations at the front or rear or include them alphabetically with the appropriate word entry.

Capitalization

Capitalize the following: proper names, titles of people, books and chapters, languages, days of the week, the months, holidays, names of organizations or groups, races and nationalities, historical events, important documents, and names of structures or vehicles. In titles of books, films, etc., capitalize the first word and all those following except articles or prepositions.

Joe Schmoe
A Tale of Two Cities
Protestant
Wednesday
the *Queen Mary*
the Statue of Liberty
April
Chicago
the Bill of Rights
the Chevrolet Vega
Russian
Labor Day
Dupont Chemical Company

> Senator John Pasteur
> France
> The War of 1812
> Daughters of the American Revolution
> The Emancipation Proclamation
> Jewish
> *Gone With the Wind*

Don't capitalize the seasons, names of college classes (*freshman, junior*), or general groups (*the younger generation,* or *the leisure class*).

Capitalize adjectives that are derived from proper nouns.

> Chaucerian English

Capitalize titles preceding a proper noun but not those following.

> State Senator Marsha Smith
> Marsha Smith, state senator

Capitalize words like *street, road, corporation, college* only when they accompany a proper name.

> Bob Jones University
> High Street
> The Rand Corporation

Capitalize *north, south, east,* and *west* when they denote specific locations, not when they are simply directions.

> the South
> the Northwest
> Turn east at the next set of lights.

Begin all sentences with capitals.

Use of Numbers

As a general rule, write out numbers that can be expressed in two words or less.

> fourteen   five
> eighty-one   two million
> ninety-nine

Use numerals for all others.

> 4,364   2,800,357
> 543   3¼

Use numerals to express decimals, precise technical figures, or any other exact measurements.

50 kilowatts	15 pounds of pressure
14.3 milligrams	4,000 rpm

Express the following in numerals: dates, census figures, addresses, page numbers, exact units of measurement, percentages, ages, and times with A.M. or P.M. designations, monetary and mileage figures.

page 14	1:15 P.M.
18.4 pounds	9 feet
115 miles	12 gallons
the 9-year-old tractor	$15
15.1 percent	

Do not begin a sentence with a numeral.

Six hundred students applied for the 102 available jobs.

If your figure consumes more than two words, revise your word order.

The 102 available jobs brought 780 applicants.

Do not use numerals to express approximate figures, time not stated with A.M. or P.M. designations, or streets named by numbers below 100.

about seven hundred fifty
four fifteen
108 East Forty-second Street

If one number immediately precedes another, spell out the first and use a numeral to express the second:

Please deliver twelve 18-foot rafters.

Only in contracts and other legal documents should a number be stated both in numerals and in words:

The tenant agrees to pay a rental fee of three hundred seventy-five dollars ($375.00) monthly.

Spelling

Spelling problems will not cure themselves. If you are bothered by certain spelling weaknesses, take the time to use your dictionary, *religiously*, for all writing assignments. As you read, note the spelling of the words that have given you trouble in the past. You might even compile a list of troublesome words. Moreover, your college may have a learning laboratory or a skills resources center where you can get professional assistance with spelling problems. If not, your instructor may suggest several self-teaching books with instructions and exercises for spelling improvement.

Appendix B

The Brainstorming Technique

Brainstorming is based on the well-proven thesis that we cannot make something from nothing. Its purpose is to get ideas down *on paper* so that they may be used and so that relationships among them may be recognized. You can hold a brainstorming session alone or with a group. All you need is a quiet room, an alarm clock, and pencil and paper. The procedure is simple: just think about a chosen subject and write down every idea that pops into your head within a given period. The technique was devised years ago by Alex Osborne as a way of attacking a problem or question from all sides. Here are his guidelines for a brainstorming session:

1. Don't criticize or evaluate any ideas during the session. Simply write down every idea that emerges. Save the criticism and evaluation until later.

2. Use your imagination for "free wheeling." The wilder the idea the better, because it might lead to some valuable insights later.

3. Strive for quantity. The more ideas, the better chance for a winner to emerge.

4. Combine and improve ideas as you proceed. As your list grows, someone may suggest ways of combining two or more ideas into a better idea, or of improving on some of the ideas already listed.[1]

The point is to write down all your ideas immediately!

The brainstorming procedure can be applied to just about any question or problem. Imagine, for instance, that you are having trouble balancing your budget. Begin by formulating the problem as a question: How can I make ends meet? Now follow these steps:

---

[1] Alex F. Osborne, *Applied Imagination* (New York: Charles Scribner's Sons, 1957), p. 84. Reprinted by permission of Charles Scribner's Sons.

1. Lock yourself in a quiet room with an alarm clock, a pencil, and a stack of paper.

2. Set the alarm to ring in twenty minutes.

3. Empty your mind of anxieties — about the bills, studies, the football team, and everything else. Sit with your eyes closed for two minutes, thinking about *absolutely nothing.*

4. Now, begin thinking about ways to solve your problem. Repeat this question: How can I make ends meet?

5. As the ideas begin to flow, write *every one* of them down. Don't make judgments about relevance or value and don't worry about complete sentences. Just *get them down on paper* — all of them!

6. Continue this pushing and sweating until the alarm goes off, thinking of nothing but the universal question.

7. And if the ideas are still flowing, reset the alarm clock and go on.

8. At the end of this time you should have an impressive mixture of gibberish, irrelevancies, and *pure gold.*

9. Take the rest of the day off!

10. On the next day confront your list. Strike out the gibberish and irrelevancies and organize related items into categories. As you work on your list, other ideas will surely come to mind; put *them* down too. Out of your list, the best solution to your problem should emerge.

Here is how your brainstorming list might read:

> Get more roommates for the apartment.
> Move to a cheaper place.
> Sell the car.
> Get a part-time job.
> Apply for a loan.
> Drink less beer.
> Do less partying.
> Find a rich girlfriend/boyfriend.
> Buy used books.
> Apply for work-study credits.
> Cook at home.
> Wash and iron my own shirts.
> Sell my stereo.
> Apply for a scholarship.
> Quit school and work full time.
> Transfer to a tuition-free institution.
> Get help from home.
> Write to my rich aunt.
> Make a strict budget.
> Sell my skis and surfboard.

Turn in my credit cards.
Lock the checkbook in the closet.
Do less dating.
Work two jobs during summer.
Live on peanut butter sandwiches.
Buy a van and live in it.
Have the telephone removed.
Don't take weekend trips.

As the next step, cross out the ideas that seem less practical and realistic than the others and add any new ideas. Now organize related items within broader categories (as discussed in Chapter 6). The six classes that seem to emerge from this list are:

Selling Things
Asking for Help
Changing Lifestyle
Making a Physical Move
Getting a Job
Managing Money

As you ponder these categories, which cover all sides of the problem, you should come up with a practical solution.

The same technique can be applied to your job-hunting campaign, especially to making up your résumé. Simply modify items 4 and 10 in the earlier list.

4. Now, reach back into your personal, educational, and work experience for specific items to answer your prospective employer's *big* question: What do you have to offer? Especially dig for items that separate you from the herd.

10. On the next day, review your list and strike out the irrelevancies. On separate sheets of paper headed Career Objectives; Educational Background; Work Experience; and Personal Interests, Activities, Awards, and Special Skills, arrange the remaining items along with any new ones that pop into your head.

Now you have the raw material for your résumé.

You will find the brainstorming technique handy for achieving the depth of concentration needed for coming up with good ideas — ideas that otherwise you might never realize you had. This technique may well be your most essential prewriting step.

Appendix C

Eliminating Sexist Language

Sexism is the stereotyping of roles for individuals of either sex on the basis of their gender. One way to avoid sexism is to eliminate those terms that suggest inequality between the sexes. Your language should reflect the fact that roughly 50 percent of American women now work outside the home and that a growing number of American males are working as homemakers — either part time or full time. Thus, *supervisor* is better than *foreman; salesperson* is better than *salesman; consumer* is better than *housewife*. Because they rightly suggest that anyone can fill these roles, *secretary, nurse,* and *homemaker* are better than *male secretary, male nurse,* and *male homemaker;* likewise, *pioneer, surgeon, athlete,* or *carpenter* is better than *female pioneer, female surgeon, female athlete,* or *female carpenter*.

The National Council of Teachers of English offers the following guidelines for avoiding sexist language:[1]

1. Although *man* in its original sense carried the dual meaning of adult human and adult male, its meaning has come to be so closely identified with adult male that the generic use of *man* and other words with masculine markers should be avoided whenever possible.

Examples	Alternatives
mankind	humanity, human beings, people
man's achievements	human achievements
the best man for the job	the best person for the job; the best man or woman for the job

[1] Reprinted by permission of the National Council of Teachers of English.

Examples	Alternatives
man-made	synthetic, manufactured, crafted, machine-made
the common man	the average person, ordinary people

2. The use of *man* in occupational terms when persons holding the jobs could be either male or female should be avoided.

Examples	Alternatives
chairman	coordinator (of a committee or department), moderator (of a meeting), presiding officer, head, chair
businessman, fireman, mailman	business executive or manager, fire fighter, mail carrier
steward and stewardess	flight attendant
policeman and policewoman	police officer

3. Because English has no generic singular — no common-sex pronoun — we have used *he, his,* and *him* in such expressions as "the student . . . he." When we constantly personify "the judge," "the critic," "the executive," "the author," etc., as male by using the pronoun *he,* we are subtly conditioning ourselves against the idea of a female judge, critic, executive, or author. There are several alternative approaches for ending the exclusion of women that results from the pervasive use of the masculine pronouns.

a. Recast into the plural.

Example	Alternative
Give each student his paper as soon as he is finished.	Give students their papers as soon as they are finished.

b. Reword to eliminate unnecessary gender problems.

Example	Alternative
The average student is worried about his grades.	The average student is worried about grades.

c. Replace the masculine pronoun with *one, you,* or (sparingly) *he or she,* as appropriate.

Example	Alternative
If the student was satisfied with his performance on the pretest, he took the posttest.	A student who was satisfied with her or his performance on the pretest took the posttest.

d. Alternate male and female examples and expressions.

Example

Let each student participate. Has he had a chance to talk? Could he feel left out?

Alternative

Let each student participate. Has she had a chance to talk? Could he feel left out?

4. Using the masculine pronoun to refer to an indefinite pronoun (*everybody, everyone, anybody, anyone*) also has the effect of excluding women. In all but strictly formal usage, plural pronouns have become acceptable substitutes for the masculine singular.

Example

Anyone who wants to go to the game should bring his money tomorrow.

Alternative

Anyone who wants to go to the game should bring their money tomorrow.

Index

Abbreviations, 350, 603–605
Abstracts. *See also* Summaries
  descriptive, 79
  informative, 67–80, 247, 251, 252–253, 457, 496, 501
  placement in reports, 79–80
Active voice, 41–43
  in instructions, 191
Adjective clause, 565
Adverb clause, 565
Agreement
  noun-pronoun, 576
  subject-verb, 574–576
Ambiguous phrasing, 40
Analytical Reports. *See also* Formal Reports
  audience for, 477–481
  audience analysis for, 496, 518
  audience questions about, 478–479
  body in, 491–493
  conclusion in, 493–495
  definition of, 477–478
  elements of, 481–487
  evidence in (*see* Evidence)
  flexible approach to, 487–488
  introduction in, 488–491
  outline for, 488–489
  purpose of, 478
  reasoning in (*see* Evidence)
  recommendations in, 483, 493–495
  revision checklist for, 539–540
  sources for, 484–485
  supplements to, 484, 495–496
typical problems, 478–481
visuals in, 484, 505, 509, 522, 525
Apostrophe, 593–595
Appendix, 17–18, 304–306, 469, 496, 515.
  See also Supplements to reports
Application letters, 384–389
Appositive, 591–592
Audience
  analysis of, 146, 163, 172, 198, 208, 333, 453, 470, 496, 518
  for analytical reports, 477–481
  background of, 11–18, 19
  definition of, 11–13
  for definitions, 88–89
  for descriptions, 156–158
  for informal reports, 401–402
  for instructions, 187–189
  for letters, 360, 363–364
  level of technical understanding, 11–22, 142
  needs of, 8, 18–20, 28
  for oral reports, 545, 547
  primary vs. secondary, 17–18
  for process analyses, 208, 215
  for process narratives, 198, 207–208
  for proposals, 431–432
  questions, 12, 69, 156, 185, 432, 435, 447, 450, 452, 478, 479, 555
  sample situations, 20–23
  for summaries, 69–70
Author-year documentation system, 328, 330–331, 496, 516–517

Bar graphs, 121, 265–268
Basis for classification, 119–120
Basis for partition, 113
Biased language, 57–58, 482. *See also*
    Objectivity; Subjectivity
Bibliographies
    for given subjects, 292, 297
    of government documents, 298
Bibliography
    card, 311–312
    as a documentation system, 328, 329
    as a report supplement, 255
    working bibliography, 320
Brackets, 597
Brainstorming, 31, 132, 133, 141, 287, 302,
    321, 375, 486, 608–610

Capitalization, 605–606
Card catalog, library, 287–292
Case of pronouns, 578–579
Cause and effect
    analysis of, 479, 486, 489–495, 496–515
    outline sequence, 140
    paragraphs of, 36
Charts, 272–276
Chronological sequence
    in mechanism descriptions, 163, 164–167
    in paragraph development, 35–36
    in process explanations, 187
Classification
    definition of, 111–112
    guidelines for, 117–123
    revision checklist for, 126
    in sentence definitions, 92
    systems in libraries, 290–292
    uses of, 111–112
    visuals in, 121–122
Clutter words, 47
Coherence in paragraphs, 33–34. *See also*
    Transitions and other connectors
Colon, 584, 587
Comma, 584, 587–593
Comma splice, 569–571
Commerce Business Daily, 430
Comparison and contrast
    as analysis, 479, 496, 519–538
    block structure in, 38–39
    in definitions, 96
    in descriptions, 160
    paragraphs of, 38–39

point-by-point structure, 39
    transitions in, 600
Complaint letters, 368–369, 370
Complex sentences, 566
Compound-complex sentences, 566
Compound sentences, 566
Computerized information retrieval, 298
Computer-Readable Data Bases, 298
Conciseness, 43–47
    in oral reports, 557
    in summaries, 68, 70–71
Concrete and specific language, 56–57
    in descriptions, 158–159
    in proposals, 442, 444
Consumer complaint reports, 417, 418
Contractions, 594–595
Coordination in sentences, 48–49, 572–573
Correction symbols, 562
Covers for reports, 239, 241

Dangling modifiers, 579–580
Dashes, 597–598
Data. *See also* Evidence
    accuracy of, 482
    analysis of, 477
    adequacy of, 482
    choosing reliable sources of, 484–485
    gathering of (*see* Research)
    interpretation of, 143, 483, 484, 485, 486,
        487, 492, 493, 494, 495, 512, 524,
        528, 529, 530, 534, 535–536
    recording of (*see* Research)
Data-based information services, 298
Definition
    of abstract and general terms, 88
    audience for, 57, 58–59
    basic properties expressed in, 89–90
    of concrete and specific terms, 87–88
    expanded, 93–94
    expansion methods, 94–97
    objectivity in, 90
    parenthetical, 90–91
    placement of, in reports, 101–102
    plain English in, 89
    purpose of, 87–89
    revision checklist for, 103–104
    in a sentence, 91–92
    visuals in, 95
Dependent clause, 565

Description
  appropriate details in, 160–161
  audience analysis for, 163, 172
  audience questions about, 156
  body in, 170–171
  conclusion in, 171–172
  definition of, 153
  elements of, 159–161
  introduction in, 169–170
  objectivity in, 156–159
  outline of, 169
  precise language in, 158–159
  purpose of, 153–156
  revision checklist for, 179–180
  sequences in, 162–168
  visuals in, 160, 161, 165, 166, 167, 170,
    174
Descriptive abstract, 79
Dewey Decimal System, 292. *See also*
    Library
Diagrams, 276–278
Diction, 51. *See also* Style
Direct object, 564
Distinguishing features, in sentence defini-
    tion, 92
Documentation. *See also* Library; Research
  bibliography, 328, 329
  choosing a system of, 324–325
  footnotes, 325
  footnotes plus bibliography, 328, 329
  footnoting articles, 326–327
  footnoting books, 325–326
  footnoting miscellaneous items, 327
  lists of references
    author-year system, 328, 330
    numerical system, 331, 332
  purpose of, 324
Dossier, employment, 389

Ellipses, 313, 596
End matter, 239
Endnotes, 255
Euphemisms, 54
Evidence. *See also* Data
  evaluation of, 485–487
  reasoning on the basis of, 483, 487
Exclamation point, 585–586
Expanded definition, 93–101, 102, 518, 519–
    520
Exploded diagram, 277

Feasibility
  analysis of, 479, 481, 496, 519–538
  of proposals, 447, 452, 466
Figures, as visuals, 265–279
Flow chart, 275–276
Footnotes, 325–328, 339, 340, 496
Footnotes plus bibliography, 328, 329, 496
Formal reports, 231–232. *See also* Analytical
    reports
Formal topic outline, 133–135
Format,
  for classifications, 121–123
  definition of, 232
  features of, 235–239, 240
  for letters, 355–356
  for memos, 403–404
  for partitions, 115–117
  for proposals, 439, 442
  purpose of, 232–234
Fragment, of a sentence, 567–569
Front matter, 239
Functional sequence
  in mechanism description, 162–163, 168
  for paragraph development, 35

Gerund phrase, 565
Glossary, 17, 232, 239, 251, 254, 496, 514
Grammar, 561–583
Grantsmanship Center News, 430
Graphs, 265–272

Headings, 232, 237–239, 240
  in letters, 350
  in memos, 439
  in outlines, 138–139
  in résumés, 375
  in tables of contents, 245–247
  as transitions, 602
Hyphen, 598–599

Imperative mood in instructions, 191
Indentation in typed reports, 236
Independent clause, 565
Indexes to periodicals, 293–297
Indirect object, 564
Infinitive phrase, 564
Inflated diction, 50–51, 359–360. *See also*
    Jargon

Informal outline, 132–133
Informal reports
  definition of, 231–232, 401
  as letters, 406, 409–412, 413–414
  as memos, 402–406, 407
  in miscellaneous forms, 418–423
  on prepared forms, 412, 415–417, 418
  purpose of, 401–402
  revision checklist for, 424
Informative abstract, 67–80, 247, 251, 252–
    253, 457, 501
Inquiry letters, 307, 322, 351–352, 365–368
Instruction, letter of, 369, 371–372
Instructions, 185, 186, 187–198, 199–206
  audience analysis for, 198
  body in, 196–197
  conclusion in, 197
  elements of, 187–193
  introduction in, 194–195
  outline for, 193–194
  revision checklist for, 226–227
  visuals in, 190, 195, 203, 204, 206
Interviews
  employment, 390–391
  informative, 300–306
Italics, 596
"It" sentence openers, 45

Jargon, needless, 52–53. *See also* Letterese
Job description, 155
Job hunting, 372–374

Lab report, 186, 207–208, 209–214
Letterese, 359–360. *See also* Inflated diction;
    Jargon
Letter reports, 406, 409–412, 413–414, 435,
    439, 440–441
Letters
  of acceptance, 391–392
  accepted form for, 356, 357–358
  clear purpose in, 364
  of complaint, 368–369, 370
  definition of, 348
  employment dossier, 389
  follow-up, 391
  format of, 355
  of inquiry, 307, 322, 351–352, 365–368
  of instruction, 369, 371–372

  introduction-body-conclusion in, 349–350
  job application, 384–389
  job interviews, 390–391
  plain English in, 356–360, 361–362
  purpose of, 348–349
  required parts, 350–354
  resume, 374–384
  revision checklist for, 393–394
  specialized parts of, 354–355
  of transmittal (*see* Supplements to reports)
  "you" perspective in, 360, 363
Library
  card catalog, 287–292
  classification systems of, 290–292
  data-based information services, 298
  periodical indexes, 293–297
  reference books, 292–293
  reference librarian, 297
  U.S. Government publications, 297–298
Library of Congress Classification System,
    290–291. *See also* Library
Limitations, statement of, in a report, 142,
    489, 491, 523
Line graph, 268–272
Location words, in description, 158

Manuscript reference, abbreviations for, 605
Margins in letters and reports, 235, 355
Mechanism, description of, 153, 156–180
Memos. *See also* Informal reports
  proposals written as, 432, 433–434, 436–
    438
Minutes, of meetings, 417, 419, 420–421
Modified block form, for letters, 356, 358
Modifiers
  dangling, 579–580
  misplaced, 580–582
  stacked, 40–41

Nonrestrictive clauses or phrases, 590–591
Nontechnical writing, 3–4
Notation
  decimal, in outlines, 135, 137
  distribution, in letters
  enclosure, in letters, 354
  Roman numeral–letter–Arabic numeral, in
    outlines, 133–134, 137
  in tables, 264

Noun addiction, 46
Noun clause, 565
NTIS (National Technical Information Service), 298. *See also* Library
Numbers
  correct use of, 606–607
  plurals of, 595
Numerical System of Documentation, 331, 332, 496, 538

Objective complement, 564
Objectivity, 3, 4, 5, 6. *See also* Biased Language
  in analytical reports, 482, 483
  in classifications, 120
  in definitions, 90
  in descriptions, 156–159
Object of a preposition, 564
Oral report
  audience for, 545–547
  audience questions about, 555
  definition of, 545
  extemporaneous delivery of, 546
  outline for, 548, 549–552
  preparation of, 547–555
  purpose of, 545
  revision checklist for, 558–559
  techniques for delivery of, 555–558
  tone in, 555, 556
  visuals in, 548, 553, 554
Organizational chart, 117, 273–274, 275
Outlines
  for analytical reports, 488–489
  construction of, 140–143
  definition of, 131
  for descrpitions, 169
  formal (topic), 133–135
  headings in, 138–139
  informal, 132–133
  for instructions, 193–194
  logical sequence in, 139–140
  notation in, 133–135, 137
  for oral reports, 548, 549–552
  parallelism in, 138
  partitioning in, 136
  for process analyses, 215–216
  for process narratives, 207
  for proposals, 447
  for a research report, 321
  revision checklist for, 149

  sentence outline, 135–136
  working outline, 321, 488
Overstatement, 54

Page numbering, 236
Paper for typed reports, 235
Paragraphs
  body in, 30–31
  coherence in, 33–34
  conclusion in, 31–32
  function of, 28–29
  introduction in, 29–30
  length of, 29, 131
  logical sequence in, 34–39
  structural variation in, 32–33
  structure of, 29–33
  transitions and connectors in, 34 (*see also* Transitions and other connectors)
  unity in, 33
Parallelism, 581–582
  in classifications, 120
  in headings, 238–239
  in instructions, 192
  in outlines, 138
  in partitions, 114
Paraphrasing in note-taking, 312–314
Parentheses, 597
Parenthetical definition, 90–91, 101
Participial phrase, 565
Partition
  in analytical reports, 491–492
  definition of, 111–112
  guidelines for, 112–117
  in outlines, 136
  parallelism in, 114
  revision checklist for, 126
  uses of, 11–112
  visuals in, 113, 115, 116, 117
Passive voice, 41–43
Period, 584, 585
Periodic activity reports, 412, 415–416, 417
Photographs, 278
Pie chart, 116, 272–273
Planning proposals, 432, 433–434
Position words, in description, 158
Possessive indicated by apostrophe, 593–594
Postscript in letters, 355
Precise language, 55–56
Predicate, 563
Prefaces, needless, 47, 71

Prepositional phrases, 564
Prepositions, excessive use of, 46
Pretentious prose, 51–52, 356, 359–360, 361
Primary reader, 17–18
Primary research, 285
Process analyses, 208, 215–216, 217–225
    elements of, 208, 215
    outlines for, 215–216
    visuals in, 208, 219, 221
Process explanations
    audience questions about, 185
    definition of, 185
    as instructions, 186 (*see also* Instructions)
    as process analyses, 187 (*see also* Process
        analyses)
    as process narratives, 186, 198 (*see also*
        Process narratives)
    purpose of, 185
    revision checklist for, 226–227
Process narratives, 186, 198, 207–208, 209–
        214
    audience analysis for, 208
    elements of, 198, 207
    outline for, 207
    visuals in, 198
Progress reports, 406, 408, 419, 422–423
Pronoun case, 578–579
Pronoun reference, 576–577
Pronouns, as connectors, 601
Proposals
    audience for, 431–432
    audience analysis for, 453, 470
    audience questions about, 432, 435, 447,
        450, 452
    body in, 450–452
    conclusion in, 452–453
    criteria for evaluating, 431
    definition of, 429
    elements of, 439–446
    external, 431–432
    internal, 431–432
    introduction in, 447–450
    outline for, 447
    planning, 432, 433–434
    process, summarized, 429–431
    purpose of, 429
    request for (RFP), 430–431
    research, 432, 435, 436–439
    revision checklist for, 471–472
    sales, 435, 439, 440–441
    solicited, 431

    supplements to, 439–442
    tone of, 432, 439, 446, 452
    unsolicited, 431
    visuals in, 444–445, 469
Punctuation, 583–599

Qualifiers, needless, 47
Question mark, 585
Questionnaires, 154, 306–307, 308
Quotation marks, 312–313, 591, 595–596

Readability. *See also* Style
    effect of word choice on, 50–58
    of paragraphs, 28–39
    of sentences, 39–50
Readers. *See* Audience
Recommendations
    in analytical reports, 143, 483, 489, 493,
        494, 495, 513, 537
    in oral reports, 551
    in proposals, 453
Redundancy, 44
Reference books, library, 292–293
Reference, letters of, 376–379. *See also* Job
        hunting; Letters
Reference, pronoun, 576–577
Repetition
    as a connector, 601
    needless, 45
Report design worksheet, 143–147
Reports
    formal, 231–232, 476–540 (*see also* Ana-
        lytical Reports)
    informal, 231–232, 400–424
Research. *See also* Data; Documentation;
        Library
    definition of, 285
    documentation practices, 324–332
    information gathering, 321, 323
    information sources, 142, 287–312
    informative interviews, 300–306
    inquiry letters, 307, 322 (*see also* Letters)
    library use, 287–299
    notecards, 311
    note-taking, 311–314
    organizational records, 310
    personal experience, 287
    personal observation, 310–311
    primary, 285

purpose of, 285–287
questionnaires, 154, 306–307, 308
secondary research, 285
Research proposals, 432, 435, 436–439
Research reports. *See also* Library; Research
audience analysis for, 333
choice of topics, 319
definition of, 319
documentation of, 324–332
focus of topics, 320
interviews, 322
outline revision, 323
questionnaires, 322
section-by-section development, 323
selective note-taking, 322 (*see also* Research)
skimming, 322
statement of purpose, 321
working outline, 321
Restrictive clauses or phrases, 590
Résumé, 374–383. *See also* Job hunting;
Letters
Revision checklists, 80–81, 103–104, 126,
149, 179–180, 226–227, 279–280,
393–394, 425, 471–472, 558–559
RFP (Request for proposal), 430–431
Rhetorical triangle, 186
Run-on sentence, 571–572

Sales proposals, 435, 439, 440–441
Secondary reader, 17–18
Secondary research, 285
Section length of various reports, 237
Semiblock form for letters, 356, 357
Semicolon, 584, 586
Sentence combining, 48–49, 72
Sentence definitions, 91–92, 101
Sentence errors, common, 567–583. *See also*
Punctuation
Sentence fragments, 567–569
Sentence length in technical writing, 39
Sentence outlines, 135–136
Sentence parts, 561–565
Sentences
clarity in, 39–43
conciseness in, 43–47
fluency in, 47–50
Sentence types, 565–566

Sentence variety, 49–50
Sexist language, elimination of, 611–613
Shifts in sentence construction, 582–583
Short sentences
for emphasis, 50
in instructions, 193
Simple sentences, 565
Site inspection reports, 406, 409–411, 412
Spacing, of typed lines, 235
Spatial sequence,
in a mechanism description, 163, 164
for paragraph development, 35
Specious reasoning, 487
Spelling, 607
Statements of purpose
for analytical reports, 481–482, 488, 490,
503, 521
for descriptions, 162, 164, 169, 170, 173
as guides to outlining, 132–133, 141, 142
for instructions, 194, 199
for oral reports, 548, 549
for proposals, 442–443, 447, 448, 458
for research reports, 321
Structure of all communication, 236. *See also*
Outlines
Style. *See also* Readability
in definitions, 89
in descriptions, 158–159
in instructions, 190–193
in letters, 356, 359–363, 366–369, 370
in process analyses, 208
in process narratives, 198
in proposals, 444, 447, 452
in summaries, 70, 71–72
Subject heading in letter reports, 406–409
Subjective complement, 564
Subjectivity, 3–4, 156, 158. *See also* Biased
language; Objectivity
Subject of a sentence, 563
Subordination in sentences, 48–49, 573–574
Summaries. *See also* Abstracts
accuracy of, 68
audience for, 69–70
audience questions about, 69
conciseness in, 68, 70–71
definition of, 67
documentation in, 72
essential message in, 68, 69–70
meaning in, 70
placement of, in reports, 79–80
purpose of, 67–68

Summaries (*cont.*)
  revision checklist for, 80–81
  steps in writing of, 71–72
  structure of, 70
  style in, 70, 71–72
Supplements to reports
  appendix, 255–256, 257 (*see also* Appendix)
  bibliography, 255 (*see also* Documentation)
  cover, 239, 241
  definition of, 232
  end matter, 239
  endnotes, 255 (*see also* Documentation)
  front matter, 239
  glossary, 251, 254 (*see also* Glossary)
  informative abstract, 247, 251, 252 (*see also* Abstracts; Summaries)
  letter of transmittal, 244–245, 246 (*see also* Transmittal, letter of)
  for proposals, 439, 442
  purpose of, 235
  table of contents, 245, 247, 248, 259 (*see also* Table of contents)
  table of illustrations, 247, 250, 500
  title page, 241–244
Survey results, reporting of, 404–406, 407
Sweeping generalizations, 54
Synonyms as connectors, 601

Table of contents, 245, 247, 248, 259, 456, 495, 499
Tables, 115, 119, 120, 122, 262–265
Technicality, level of, 11–18, 19, 21–22
  in definitions, 89
  in descriptions, 160–161
  in instructions, 188–189
  in process analyses, 208, 215
  in process narratives, 198, 207
  in proposals, 445–446
  in summaries, 70
Technical writing. *See also* Nontechnical writing
  definition of, 3
  example of, 4–5
  uses of, 6–7
  value of, 7–8
"There" sentence openers, 45
Title page, 242–244, 454, 495, 497

Titles, 241–242
  for analytical reports, 482
  for descriptions, 159–160
  for classifications, 118–119
  for instructions, 187
  for partitions, 113
  for process analyses, 208
  for process narratives, 198
  for proposals, 442
  for research reports, 320
Tone
  avoiding bias in, 57–58, 482
  of letters, 348, 356, 359–363, 366–369, 370, 384, 385–388
  of oral reports, 555, 556
  of proposals, 432, 439, 446, 452
Topic outline, formal, 133–135
Topic sentence, 29–30
Transmittal, letter of, 244–245, 246, 455, 495, 498
Transitional paragraphs, 602
Transitional sentences, 602
Transitions and connectors, 599–602. *See also* Coherence in paragraphs
  in instructions, 191–192
  in paragraphs, 33–34
  in summaries, 72, 77–78
Triteness, 53, 359–360

Unity, in paragraphs, 33
U.S. Government publications, 297–298

Verbal phrase, 565
Visual aids
  in analytical reports, 484, 505, 509, 522, 525
  in classifications, 118–119, 120, 121, 122
  in definitions, 95, 98, 99, 100
  definition of, 261
  in descriptions, 160, 161, 165, 166, 167, 170, 174
  in instructions, 190, 195, 203, 204, 206
  in oral reports, 548, 553, 554
  in partitions, 113, 115, 116, 117
  in process analyses, 208, 219, 221
  in process narratives, 198
  in proposals, 444–445, 469

purpose of, 261–262, 263
revision checklist for, 279–280

Warnings in instructions, 190, 200, 202, 205
Word choice, 50–58. *See also* Style; Tone

Wordiness, 43–47
Work estimates, 412, 413–414
Working definition, 92, 142, 195, 491

"You" perspective, in letters, 360–363